Marlies Pirner

Kinetic modelling of gas mixtures

Marlies Pirner

Kinetic modelling of gas mixtures

Würzburg
University Press

Dissertation, Julius-Maximilians-Universität Würzburg
Fakultät für Mathematik und Informatik, 2018
Gutachter: Prof. Dr. Christian Klingenberg, Prof. Dr. Stéphane Brull, Prof. Dr. Seok-Bae Yun

Impressum

Julius-Maximilians-Universität Würzburg
Würzburg University Press
Universitätsbibliothek Würzburg
Am Hubland
D-97074 Würzburg
www.wup.uni-wuerzburg.de

© 2018 Würzburg University Press
Print on Demand

Coverdesign: Julia Bauer

ISBN 978-3-95826-080-1 (print)
ISBN 978-3-95826-081-8 (online)
URN urn:nbn:de:bvb:20-opus-161077

Acknowledgments

I would particularly like to thank my advisor Prof. Christian Klingenberg for the continuous support of my PhD study and related research. I want to thank him for his patience, his friendly nature, his motivation and his constant feedback under his supervision. I appreciate his encouragement and his suggestions during our many discussions to find issues to work on. His guidance helped me in all the time of research, writing articles and giving presentations. I also want to thank him for the creation of collaborative projects.

Besides my advisor, I would like to thank Prof. Gabriella Puppo at the University of Como for suggesting interesting topics and for all the useful scientific discussions. I received generous financial support from the DAAD-MIUR Joint Mobility Programme which enabled us to visit each other during the time of my PhD study.

My thanks also go to Anaïs Crestetto at the University of Nantes for her fruitful collaboration during her visits in Wuerzburg and Prof. Qin Li for inviting me to the department of mathematics at the University of Madison.

Special thanks to Benedikt and Gero for reading this thesis and for all their good suggestions and comments. I thank Martin for his critical reading of my papers and preprints.

I greatly appreciate the German Priority Programme 1648 for financial support and the University of Wuerzburg for the support to take up the position of a teaching assistant.

Last but not least I would like to thank my family and my friends for supporting me at any time.

Marlies Pirner

Abstract

The present thesis considers the modelling of gas mixtures via a kinetic description. Fundamentals about the Boltzmann equation for gas mixtures and the BGK approximation are presented. Especially, issues in extending these models to gas mixtures are discussed. A non-reactive two component gas mixture is considered. The two species mixture is modelled by a system of kinetic BGK equations featuring two interaction terms to account for momentum and energy transfer between the two species. The model presented here contains several models from physicists and engineers as special cases. Consistency of this model is proven: conservation properties, positivity of all temperatures and the H-theorem. The form in global equilibrium as Maxwell distributions is specified. Moreover, the usual macroscopic conservation laws can be derived.

In the literature, there is another type of BGK model for gas mixtures developed by Andries, Aoki and Perthame, which contains only one interaction term. In this thesis, the advantages of these two types of models are discussed and the usefulness of the model presented here is shown by using this model to determine an unknown function in the energy exchange of the macroscopic equations for gas mixtures described in the literature by Dellacherie. In addition, for each of the two models existence and uniqueness of mild solutions is shown. Moreover, positivity of classical solutions is proven.

Then, the model presented here is applied to three physical applications: a plasma consisting of ions and electrons, a gas mixture which deviates from equilibrium and a gas mixture consisting of polyatomic molecules.

First, the model is extended to a model for charged particles. Then, the equations of magnetohydrodynamics are derived from this model. Next, we want to apply this extended model to a mixture of ions and electrons in a special physical constellation which can be found for example in a Tokamak. The mixture is partly in equilibrium in some regions, in some regions it deviates from equilibrium. The model presented in this thesis is taken for this purpose, since it has the advantage to separate the intra and interspecies interactions. Then, a new model based on a micro-macro decomposition is proposed in order to capture the physical regime of being partly in equilibrium, partly not. Theoretical results are presented, convergence rates to equilibrium in the space-homogeneous case and the Landau damping for mixtures, in order to compare it with numerical results.

Second, the model presented here is applied to a gas mixture which deviates from equilibrium such that it is described by Navier-Stokes equations on the macroscopic level. In this macroscopic description it is expected that four physical coefficients will show up, characterizing the physical behaviour of the gases, namely the diffusion coefficient, the viscosity coefficient, the heat conductivity and the thermal diffusion parameter. A Chapman-Enskog expansion of the model presented here is performed in order to capture three of these four physical coefficients. In addition, several possible extensions to an ellipsoidal statistical model for gas mixtures are proposed in order to capture the fourth coefficient. Three extensions are proposed: An extension which is as simple as possible, an intuitive extension copying the one species case and an extension which takes into account the physical motivation of the physicist Holway who invented the ellipsoidal statistical model for one species. Consistency of the extended models like conservation properties, positivity of all temperatures and the H-theorem are proven. The shape of global Maxwell distributions in equilibrium are specified.

Third, the model presented here is applied to polyatomic molecules. A multi component gas mixture with translational and internal energy degrees of freedom is considered. The two species are allowed to have different degrees of freedom in internal energy and are modelled by a system of kinetic ellipsoidal statistical equations. Consistency of this model is shown: conservation properties, positivity of the temperature, H-theorem and the form of Maxwell distributions in equilibrium. For numerical purposes the Chu reduction is applied to the developed model for polyatomic gases to reduce the complexity of the model and an application for a gas consisting of a mono-atomic and a diatomic gas is given.

Last, the limit from the model presented here to the dissipative Euler equations for gas mixtures is proven.

Contents

Introduction

In 1872, the physicist Boltzmann developed a partial differential equation which describes the time evolution of a density distribution of particles in a rarefied mono-atomic gas of one species. This partial differential equation is called Boltzmann equation. The physical theory based on this equation is called kinetic theory and models phenomena in the statistical physics. The kinetic theory was developed among other physicists by Cercignani, Maxwell, Chapman, Cowling, Esposito and Pulvirenti, see for example [24, 26, 39]. The properties of the Boltzmann equation namely conservation properties of mass, momentum and energy were considered more rigorously for example by Golse [47] or Villani [83]. There are a lot of articles dealing with the existence of solutions to the Boltzmann equation, see for example [4, 36, 37, 45].

In 1954, Bathnagar, Gross and Krook [16] proposed a simplified model with the same main properties as the Boltzmann equation for one species. It is called BGK model and is less complicated than the full Boltzmann equation. Moreover, it leads to efficient numerical simulations, see for example [13, 34, 35, 41, 72]. In 1989, Perthame proved existence of global solutions to the BGK equation for one species in [69] and in 1993, Perthame and Pulvirenti established the existence and uniqueness of mild solutions on a bounded domain in space with periodic boundary conditions in [70].

Several physicists and engineers developed extensions of the BGK model for one species to a BGK model for gas mixtures, for example Gross and Krook [49], Hamel [51], Garzo, Santos and Brey [43], Sofonea and Sekerka [79] and Asinari [6]. All this models have one thing in common. The interactions are described by a sum of relaxation operators, one for interactions of a species with itself and one for the interaction of a species with the other one. In 2002, Andries, Aoki and Perthame proposed a BGK model for gas mixtures [1] with only one relaxation operator describing all interactions in one relaxation operator. For this model they proved consistency: conservation properties, positivity of all temperatures and the H-theorem. Then, this model was used by several applied mathematicians for example in [22, 48, 18].

There are three main applications where a gas mixture is modelled via a BGK approach for gas mixtures. One main application is the physical regime of a plasma, a mixture of ions and electrons. Here, the BGK equation was extended to the Vlasov-BGK equation which has an additional force term taking into account the magnetic and the electric fields generated by the charged particles [80]. In [75] the existence of mild solutions for one species is established using estimates from [82]. Moreover, in the collision less limit Landau discovered a damping property of the Vlasov equation called Landau damping [63]. This was made more rigorous by Villani in [84] and is now often used as a test case in numeric simulations concerning a plasma. One specific application of a plasma is fusion in a Tokamak. In this case, the plasma is in equilibrium near the wall of the chamber of a fusion reactor, in the core plasma it is not. For this regime, a micro-macro decomposition is used in the literature, see [12, 28, 30].

Another application is to describe a gas which deviates from equilibrium such that it is described by the Navier-Stokes equations on the macroscopic level. In this case, it turns out that the BGK model is not an accurate description. Therefore, Holway suggested a correction called ellipsoidal statistical model for one species (ES-BGK model) in [54] in 1966. It is based on a physical theory called persistence of velocities discovered by Jeans [55] in 1904. In [3], Andries and Perthame proved that this model satisfies an H-theorem. In [21], Brull showed that this model is linked to a minimization problem and Yun established existence and uniqueness of mild solutions in [87].

The last main application is the modelling of polyatomic molecules. A lot of models are proposed by physicists [76, 68, 54]. Andries, LeTallec, Perlat and Perthame proved the H-theorem of one suggestion in [54] for one species in [2]. An alternative model for one species was suggested by Bernard, Iollo and Puppo in [14]. A further attempt in the area of polyatomic molecules is to take into account chemical reactions, see for example [18].

Another recent area of research is to prove rigorously limits from kinetic to macroscopic equations in the case of one species, see [46, 47, 78, 77].

The outline of this thesis is the following. In chapter 1, we introduce all fundamental things like the Boltzmann equation, link of the distribution function to macroscopic quantities and the BGK approximation. In chapter 2, we present a model for a gas mixture which contains well-known models from physicists and engineers as special cases and prove consistency of this model, meaning conservation properties, positivity of all temperatures and the H-theorem. In chapter 3, we illustrate the usefulness of the model presented in chapter 2 by using it to determine an unknown function in the energy exchange in a macroscopic model of Dellacherie [32]. In chapter 4, we prove existence, uniqueness and positivity of mild solutions for the model presented in chapter 2 and for the other type of model for gas mixtures [1]. In chapter 5, we extend the model presented in chapter 2 to charged particles. In chapter 6, we use this extended model to model a gas mixture consisting of ions and electrons. We perform a micro-macro decomposition and establish theoretical estimates such that one can perform a simulation for a plasma and compare it with these theoretical results. In chapter 7, we extend the model from chapter 2 to an ES-BGK model for gas mixtures and perform a Chapman-Enskog expansion in order to see that we can capture the right hydrodynamic regime on the level of the Navier-Stokes equations. In chapter 8, we extend the model presented in chapter 2 to polyatomic molecules. In chapter 9, we prove convergence of the model presented in chapter 2 to the incompressible Euler equations for gas mixtures.

Chapter 1

Fundamentals

In this chapter we give a brief introduction into the fundamental differential equation in the kinetic theory of gases, the Boltzmann equation. Section 1.1 deals with the mathematical foundation of the Boltzmann equation, the homogeneous transport equation. In section 1.2 we give an introduction into the basic physical treatment of a gas, Newton's laws of motion. In section 1.3 we give a brief introduction into the Boltzmann equation itself whereas section 1.4 deals with a well-known approximation of the Boltzmann equation, the BGK equation.

1.1 The transport equation and characteristic curves

In this section we consider the homogeneous transport equation. Literature on the homogeneous transport equation and characteristic curves can be found for example in chapter 1.2 of Evans [40] or in chapter 1.3 and 1.4 of John [57]. The homogeneous transport equation is given by

$$\partial_t f^{hom} + b \cdot \nabla_x f^{hom} + c \cdot \nabla_v f^{hom} = 0 \quad \text{in} \quad \mathbb{R}^3 \times \mathbb{R}^3 \times \mathbb{R}^+,$$
$$f^{hom}(x, v, 0) = f^{hom,0}(x, v), \tag{1.1}$$

where $b, c : \mathbb{R}^3 \times \mathbb{R}^3 \times \mathbb{R}_0^+ \to \mathbb{R}^3$ are given functions, $b = b(x, v, t)$, $c = c(x, v, t)$, the function $f^{hom} : \mathbb{R}^3 \times \mathbb{R}^3 \times \mathbb{R}^+ \to \mathbb{R}$ is the unknown, $f^{hom} = f^{hom}(x, v, t)$ and $f^{hom,0} : \mathbb{R}^3 \times \mathbb{R}^3 \to \mathbb{R}$ the initial value. Here $(x, v) \in \mathbb{R}^3 \times \mathbb{R}^3$ denotes a point in the position-velocity space called phase space and $t \geq 0$ denotes the time.

For later purposes, we want to introduce an approach how to construct solutions to (1.1). The idea is to reduce the problem of finding a solution to a partial differential equation to a problem of finding a solution to a system of ordinary differential equations. This approach is described in the following. We can show that any classical solution to (1.1) is constant along certain curves in the t-(x, v)-space. Let $\gamma : [0, 1] \to \mathbb{R}^3 \times \mathbb{R}^3 \times \mathbb{R}^+, s \mapsto \gamma(s) = (v(s), x(s), t(s))$ be a smooth parametrisation of a curve in $\mathbb{R}^3 \times \mathbb{R}^3 \times \mathbb{R}^+$. Now, consider the function $z(s) = f^{hom}(x(s), v(s), t(s))$. Then the derivative of z with respect to s is given by

$$\frac{d}{ds} z(s) = \frac{dt(s)}{ds} \partial_t f^{hom}(x(s), v(s), t(s)) + \frac{dx(s)}{ds} \cdot \nabla_x f^{hom}(x(s), v(s), t(s))$$
$$+ \frac{dv(s)}{ds} \cdot \nabla_v f^{hom}(x(s), v(s), t(s)). \tag{1.2}$$

Comparing (1.2) with (1.1), we see that f^{hom} is constant along the curve parametrized by γ, if γ satisfies the following system of ordinary differential equations

$$\frac{dt(s)}{ds} = 1,$$

$$\frac{dx(s)}{ds} = b(x(s), v(s), t(s)), \tag{1.3}$$

$$\frac{dv(s)}{ds} = c(x(s), v(s), t(s)).$$

So we reduced the problem (1.1) to a system of ordinary differential equations by introducing this specific curve γ.

Definition 1.1.1. The curve γ, on which the solution of the partial differential equation (1.1) is constant, is called a characteristic curve or characteristic line, and the corresponding ordinary differential equations are called characteristic equations.

If we can solve the characteristic equations, we can solve (1.1) for f^{hom} by going back to the initial data along the characteristic line, which has the same value there, since we showed that the solution is constant along characteristic lines. We illustrate this in the following example.

Example 1.1.1. Let $b(x, v, t) = v$ and $c(x, v, t) = 0$ in (1.1). In this case, we can solve the characteristic equations (1.3) and obtain

$$t(s) = s + t_0, \quad x(s) = vs + x_0, \quad v(s) = v_0, \tag{1.4}$$

for some $t_0 \geq 0$, $x_0, v_0 \in \mathbb{R}^3$. Since our initial data on f^{hom} is given at time $t = 0$, we choose $t_0 = 0$. The solution is unique since in this case the right-hand side of (1.3) is Lipschitz continuous in (x, v) and continuous in the variables (x, v, t). This guarantees that the solution is unique according to the theorem of Picard-Lindelöf, see for example theorem 16.1 in volume 1 of [33]. Since we know that the solution of (1.1) is constant along the characteristic lines, we can solve the equation for a given initial data $f^{hom}(x, v, 0) = f^{hom,0}(x, v)$. The solution f^{hom} is equal to the function z if we choose $s = t$, so $f^{hom}(x, v, t) = z(t)$. The function z is constant along characteristic lines, therefore $z(t) = z(0)$ along solutions to (1.4). By the definition of z, $z(0)$ is nothing else than the function f^{hom} evaluated at $(x_0, v_0, 0)$, so $z(0) = f^{hom}(x_0, v_0, 0)$. Now, we can invert the equations (1.4) for (x_0, v_0) and write (x_0, v_0) in terms of (x, v). All in all, we obtain

$$f^{hom}(x, v, t) = f^{hom,0}(x - vt, v),$$

which is a solution to the initial value problem (1.1).

1.2 Newton's fundamental laws of motion

We consider a gas consisting of particles. According to physical axioms, the time evolution of the individual particles is described by the fundamental laws of Newton.

In this section we introduce these two fundamental laws and apply them to a situation in which the gas particles are modelled as balls and can collide with each other. Note, that parts of the following are also mentioned in my Bachelor thesis [73].

1.2.1 Newton's equations and Newton's third law

We assume that our gas consisting of one species of particles can be modelled as a collection of N identical particles with masses m interacting via a force. We number the particles by the index $s \in \{1, ..., N\}$. The state of each particle is described by their positions of the centre $x_s \in \mathbb{R}^3$ and their velocities $v_s \in \mathbb{R}^3$ at any time $t \in \mathbb{R}_0^+$ for $s = 1, ..., N$. Consider a particle with position $x_s^0 \in \mathbb{R}^3$ and velocity $v_s^0 \in \mathbb{R}^3$ at a time $t_0 \in \mathbb{R}_0^+$. Then, it is an axiom of the classical mechanics that the time evolution of the position $x_s(t)$ and the velocity $v_s(t)$ for $t \geq t_0$ is given by the following system of ordinary differential equations called Newton's equations:

Axiom 1.2.1 (Newton's equations). Let $x_s^0 \in \mathbb{R}^3$ and $v_s^0 \in \mathbb{R}^3$ be the position and the velocity of a particle $s \in \{1, ..., N\}$ at a time $t_0 \in \mathbb{R}_0^+$. Then the time evolution of the position $x_s(t)$ and the velocity $v_s(t)$ of this particle are determined by Newton's equations given by

$$\frac{d}{dt}x_s(t) = v_s(t),$$

$$m\frac{d}{dt}v_s(t) = \sum_{\substack{j=1 \\ j \neq s}}^{N} F_{s,int,j}(x_s(t), v_s(t), x_j(t), v_j(t), t) + F_{ext}(x_s(t), v_s(t), t), \quad (1.5)$$

$$x_s(t_0) = x_s^0, \; v_s(t_0) = v_s^0,$$

for all $t > t_0$, where $\sum_{\substack{j=1 \\ j \neq s}}^{N} F_{s,int,j}(x_s(t), v_s(t), x_j(t), v_j(t), t) + F_{ext}(x_s(t), v_s(t), t)$ denotes the force on the particle s and can be split into an external force $F_{ext}(x_s(t), v_s(t), t)$ and a force describing the interactions with the other particles

$$\sum_{\substack{j=1 \\ j \neq s}}^{N} F_{s,int,j}(x_s(t), v_s(t), x_j(t), v_j(t), t).$$

The meaning of the equations is the following. If the particle s has a velocity v_s, the position x_s of the particle s will change in time. And if a force is acting on this particle s, the velocity v_s will change in time.

In addition, there is another law of Newton concerning the forces in an interaction of two particles. It states the following:

Axiom 1.2.2 (Newton's third law). If a particle exerts a force on another particle, the other particle always inserts a force on the first particle with the same absolute value in the opposite direction.

1.2.2 Conservation laws and elastic interactions

From the dynamic of Newton's laws one can deduce some fundamental physical properties, the conservation of momentum and energy in interactions. This properties allow to determine the velocities of two particles after an interaction knowing the velocities before an interaction in the case of particles interacting via elastic interactions. We start with the conservation of momentum.

Consequence 1.2.1 (Conservation of momentum). Let m_1, m_2 be the masses, $x_1(t)$, $x_2(t)$ the positions and $v_1(t), v_2(t)$ the velocities determined by Newton's laws (1.5) of two particles. Assume that $F_{ext} = 0$. Then, under the hypothesis of axiom 1.2.2, we have

$$m_1 v_1(t) + m_2 v_2(t) = \text{const},$$

for all $t \in I$ where $I \subset [t_0, \infty)$ is an interval where the two particles do not interact with the other $N - 2$ particles.

Proof. If $F_{ext} = 0$, Newton's equations are given by

$$m_1 \frac{d}{dt} v_1(t) = F_{1,int,2}(x_1(t), v_1(t), x_2(t), v_2(t), t), \tag{1.6}$$

$$m_2 \frac{d}{dt} v_2(t) = F_{2,int,1}(x_1(t), v_1(t), x_2(t), v_2(t), t). \tag{1.7}$$

Under the hypothesis of axiom 1.2.2, we have

$$F_{1,int,2}(x_1(t), v_1(t), x_2(t), v_2(t), t) = -F_{2,int,1}(x_1(t), v_1(t), x_2(t), v_2(t), t),$$

for all $t \in I$. If we then add (1.6) and (1.7) and integrate with respect to t, this leads to

$$m_1 v_1(t) + m_2 v_2(t) = \text{const} \quad \text{for all} \quad t \in I.$$

$\square$

Another consequence from Newton's laws is the conservation of energy.

Consequence 1.2.2 (Conservation of energy). Let m_1, m_2 be the masses, $x_1(t), x_2(t)$ the positions and $v_1(t), v_2(t)$ the velocities determined by Newton's laws (1.5) of two particles. Assume that $F_{ext} = 0$; and that $F_{1,int,2}$ depends only on $x_1(t) - x_2(t)$, and assume that there exists a function $\Phi(x)$ such that $F_{1,int,2} = -\nabla_x \Phi$. Then, under the hypothesis of axiom 1.2.2, we have

$$\frac{1}{2} m_1 |v_1(t)|^2 + \frac{1}{2} m_2 |v_2(t)|^2 + \Phi = \text{const},$$

for all $t \in I$ where $I \subset [t_0, \infty)$ is an interval where the two particles do not interact with the other $N - 2$ particles.

Proof. If $F_{ext} = 0$ and under the assumption that $F_{1,int,2}$ depends only on $x_1(t) - x_2(t)$, Newton's equations are given by

$$m_1 \frac{d}{dt} v_1(t) = F_{1,int,2}(x_1(t) - x_2(t)), \tag{1.8}$$

$$m_2 \frac{d}{dt} v_2(t) = F_{2,int,1}(x_1(t), v_1(t), x_2(t), v_2(t), t). \tag{1.9}$$

Under the hypothesis of axiom 1.2.2, we have

$$F_{1,int,2}(x_1(t) - x_2(t)) = -F_{2,int,1}(x_1(t), v_1(t), x_2(t), v_2(t), t) \quad \text{for all} \quad t \in I.$$

We multiply (1.8) by $v_1(t)$, (1.9) by $v_2(t)$ and add them. Under the assumption that $F_{1,int,2}$ can be written as minus a gradient of a potential Φ, we obtain

$$\frac{d}{dt}\left(\frac{1}{2}(m_1|v_1(t)|^2 + m_2|v_2(t)|^2\right) = F_{1,int,2}(x_1(t) - x_2(t)) \cdot (v_1(t) - v_2(t))$$

$$= -\nabla_x \Phi(x_1(t) - x_2(t)) \cdot (v_1(t) - v_2(t))$$

$$= -\frac{d}{dt}\Phi(x_1(t) - x_2(t)) \quad \text{for all} \quad t \in I.$$

The last equality follows by chain rule and the first equation in Newton's equations (1.6). If we integrate the obtained equation with respect to t from t_0 to t we obtain the result. $\qquad\square$

In consequence 1.2.2 we assumed the existence of a scalar potential Φ such that we can write $F_{1,int,2}$ as minus the gradient of the scalar function Φ. The physical meaning of this is the following. Let $\tilde{\gamma} : [0,1] \to \Gamma \subset \mathbb{R}^3$ be a smooth parametrization of a curve in $\mathbb{R}^3$ and $F : \Gamma \to \mathbb{R}^3$ be a smooth vector field representing the force exerting on a particle which moves along Γ. Then the physical work of this particle is defined by

$$\int_{\tilde{\gamma}} F \cdot ds = \int_0^1 F(\tilde{\gamma}(t))\tilde{\gamma}'(t)dt.$$

One can prove the following lemma.

Lemma 1.2.3. *Let $\Omega \subset \mathbb{R}^3$ be an open and connected subset and F be a smooth force field on Ω. Then we have: It exists a smooth function Φ such that $F = -\nabla_x \Phi$ in Ω if and only if F is independent of the path, that means*

$$\int_{\Gamma_1} F \cdot ds = \int_{\Gamma_2} F \cdot ds$$

for all curves Γ_1, Γ_2 in Ω which coincide in the start and end points.

The proof is given in theorem 13.50 in volume 1 of [33]. This lemma has the following physical consequence. The work which is needed to move a particle in this field F is independent of the path. That means the energy of the particle is conserved in a closed path, it neither gains nor loses energy there.

We conclude with the last consequence from Newton's laws which is the conservation of angular momentum.

Consequence 1.2.4 (Conservation of angular momentum). Let $\tilde{b} \in \mathbb{R}^3$ be a fixed point in space. Let m_1, m_2 be the masses, $x_1(t), x_2(t)$ the positions and $v_1(t), v_2(t)$ the velocities determined by Newton's laws (1.5) of two particles. Assume that $F_{ext} = 0$, $F_{1,int,2}$ is parallel to $x_1(t) - \tilde{b}$ and $F_{2,int,1}$ is parallel to $x_2(t) - \tilde{b}$. Then, under the hypothesis of axiom 1.2.2, we have

$$m_k(x_k(t) - \tilde{b}) \times v_k(t) = c \quad \text{for} \quad k = 1, 2, \quad \text{and a constant} \quad c \in \mathbb{R},$$

and for all $t \in I$ where $I \subset [t_0, \infty)$ is an interval where the two particles do not interact with the other $N - 2$ particles. Furthermore, the map $c(\tilde{b})$ is continuous in $\tilde{b}$.

Proof. If we use the product rule for a cross product and Newton's equations (1.5) for $F_{ext} = 0$, we obtain

$$\frac{d}{dt}[m_k(x_k(t) - \tilde{b}) \times v_k(t)] = m_k(x_k(t) - \tilde{b}) \times \frac{d}{dt}v_k(t)$$

$$= (x_k(t) - \tilde{b}) \times F_{k,int,j}(x_k(t), v_k(t), x_j(t), v_j(t), t) = 0,$$

for all $k, j = 1, 2$, $k \neq j$ and for all $t \in I$. The last equality is satisfied since we assumed that $F_{1,int,2}$ is parallel to $x_1(t) - \tilde{b}$ and $F_{2,int,1}$ is parallel to $x_2(t) - \tilde{b}$. This shows the first statement. It remains to prove that the map $c(\tilde{b})$ is continuous in $\tilde{b}$. Let $\tilde{b}_1, \tilde{b}_2 \in \mathbb{R}^3$ be arbitrary. Then we have

$$|c(\tilde{b}_1) - c(\tilde{b}_2)| = |m_k(x_k - \tilde{b}_1) \times v_k - m(x_k - \tilde{b}_2) \times v_k| = |m(\tilde{b}_2 - \tilde{b}_1) \times v_k| \leq m|v_k||\tilde{b}_2 - \tilde{b}_1|.$$

Therefore $c(\tilde{b})$ is Lipschitz continuous in $\tilde{b}$, especially continuous. $\qquad \square$

The physical meaning is the following. In the proof of the next corollary, we will see that the conservation of angular momentum means that the particles always stay in the same plane.

In the rest of this section we want to apply the conservation laws to a specific physical example. We now want to apply the conservation laws in order to determine the velocities of the particles after the interaction depending on the velocities before the interaction.

Corollary 1.2.5 (Velocities after an interaction). *Let m_1, m_2 be the masses of two particles. Let v_1, v_2 be two velocities before a collision of the two particles. Assume that $F_{ext} = 0$, $F_{1,int,2}$ depends only on $x_1(t) - x_2(t)$ and is parallel to $x_1(t) - x_2(t)$ and $F_{2,int,1}$ is parallel to $x_2(t) - x_1(t)$, and that during the interaction the two particles do not interact with the other $N - 2$ particles. We assume that we can write the force $F_{1,int,2}$ as minus a gradient of a scalar potential Φ with compact support. Let ω be the unit vector along the line with the minimal distance of the two particles during the interaction in the direction of particle 2, see figure 1.1. Then under the hypothesis of axiom 1.2.2, we can derive the following conservation laws during a collision*

$$\begin{aligned} m_1 v_1 + m_2 v_2 &= m_1 v_1' + m_2 v_2', \\ m_1|v_1|^2 + m_2|v_2|^2 &= m_1|v_1'|^2 + m_2|v_2'|^2, \end{aligned} \tag{1.10}$$

and the velocities of the two particles v_1', v_2' after the interaction are given by

$$v_1' = v_1 - \frac{2m_2}{m_1 + m_2}\omega[(v_1 - v_2) \cdot \omega],$$

$$v_2' = v_2 + \frac{2m_1}{m_1 + m_2}\omega[(v_1 - v_2) \cdot \omega],$$

provided that the particles change its velocities instantaneously in time into v_1', v_2' during the interaction at the time t^* when the two particles have their minimal distance during the interaction. Then ω can be written as $\omega = \frac{x_1(t^*) - x_2(t^*)}{|x_1(t^*) - x_2(t^*)|}$.

Proof. Since $F_{1,int,2}$ is parallel to $x_1(t^* - \varepsilon) - x_2(t^* - \varepsilon)$ at time $t^* - \varepsilon$ and $F_{1,int,2}$ is parallel to $x_1(t^* + \varepsilon) - x_2(t^* + \varepsilon)$ at time $t^* + \varepsilon$, we have

$$m_1(x_1(t^* - \varepsilon) - x_2(t^* - \varepsilon)) \times v_1(t^* - \varepsilon) = c(x_2(t^* - \varepsilon)),$$

$$m_1(x_1(t^* + \varepsilon) - x_2(t^* + \varepsilon)) \times v_1'(t^* + \varepsilon) = c(x_2(t^* + \varepsilon)),$$

according to consequence 1.2.4 with $\tilde{b} = x_2(t^* - \varepsilon)$ and $\tilde{b} = x_2(t^* + \varepsilon)$, respectively, where ε denotes a positive constant such that $[t^* - \varepsilon, t^* + \varepsilon] \subset I$ where $I \subset [t_0, \infty)$ is an interval where the two particles do not interact with the other $N - 2$ particles and outside the compact support of Φ. Since x_1, x_2, v_1, v_1' are continuous as classical solutions and c is continuous in $\tilde{b}$, we obtain in the limit $\varepsilon \to 0$

$$m_1(x_1(t^*) - x_2(t^*)) \times v_1(t^*) = m_1(x_1(t^*) - x_2(t^*)) \times v_1'(t^*),$$

under the assumption that particle 1 changes its velocity v_1 instantaneously at t^* into v_1'; and v_1 and v_1' are assumed to go forwards and backwards, respectively, in time as there were no interaction at this point. This type of interaction is called localized in time. This means that $v_1'(t^*) - v_1(t^*)$ is parallel to ω. Therefore, there exists $\alpha \in \mathbb{R}$ such that

$$v_1' - v_1 = \alpha\omega,$$

which is equivalent to

$$v_1' = \alpha\omega + v_1. \tag{1.11}$$

Here, and in the following, we will omit the argument t^*. Analogously, one obtains

$$v_2' = \beta\omega + v_2, \tag{1.12}$$

choosing $\tilde{b} = x_2(t^* \pm \varepsilon)$ for some $\beta \in \mathbb{R}$. From consequence 1.2.1, we get

$$m_1 v_1' + m_2 v_2' = m_1\alpha\omega + m_1 v_1 + m_2\beta\omega + m_2 v_2 \overset{!}{=} m_1 v_1 + m_2 v_2.$$

This leads to $\alpha = -\frac{m_2}{m_1}\beta$. From conservation of energy (consequence 1.2.1), we obtain

$$m_1|v_1'|^2 + m_2|v_2'|^2 = m_1|\alpha\omega + v_1|^2 + m_2\left| -\frac{m_1}{m_2}\alpha\omega + v_2\right|^2$$

$$= m_1\alpha^2 + m_1|v_1|^2 + 2m_1\alpha\omega \cdot v_1 + m_2|v_2|^2 + \frac{m_1^2}{m_2}\alpha^2 - 2m_1\alpha\omega \cdot v_2$$

$$\overset{!}{=} m_1|v_1|^2 + m_2|v_2|^2.$$

This leads to

$$\alpha m_1 \left(\alpha + 2\omega \cdot v_1 + \frac{m_1}{m_2} \alpha - 2\omega \cdot v_2 \right) = 0.$$

If $\alpha = 0$, we would have $v_1 = v_1'$ and $v_2 = v_2'$ and there would be no interaction. Therefore $\alpha \neq 0$ and we have

$$\alpha + 2\omega \cdot v_1 + \frac{m_1}{m_2} \alpha - 2\omega \cdot v_2 = 0.$$

Thus

$$\alpha = -\frac{2m_2}{m_1 + m_2} \omega \cdot (v_1 - v_2),$$

and therefore

$$\beta = \frac{2m_1}{m_1 + m_2} \omega \cdot (v_1 - v_2).$$

Inserting this into (1.11) and (1.12), we obtain the expressions for the velocities after the interaction claimed in corollary 1.2.5. $\qquad \square$

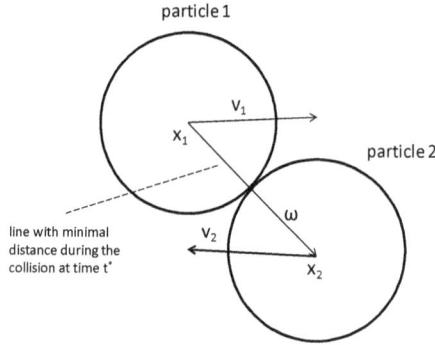

Figure 1.1: Collision of two particles. Particle 2 has the minimal distance to particle 1. The vector ω represents the direction of the line connecting the two positions x_1 and x_2 during they are at the minimal distance.

Remark 1.2.1. Note that the operator $T_\omega : (v_1, v_2) \mapsto (v_1', v_2')$ has the following properties:

   i) T_ω is invertible with $\qquad\qquad T_\omega \circ T_\omega = \mathbf{1}_{6\times 6},$

   ii) T_ω has unit Jacobian,

   iii) We have $\qquad\qquad\qquad \omega \cdot (v_1 - v_2) = -\omega \cdot (v_1' - v_2'),$

$$|v_1 - v_2| = |v_1' - v_2'|, \qquad (1.13)$$

   for every $v_1, v_2 \in \mathbb{R}^3$, $\omega \in S^2$.

These properties can easily be verified and are also mentioned in [25, 24, 83, 1].

The physical meaning of the first part of the remark is the following. The fact that T_ω is self-invertible reflects the physical principle that for elastic collisions the collision is reversible.

1.3 The Boltzmann equation for gas mixtures

In this section we want to give an overview of the fundamental equation in gas dynamics, the Boltzmann equation, and its fundamental properties. For one species of particles, there are a lot of introductory books and lecture notes in the literature concerning the Boltzmann equation and its fundamental properties, see [25, 24, 39, 46, 47, 81, 83]. Since we attempt to extend the subject to gas mixtures, the following overview is an extension of the properties of the Boltzmann equation found in the literature for one species to gas mixtures. For simplicity, we only consider two species. Furthermore, we assume that we have no chemical reactions and the number of particles of each species remains constant. Note that parts of the following are also given in my Master thesis [74].

1.3.1 Fundamental definitions

The most natural way to describe a gas with one species of N identical particles is to model the particles as balls interacting via a force. We number the particles by the index $s \in \{1, ..., N\}$. The state of each particle is described by their positions of the centre $x_s \in \mathbb{R}^3$ and velocities $v_s \in \mathbb{R}^3$ at any time $t \in \mathbb{R}_0^+$. We describe the time evolution of these positions and velocities by Newtons equations (1.5). So knowing all initial data, we are able to compute the time evolution of this gas. But a gas consists of a number of interacting particles in the order of 10^{23}, so one would have to solve a set of about 10^{23} coupled equations. This is not possible, among others there is no computer which has a high enough capacity to solve this. This fact was realized by Maxwell and Boltzmann. Therefore they started to work with a distribution function $f(x, v, t)$ where $x \in \mathbb{R}^3$ and $v \in \mathbb{R}^3$ are the phase space variables and $t \geq 0$ the time. The meaning of f is as follows: any infinitesimal volume $dxdv$ centred at (x, v) contains at time t about $f(x, v, t)dxdv$ particles. So we do not consider the positions and velocities of each single particle, we only consider the distribution of the positions and velocities in phase-space. We state this in our first definition.

Definition 1.3.1 (Distribution function). A function $f : \mathbb{R}^3 \times \mathbb{R}^3 \times \mathbb{R}_0^+ \to \mathbb{R}, (x, v, t) \mapsto f(x, v, t)$ is called a distribution function if and only if $f(x, v, t)dxdv$ is the number of particles with velocities in $(v, v + dv)$ located in the interval $(x, x + dx)$ at time t.

Since we consider a mixture composed of two different species, our kinetic model has two distribution functions $f_1(x, v, t) > 0$ and $f_2(x, v, t) > 0$, one for each species. The distribution function gives a detailed picture of the state of the gas, but it is not measurable in experiments. Measurable in experiments are quantities as the number density, the mean velocity and the temperature. For any $f_1, f_2 : \Lambda \times \mathbb{R}^3 \times \mathbb{R}_0^+ \to \mathbb{R}$, $\Lambda \subset \mathbb{R}^3$ with $(1 + |v|^2)f_1, (1 + |v|^2)f_2 \in L^1(dv), f_1, f_2 \geq 0$, we can relate them to our distribution functions in terms of microscopic averages of $f_k, k = 1, 2$ with respect to the velocity v illustrated in the following definitions.

From definition 1.3.1 follows that the total number of particles N_k in a volume $V \subset \Lambda \times \mathbb{R}^3$ is given by

$$N_k = \int_V f_k \, dv \, dx.$$

When we integrate only with respect to the velocity, we obtain a density only in x-space.

Definition 1.3.2 (Number density). Let $f_1, f_2 : \Lambda \times \mathbb{R}^3 \times \mathbb{R}_0^+ \to \mathbb{R}$, $\Lambda \subset \mathbb{R}^3$ with $f_1, f_2 \in L^1(dv)$, $f_1, f_2 \geq 0$ be two distribution functions. Then, the function

$$n_k : \mathbb{R}^3 \times \mathbb{R}_0^+ \to \mathbb{R}; \quad (x, t) \mapsto n_k(x, t) = \int_{\mathbb{R}^3} f_k(x, v, t) \, dv,$$

is called the number density of species k for $k = 1, 2$.

Definition 1.3.3 (Mean velocity). Let $f_1, f_2 : \Lambda \times \mathbb{R}^3 \times \mathbb{R}_0^+ \to \mathbb{R}$, $\Lambda \subset \mathbb{R}^3$ with $(1 + |v|^2) f_1, (1 + |v|^2) f_2 \in L^1(dv)$, $f_1, f_2 \geq 0$ be two distribution functions. Then, we define the function

$$n_k u_k : \mathbb{R}^3 \times \mathbb{R}_0^+ \to \mathbb{R}^3; \quad (x, t) \mapsto (n_k u_k)(x, t) = \int_{\mathbb{R}^3} f_k(x, v, t) v \, dv.$$

If $n_k > 0$, the function $u_k = \frac{n_k u_k}{n_k}$ is called the mean velocity of species k for $k = 1, 2$.

The kinetic energy of a particle is given by $\frac{m_k}{2} |v|^2$, where m_k denotes the mass of the particles of species k. By averaging over all microscopic energies we obtain a macroscopic energy density.

Definition 1.3.4 (Energy density). Let $f_1, f_2 : \Lambda \times \mathbb{R}^3 \times \mathbb{R}_0^+ \to \mathbb{R}$, $\Lambda \subset \mathbb{R}^3$ with $|v|^2 f_1, |v|^2 f_2 \in L^1(dv)$, $f_1, f_2 \geq 0$ be two distribution functions. Then, the function

$$E_k : \mathbb{R}^3 \times \mathbb{R}_0^+ \to \mathbb{R}; \quad (x, t) \mapsto E_k(x, t) = \frac{1}{2} m_k \int_{\mathbb{R}^3} f_k(x, v, t) |v|^2 \, dv,$$

is called the energy density of species k for $k = 1, 2$.

We split the energy E_k into the kinetic energy $\frac{1}{2} m_k n_k |u_k|^2$ and the remainder which has the physical meaning of the internal energy. This defines the internal energy $e_k := E_k - \frac{1}{2} m_k n_k |u_k|^2$. One can compute that e_k can be rearranged to $e_k = \frac{1}{2} m_k \int_{\mathbb{R}^3} |v - u_k|^2 f_k \, dv$. This is true since $\frac{1}{2} |v|^2 - \frac{1}{2} |u_k|^2 = \frac{1}{2} |v - u_k|^2 - |u_k|^2 + v \cdot u_k$, and we see that $\int_{\mathbb{R}^3} |u_k|^2 f_k \, dv = \int_{\mathbb{R}^3} v \cdot u_k f_k \, dv$ using the definitions 1.3.2 and 1.3.3. Therefore

$$\int_{\mathbb{R}^3} \frac{1}{2} |v|^2 f_k \, dv - \int_{\mathbb{R}^3} \frac{1}{2} |u_k|^2 f_k \, dv = \int_{\mathbb{R}^3} \frac{1}{2} |v - u_k|^2 f_k \, dv,$$

which means

$$E_k - \frac{1}{2} m_k n_k |u_k|^2 = \frac{1}{2} m_k \int_{\mathbb{R}^3} |v - u_k|^2 f_k \, dv.$$

So e_k is equal to $\int_{\mathbb{R}^3} \frac{1}{2} m_k |v - u_k|^2 f_k \, dv$.

Moreover, we assume that we are in an ideal gas meaning that the gas is imagined as consisting of moving balls or point particles whose only interactions are perfectly elastic collisions. Furthermore, the gas is dilute enough such that we have only dual interactions. In this case, one observes in experiments that the internal energy is proportional to its temperature. We summarize both facts in the next definition.

Definition 1.3.5 (Internal energy and temperature). Let $f_1, f_2 : \Lambda \times \mathbb{R}^3 \times \mathbb{R}_0^+ \to \mathbb{R}$, $\Lambda \subset \mathbb{R}^3$ with $(1+|v|^2)f_1, (1+|v|^2)f_2 \in L^1(dv), f_1, f_2 \geq 0$ be two distribution functions. Then, the function

$$
e_k : \mathbb{R}^3 \times \mathbb{R}_0^+ \to \mathbb{R}; \quad (x, t) \mapsto e_k(x, t) = E_k(x, t) - \frac{1}{2} m_k n_k |u_k|^2
$$
$$
= \frac{1}{2} m_k \int_{\mathbb{R}^3} f_k(x, v, t)|v - u_k(x, t)|^2 dv,
$$

is called the internal energy of species k for $k = 1, 2$. If we are in an ideal gas and $n_k > 0$, the function $T_k = \frac{2}{3} \frac{e_k}{n_k}$ is called the temperature of species k for $k = 1, 2$.

The integral $e_k = \frac{1}{2} m_k \int_{\mathbb{R}^3} f_k |v - u_k|^2 dv$ can be motivated as follows. The integral has a value different from zero if the distribution of the microscopic velocities has a deviation from the mean velocity. If the gas has internal energy, the particles can use this energy to deviate from the macroscopic velocity. So the internal energy is a measure of the deviation from the mean velocity. If the internal energy of a gas is high, the deviation is high and the other way round.

There are two more moments which have a meaningful physical interpretation being the energy flux Q_k and the pressure tensor $\mathbb{P}_k$.

Definition 1.3.6 (Energy flux). Let $f_1, f_2 : \Lambda \times \mathbb{R}^3 \times \mathbb{R}_0^+ \to \mathbb{R}, \Lambda \subset \mathbb{R}^3$ with $(1 + |v|^2)f_1, (1 + |v|^2)f_2 \in L^1(dv), f_1, f_2 \geq 0$ be two distribution functions. Then, the function

$$
Q_k : \mathbb{R}^3 \times \mathbb{R}_0^+ \to \mathbb{R}^3; \quad (x, t) \mapsto Q_k(x, t) = \frac{1}{2} m_k \int_{\mathbb{R}^3} f_k(x, v, t)|v|^2 v dv,
$$

is called the energy flux of species k for $k = 1, 2$.

Definition 1.3.7 (Pressure tensor). Let $f_1, f_2 : \Lambda \times \mathbb{R}^3 \times \mathbb{R}_0^+ \to \mathbb{R}, \Lambda \subset \mathbb{R}^3$ with $(1 + |v|^2)f_1, (1 + |v|^2)f_2 \in L^1(dv), f_1, f_2 \geq 0$ be two distribution functions. Then, the function

$$
\mathbb{P}_k : \mathbb{R}^3 \times \mathbb{R}_0^+ \to \mathbb{R}^{3\times 3}; (x, t) \mapsto \mathbb{P}_k(x, t) = m_k \int_{\mathbb{R}^3} (v - u_k(x, t)) \otimes (v - u_k(x, t)) f_k(x, v, t) dv,
$$

is called the pressure tensor of species k for $k = 1, 2$.

The pressure tensor contains the friction of a gas. On particles, which are faster than the mean velocity u_k, acts a force, that decelerate the particles and on particles, which are slower than the mean velocity, u_k acts a force which accelerates the particles.

1.3.2 The Boltzmann model

Now, we want to describe the time evolution of the distribution functions f_1 and f_2. Since we have two distribution functions, we expect to need two equations to describe their time evolution. Furthermore, the particles of one species can interact with particles of the same species or with particles of the other species. The following Boltzmann model for two species is also mentioned in [25, 24, 80]. We assume that we have no external forces and only binary interactions. Then the Boltzmann equation governs the time evolution of f_1 and f_2. It reads

$$\partial_t f_1 + v \cdot \nabla_x f_1 = Q_{11}(f_1, f_1) + Q_{12}(f_1, f_2),$$
$$\partial_t f_2 + v \cdot \nabla_x f_2 = Q_{22}(f_2, f_2) + Q_{21}(f_2, f_1),$$

(1.14)

with

$$Q_{11}(f_1, f_1) = \frac{1}{m_1} \int_{\mathbb{R}^3} \int_{S^2} B_{11}(|v - v_1|, \omega)$$
$$\cdot (f_1(x, v', t) f_1(x, v_1', t) - f_1(x, v, t) f_1(x, v_1, t)) d\omega dv_1,$$

$$Q_{22}(f_2, f_2) = \frac{1}{m_2} \int_{\mathbb{R}^3} \int_{S^2} B_{22}(|v - v_1|, \omega)$$
$$\cdot (f_2(x, v', t) f_2(x, v_1', t) - f_2(x, v, t) f_2(x, v_1, t)) d\omega dv_1,$$

$$Q_{12}(f_1, f_2) = \frac{1}{m_1} \int_{\mathbb{R}^3} \int_{S^2} B_{12}(|v - v_1|, \omega)$$
$$\cdot (f_1(x, v', t) f_2(x, v_1', t) - f_1(x, v, t) f_2(x, v_1, t)) d\omega dv_1,$$

and

$$Q_{21}(f_2, f_1) = \frac{1}{m_2} \int_{\mathbb{R}^3} \int_{S^2} B_{21}(|v - v_1|, \omega)$$
$$\cdot (f_2(x, v', t) f_1(x, v_1', t) - f_2(x, v, t) f_1(x, v_1, t)) d\omega dv_1.$$

The vector ω is the unit vector on S^2 defined in corollary 1.2.5, v_1' and v' are the velocities of two particles after an interaction given by the formula proven in corollary 1.2.5 with velocities v_1 and v in the beginning. The non-negative functions $B_{jk}(|v - v_1|, \omega)$ for $j, k = 1, 2$ are called the collision kernels and contain the properties of the interaction between the particles of the gas. Precisely it is the norm of the relative velocity of v and v_1 times the differential cross section which will be explained in a moment.

Let us motivate the physical meaning of this equation. First, let us motivate the transport part on the left-hand side of the Boltzmann equation. For this, neglect for the moment the interaction between particles. We consider the time evolution of the individual particles. According to Newton's equations (1.5), the particles travel at a constant velocity, along a straight line in the absence of interactions, and therefore the distribution function is constant along the characteristic lines

$$\frac{dx(t)}{dt} = v(t), \quad \frac{dv(t)}{dt} = 0.$$

The corresponding partial differential equation to this characteristic lines is the transport equation

$$\partial_t f_k + v \cdot \nabla_x f_k = 0,$$

see example 1.1.1 in section 1.1. This motivates the left-hand side of the Boltzmann equations. Next, let us motivate the collision operators on the right-hand side.

The operators $Q_{kj}, k, j = 1, 2$ represent the change of the distribution functions due to interactions of the particles. In interactions the distribution functions change due to a change of the velocities in an interaction according to the formula proven in corollary 1.2.5. So Q_{kj} are operators acting on the velocity variables only. The operators Q_{11} and Q_{22} describe the interactions of particles of a species with itself whereas Q_{12} and Q_{21} describe the interaction of particles of a species with the other one.

We assume that the particles interact via binary interactions. From the physical point of view this is true if the gas is dilute enough that the effect of interactions involving more than two particles can be neglected. This is reflected in the quadratic structure of the collision term.

We assume that our collisions are elastic which means that the conservation properties 1.2.1, 1.2.2 and 1.2.4 are satisfied. Therefore we describe the change of the velocities with the help of the formula given in corollary 1.2.5.

Inside the integrals in Q_{kj} we describe the influence of an interaction by the non-negative collision kernel B_{kj}. It is the norm of the relative velocity of v and v_1 times the differential cross section. The differential cross section describes the probability that a collision occurs, see section 1 in [20]. Since the probability that a particle of species 1 interacts with a particle of species 2 and the probability that a particle of species 2 interacts with a particle of species 1 is the same, we have the equality $B_{12} = B_{21}$.

The operator Q_{kj} can be split into a gain and a loss term. The loss term counts all interactions in which a given particle with velocity v will meet another particle with velocity v_1. As a result of such an interactions, this particle will change its velocity and this will make less particles with velocity v. On the other hand, each time particles interact with velocities v' and v'_1, the particle with velocity v' will get v as a new velocity after the collision, and this will make more particles with velocity v: this is the meaning of the gain term. This is illustrated in figure 1.2.

The appearance of the products $f_k(x, v, t) f_j(x, v_1, t)$ and $f_k(x, v', t) f_j(x, v'_1, t)$ is a consequence of the so-called chaos assumption: First of all, we expect that the probability of a collision depends on a joint distribution function of two particles $f_{12}(x, v, v_1, t)$. Then the chaos assumption is the following. We assume that velocities of the two particles which are about to collide are uncorrelated. This means that if we randomly take two particles at position x, which have not collided yet, then the joint distribution of their velocities will be given by the tensor product $f_1(x, v, t) f_2(x, v_1, t)$.

Last, the fact that the variables t, x appear only as parameters reflects the assumption that collisions are localized in space and time.

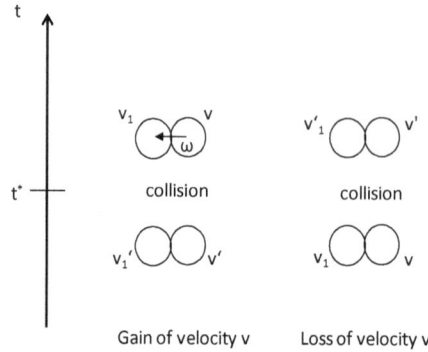

Figure 1.2: Gain and loss of velocity. On the left, two particles with velocities v'_1 and v' collide at time t^* and change its velocities into v_1 and v. On the right, two particles with velocities v and v_1 collide at time t^* and change its velocities into v' and v'_1. The right case describes a loss of velocity v, the left one a gain of velocity v.

1.3.3 Fundamental properties of the Boltzmann equation

The Boltzmann equation has two main properties. The first one is the following. In interactions the number of particles, the total momentum and the total energy are conserved. The second property is that one can show that the Boltzmann equation admits an entropy. An entropy is a function which always decreases in time. In this case it will turn out to correspond to the notion of entropy used in physics. The proofs for the collision operators Q_{11} and Q_{22} describing the interactions of a species with itself can be found in [46, 47, 83]. In this case we just state the results and show only the proofs for the collision operators describing the interactions of a species with the other one. These proofs are extended versions of the proofs for the single collision operators given in [46, 47, 83].

Conservation properties

We start with the conservation properties namely conservation of the number of particles, conservation of momentum and conservation of energy in interactions of a species with itself. This concerns the collision operators Q_{11} and Q_{22}.

Theorem 1.3.1 (Conservation properties in interactions of a species with itself). *For all $f \in L^\infty(dv)$ with compact support or decaying fast enough at infinity and all $i = 1, 2, 3,\ k = 1, 2$, we have*

$$\int_{\mathbb{R}^3} Q_{kk}(f, f)\, dv = \int_{\mathbb{R}^3} Q_{kk}(f, f) v_i\, dv = \int_{\mathbb{R}^3} Q_{kk}(f, f)|v|^2 dv = 0.$$

The proof is given in [46, 47, 83] and is a special case of the proof shown next for the collision operators describing collisions of a species with the other one.

Theorem 1.3.2 (Conservation properties in interactions of a species with the other one). *For all $f_1, f_2 \in L^\infty(dv)$ with compact support or decaying fast enough at infinity, we have the following conservation properties.*
We have conservation of the number of particles in every interaction

$$\int_{\mathbb{R}^3} Q_{12}(f_1, f_2)dv = \int_{\mathbb{R}^3} Q_{21}(f_2, f_1)dv = 0. \tag{1.15}$$

We have conservation of total momentum

$$\int_{\mathbb{R}^3} (m_1 v Q_{12}(f_1, f_2) + m_2 v Q_{21}(f_2, f_1))dv = 0. \tag{1.16}$$

We have conservation of total energy

$$\int_{\mathbb{R}^3} (m_1 |v|^2 Q_{12}(f_1, f_2) + m_2 |v|^2 Q_{21}(f_2, f_1))dv = 0. \tag{1.17}$$

Proof. Let h and g be arbitrary continuous functions of the velocity v.
First, we consider the term $\int_{\mathbb{R}^3} h(v) Q_{12}(f_1, f_2)(v)dv$. Then, we will consider the term $\int_{\mathbb{R}^3} g(v) Q_{21}(f_2, f_1)(v)dv$. The idea is to rewrite these two expressions in another way such that we can directly see that the integral is zero if we choose h and g as 1, the microscopic momentum $m_1 v$ and $m_2 v$, respectively and the microscopic energies $\frac{1}{2}m_1|v|^2$ and $\frac{1}{2}m_2|v|^2$, respectively. We start with $\int_{\mathbb{R}^3} h(v)Q_{12}(f_1, f_2)(v)dv$. From the definition of $Q_{12}(f_1, f_2)$, we see

$$\int_{\mathbb{R}^3} h(v)Q_{12}(f_1, f_2)(v)dv = \int_{\mathbb{R}^3 \times \mathbb{R}^3 \times S^2} \frac{1}{m_1} B_{12}(|v - v_1|, \omega)h(v)$$
$$\cdot [f_1(v')f_2(v_1') - f_1(v)f_2(v_1)]d\omega dv dv_1. \tag{1.18}$$

Now we exchange the notation (v, v') and (v_1, v_1') and m_1 and m_2 on the right-hand side, so we get

$$\int_{\mathbb{R}^3} h(v)Q_{12}(f_1, f_2)(v)dv = \int_{\mathbb{R}^3 \times \mathbb{R}^3 \times S^2} \frac{1}{m_2} B_{21}(|v - v_1|, \omega)h(v_1)$$
$$\cdot [f_1(v_1')f_2(v') - f_2(v)f_1(v_1)]d\omega dv dv_1. \tag{1.19}$$

This is possible due to the following reason. We can say that we take a particle with velocity v and mass m_1 which interacts with a particle with velocity v_1 and mass m_2. It is equivalent if we say that we take a particle with velocity v_1 and mass m_2 which interacts with a particle with velocity v and mass m_1. Now we exchange (v, v_1) and (v', v_1') on the right-hand side of (1.18) and (1.19). This is possible since the operator T_ω is self-inverse as stated in remark 1.2.1.

$$\int_{\mathbb{R}^3} h(v)Q_{12}(f_1, f_2)(v)dv = \int_{\mathbb{R}^3 \times \mathbb{R}^3 \times S^2} \frac{1}{m_1} B_{12}(|v' - v_1'|, \omega)(-h(v'))$$
$$\cdot [f_1(v')f_2(v_1') - f_1(v)f_2(v_1)]d\omega dv' dv_1', \tag{1.20}$$

$$\int_{\mathbb{R}^3} h(v)Q_{12}(f_1, f_2)(v)dv = \int_{\mathbb{R}^3 \times \mathbb{R}^3 \times S^2} \frac{1}{m_2} B_{21}(|v' - v_1'|, \omega)(-h(v_1'))$$
$$\cdot [f_2(v')f_1(v_1') - f_2(v)f_1(v_1)]d\omega dv' dv_1'. \tag{1.21}$$

Using the property (1.13) stated in remark 1.2.1, we obtain

$$\int_{\mathbb{R}^3} h(v)Q_{12}(f_1, f_2)(v)dv = \int_{\mathbb{R}^3 \times \mathbb{R}^3 \times S^2} \frac{1}{m_1} B_{12}(|v - v_1|, \omega)(-h(v'))$$
$$\cdot [f_1(v')f_2(v_1') - f_1(v)f_2(v_1)]d\omega dv' dv_1', \tag{1.22}$$

$$\int_{\mathbb{R}^3} h(v)Q_{12}(f_1, f_2)(v)dv = \int_{\mathbb{R}^3 \times \mathbb{R}^3 \times S^2} \frac{1}{m_2} B_{21}(|v - v_1|, \omega)(-h(v_1'))$$
$$\cdot [f_2(v')f_1(v_1') - f_2(v)f_1(v_1)]d\omega dv' dv_1'. \tag{1.23}$$

According to remark 1.2.5, the Jacobian of the transformation T_ω is 1, so we can replace $dv'dv_1'$ by $dvdv_1$ and obtain

$$\int_{\mathbb{R}^3} h(v)Q_{12}(f_1, f_2)(v)dv = \int_{\mathbb{R}^3 \times \mathbb{R}^3 \times S^2} \frac{1}{m_1} B_{12}(|v - v_1|, \omega)(-h(v'))$$
$$\cdot [f_1(v')f_2(v_1') - f_1(v)f_2(v_1)]d\omega dvdv_1, \tag{1.24}$$

$$\int_{\mathbb{R}^3} h(v)Q_{12}(f_1, f_2)(v)dv = \int_{\mathbb{R}^3 \times \mathbb{R}^3 \times S^2} \frac{1}{m_2} B_{21}(|v - v_1|, \omega)(-h(v_1'))$$
$$\cdot [f_2(v')f_1(v_1') - f_2(v)f_1(v_1)]d\omega dvdv_1. \tag{1.25}$$

We add (1.18), (1.19), (1.24) and (1.25) and use $B_{12} = B_{21}$.

$$\int_{\mathbb{R}^3} h(v)Q_{12}(f_1, f_2)(v)dv$$
$$= \frac{1}{4} \int_{\mathbb{R}^3} ([(h(v) - h(v'))Q_{12}(f_1, f_2)] + [(h(v_1) - h(v_1'))Q_{21}(f_2, f_1)])dv. \tag{1.26}$$

If we do the same for the integral $\int g(v)Q_{21}(f_2, f_1)dv$, we obtain

$$\int_{\mathbb{R}^3} g(v)Q_{21}(f_2, f_1)(v)dv$$
$$= \frac{1}{4} \int_{\mathbb{R}^3} ([(g(v_1) - g(v_1'))Q_{12}(f_1, f_2)] + [(g(v) - g(v'))Q_{21}(f_2, f_1)])dv. \tag{1.27}$$

Again an additional exchange of primed and unprimed variables in the parts with primed velocities in the arguments of f_1, f_2 in Q_{12}, Q_{21} leads to

$$\int_{\mathbb{R}^3} h(v)Q_{12}(f_1, f_2)(v)dv = \frac{1}{2} \int_{\mathbb{R}^3 \times \mathbb{R}^3 \times S^2} B_{12}(|v - v_1|, \omega)$$
$$\cdot [\frac{1}{m_1} f_1(v)f_2(v_1)(h(v') - h(v)) + \frac{1}{m_2} f_2(v)f_1(v_1)(h(v_1') - h(v_1))]d\omega dvdv_1. \tag{1.28}$$

As a consequence we see that $\int_{\mathbb{R}^3} h(v)Q_{12}(f_1, f_2)(v)dv = 0$ if h satisfies the equation

$$h(v) = h(v') \quad \text{and} \quad h(v_1) = h(v_1').$$

Obviously the function $h(v) = 1$ is a solution. So we have conservation of the number of particles of species 1 in interactions with species 2.
We can do the same for $\int_{\mathbb{R}^3} g(v)Q_{21}(f_2, f_1)dv$ and get

$$\int_{\mathbb{R}^3} g(v)Q_{21}(f_2, f_1)(v)dv = \frac{1}{2} \int_{\mathbb{R}^3 \times \mathbb{R}^3 \times S^2} B_{12}(|v - v_1|, \omega)$$

$$\cdot [\frac{1}{m_2} f_2(v)f_1(v_1)(g(v') - g(v)) + \frac{1}{m_1} f_1(v)f_2(v)(g(v_1') - g(v_1))dvdv_1 d\omega. \tag{1.29}$$

If we add (1.28) and (1.29), we see

$$\int_{\mathbb{R}^3} (h(v)Q_{12}(f_1, f_2) + g(v)Q_{21}(f_2, f_1)(v))dv = \frac{1}{2} \int_{\mathbb{R}^3 \times \mathbb{R}^3 \times S^2} B_{12}(|v - v_1|, \omega)$$

$$\cdot [\frac{1}{m_2} f_2(v)f_1(v_1)(h(v_1') - h(v_1) + g(v') - g(v))$$

$$+ \frac{1}{m_1} f_1(v)f_2(v_1)(h(v') - h(v) + g(v_1') - g(v_1))]dvdv_1 d\omega, \tag{1.30}$$

and observe that the choice of $h(v) = m_1 v$ or $h(v) = m_1|v|^2$ and $g(v) = m_2 v$ or $g(v) = m_2|v|^2$ leads to conservation of momentum (consequence 1.2.1) and conservation of energy (consequence 1.2.2). So the properties of conservation of total momentum and total energy are also satisfied. $\qquad \square$

The names conservation of the number of particles, conservation of total momentum and conservation of total energy will be motivated in the following theorem.

Theorem 1.3.3 (Macroscopic equations). *If $f_1, f_2 \in L^\infty(dv)$ decay fast enough to zero in the v variable and are a solution to (1.14) in the sense of distributions, they satisfy the following local macroscopic equations.*

$$\partial_t n_1 + \nabla_x \cdot (n_1 u_1) = 0,$$

$$\partial_t n_2 + \nabla_x \cdot (n_2 u_2) = 0,$$

$$\partial_t(m_1 n_1 u_1) + \nabla_x \cdot \mathbb{P}_1 + \nabla_x \cdot (m_1 n_1 u_1 \otimes u_1) = \int m_1 Q_{12}(f_1, f_2)vdv,$$

$$\partial_t(m_2 n_2 u_2) + \nabla_x \cdot \mathbb{P}_2 + \nabla_x \cdot (m_2 n_2 u_2 \otimes u_2) = \int m_2 Q_{21}(f_2, f_1)vdv,$$

$$\partial_t \left(\frac{m_1}{2} n_1 |u_1|^2 + \frac{3}{2} n_1 T_1 \right) + \nabla_x \cdot Q_1 = \int Q_{12}(f_1, f_2) \frac{m_1}{2} |v|^2 dv,$$

$$\partial_t \left(\frac{m_2}{2} n_2 |u_2|^2 + \frac{3}{2} n_2 T_2 \right) + \nabla_x \cdot Q_2 = \int Q_{21}(f_2, f_1) \frac{m_2}{2} |v|^2 dv.$$

Proof. If we integrate the equation (1.14) for species 1 with respect to v and use the conservation property (1.15) and theorem 1.3.1, we get:

$$\int \partial_t f_1(x, v, t) dv + \int \nabla_x \cdot (v f_1) dv = 0.$$

This is equivalent to

$$\partial_t n_1 + \nabla_x \cdot (n_1 u_1) = 0,$$

if we use the definitions of the number density and the mean velocity given by the definitions 1.3.2 and 1.3.3, respectively. We can do the same with the equation for f_2. In this case we get

$$\partial_t n_2 + \nabla_x \cdot (n_2 u_2) = 0.$$

Multiplying the equation (1.14) for species 1 by $m_1 v$ and integrating it with respect to the velocity v, leads to

$$m_1 \int v \partial_t f_1 dv + m_1 \int v \nabla_x \cdot (v f_1) dv = \int m_1 Q_{12}(f_1, f_2) v dv.$$

In the first and in the second term we formally exchange derivative and integration and obtain

$$m_1 \partial_t (n_1 u_1) + \nabla_x \cdot \int m_1 v \otimes v f_1 dv.$$

So the equation is equivalent to

$$m_1 \partial_t (n_1 u_2) + \nabla_x \cdot \int m_1 v \otimes v f_1 dv = \int Q_{12}(f_1, f_2) v dv.$$

With the definition of the pressure tensor given by definition 1.3.7, the second term turns into

$$\nabla_x \cdot \mathbb{P}_1 + \nabla_x \cdot (m_1 n_1 u_1 \otimes u_1).$$

So all in all, we get

$$\partial_t (m_1 n_1 u_1) + \nabla_x \cdot \mathbb{P}_1 + \nabla_x \cdot (m_1 n_1 u_1 \otimes u_1) = \int m_1 Q_{12}(f_1, f_2) v dv.$$

We can do the same with f_2 and obtain

$$\partial_t (m_2 n_2 u_2) + \nabla_x \cdot \mathbb{P}_2 + \nabla_x \cdot (m_2 n_2 u_2 \otimes u_2) = \int m_2 Q_{21}(f_2, f_1) v dv.$$

Multiplying the equation (1.14) for species 1 by $\frac{m_1}{2}|v|^2$ and integrating it with respect to v leads to

$$\frac{m_1}{2} \int |v|^2 \partial_t f_1 dv + \frac{m_1}{2} \int |v|^2 \nabla_x \cdot (v f_1) dv = \int Q_{12}(f_1, f_2) \frac{m_2}{2} |v|^2 dv. \qquad (1.31)$$

In the first two terms we formally exchange derivative and integration to obtain

$$\partial_t \left(\frac{m_1}{2} n_1 |u_1|^2 + \frac{3}{2} n_1 T_1 \right) + \nabla_x \cdot \int m_1 v |v|^2 f_1 dv.$$

So the equation (1.31) is equivalent to

$$\partial_t \left(\frac{m_1}{2} n_1 |u_1|^2 + \frac{3}{2} n_1 T_1 \right) + \nabla_x \cdot \int m_1 v |v|^2 f_1 dv = \int Q_{12}(f_1, f_2) \frac{m_1}{2} |v|^2 dv.$$

With the definition of the energy flux given by the definition 1.3.6 we get

$$\partial_t \left(\frac{m_1}{2} n_1 |u_1|^2 + \frac{3}{2} n_1 T_1 \right) + \nabla_x \cdot Q_1 = \int Q_{12}(f_1, f_2) \frac{m_1}{2} |v|^2 dv.$$

So all in all, we get the system of partial differential equations from theorem 1.3.3. □

Corollary 1.3.4. *If $f_1, f_2 \in L^\infty(dv)$ decay fast enough to zero in the v variable and are a solution to (1.14) in the sense of distributions, they satisfy the following local macroscopic conservation laws.*

$$\partial_t n_1 + \nabla_x \cdot (n_1 u_1) = 0,$$
$$\partial_t n_2 + \nabla_x \cdot (n_2 u_2) = 0,$$
$$\partial_t (m_1 n_1 u_1 + m_2 n_2 u_2) + \nabla_x \cdot (\mathbb{P}_1 + \mathbb{P}_2) + \nabla_x \cdot (m_1 n_1 u_1 \otimes u_1 + m_2 n_2 u_2 \otimes u_2) = 0,$$
$$\partial_t (\frac{m_1}{2} n_1 |u_1|^2 + \frac{m_2}{2} n_2 |u_2|^2 + \frac{3}{2} n_1 T_1 + \frac{3}{2} n_2 T_2) + \nabla_x \cdot (Q_1 + Q_2) = 0.$$

Proof. We take the equations from theorem 1.3.3 and add the third and the fourth one, and the fifth and the sixth one. Then we use the conservation properties (1.16) and (1.17). □

The system in theorem 1.3.3 describes conservation of the number of particles and balance laws for the momentum and the energy. The system in corollary 1.3.4 describes conservation of the number of particles, total momentum and total energy. Note that both systems of macroscopic equations are not closed since we have more unknowns than equations.

The H-theorem, entropy inequality and equilibrium

Another property of the Boltzmann equation is that it admits an entropy. In order to prove this we need the following three lemmas.

Lemma 1.3.5. *Let $\phi > 0$ be a measurable function such that $\int_{\mathbb{R}^3} (1 + |v|^2) \phi(v) dv < \infty$. If*

$$\phi(v) \phi(v_1) = \phi(v') \phi(v_1'), \tag{1.32}$$

for a.e. $(v, v_1, \omega) \in \mathbb{R}^3 \times \mathbb{R}^3 \times S^2$, then ϕ is a Maxwell distribution.

The proof is given in [46, 47] and is a special case of the proof shown next with two different functions ϕ_1 and ϕ_2.

Lemma 1.3.6. *Let $\phi_1, \phi_2 > 0$ be two measurable functions such that $\int_{\mathbb{R}^3} (1+|v|^2)\phi_1(v)dv < \infty$ and $\int_{\mathbb{R}^3}(1+|v|^2)\phi_2(v)dv < \infty$. If*

$$\phi_1(v)\phi_2(v_1) = \phi_1(v')\phi_2(v_1'), \tag{1.33}$$

for a.e. $(v, v_1, \omega) \in \mathbb{R}^3 \times \mathbb{R}^3 \times S^2$, then ϕ_1 and ϕ_2 are Maxwell distributions with equal mean velocity and temperature.

Before we prove this lemma, let us illustrate the shape of a Maxwell distribution (see figure 1.3).

Definition 1.3.8 (Maxwell distribution). A Maxwell distribution is a distribution of the form

$$M(v) = C \exp\left(-\frac{|v-U|^2}{A}\right),$$

where $C, A \in \mathbb{R}^+, U \in \mathbb{R}^3$. If C, A and U are functions of x and t, $M(x, v, t)$ is called a local Maxwell distribution.

If the integrals from definition 1.3.2, 1.3.3 and 1.3.4 of the Maxwell distribution coincide with the integrals from 1.3.2, 1.3.3 and 1.3.4 of the distribution function itself, we obtain the form

$$M_k(v) = \frac{n_k}{\sqrt{2\pi T_k/m_k}^3} \exp\left(-\frac{|v-u_k|^2}{2T_k/m_k}\right). \tag{1.34}$$

So lemma 1.3.6 tells that if the condition (1.33) for a.e. $(v, v_1, \omega) \in \mathbb{R}^3 \times \mathbb{R}^3 \times S^2$ is satisfied then ϕ_1, ϕ_2 are of the form

$$\phi_1(v) = \frac{n_1}{\sqrt{2\pi T/m_1}^3} \exp\left(-\frac{|v-u|^2}{2T/m_1}\right), \quad \phi_2(v) = \frac{n_2}{\sqrt{2\pi T/m_2}^3} \exp\left(-\frac{|v-u|^2}{2T/m_2}\right),$$

with a common value u and T still depending on x and t. According to the definitions 1.3.3 and 1.3.4, this has the physical meaning of a common mean velocity and temperature. First, we will prove lemma 1.3.6, and then we will conclude that the condition (1.33) characterizes the equilibrium. So finally, we will see that Maxwell distributions with common velocity and temperature are the expected distributions in thermodynamic equilibrium.

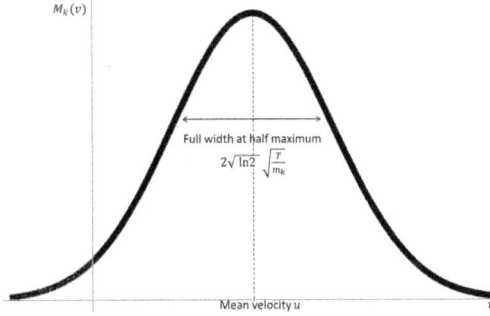

Figure 1.3: Maxwell distribution. In this figure you can see a Maxwell distribution of the velocities for fixed x and t. The centre of this distribution is at the mean velocity, meaning $v = u(x,t)$ and the width of the distribution is related to the temperature $T(x,t)$.

Proof of lemma 1.3.6. After a translation and a multiplication by a constant, ϕ_2 is supposed to satisfy

$$\int_{\mathbb{R}^3} \phi_2(v)dv = 1, \int_{\mathbb{R}^3} v\phi_2(v)dv = 0. \tag{1.35}$$

Now we take the Fourier transform of

$$\phi_1(v)\phi_2(v_1) = \phi_1(v')\phi_2(v_1'),$$

which leads for almost every $\omega \in S^2$ to

$$\hat{\phi}_1(\xi)\hat{\phi}_2(\xi_1) = \int_{\mathbb{R}^3}\int_{\mathbb{R}^3} \phi_1(v')\phi_2(v_1')e^{-i\xi\cdot v - i\xi_1\cdot v_1}dvdv_1.$$

Now we change coordinates by $(v,v_1) \mapsto (v',v_1')$ as in the proof of the conservation properties 1.15, 1.16 and 1.17. For this, we take the transformation in corollary 1.2.5 and obtain

$$\hat{\phi}_1(\xi)\hat{\phi}_2(\xi_1) = \int_{\mathbb{R}^3}\int_{\mathbb{R}^3} \phi_1(v)\phi_2(v_1)e^{-i\xi\cdot v - i\xi_1\cdot v_1}e^{i\left(\frac{2m_2}{m_1+m_2}\xi - \frac{2m_1}{m_1+m_2}\xi_1\right)\cdot\omega[(v-v_1)\cdot\omega]}dvdv_1.$$

$$\tag{1.36}$$

The change of the coordinates is possible due to remark 1.2.5. Let ξ, ξ_1 be fixed. Now we differentiate the equality (1.36) with respect to ω at any ω_* orthogonal to $m_2\xi - m_1\xi_1$. This leads to

$$0 = \int_{\mathbb{R}^3}\int_{\mathbb{R}^3} (m_2\xi - m_1\xi_1)\phi_1(v)\phi_2(v_1)e^{-i\xi\cdot v - i\xi_1\cdot v_1}(v-v_1)\cdot\omega_*dvdv_1, \tag{1.37}$$

using $(m_2\xi - m_1\xi_1)\cdot\omega_* = 0$ in the exponential function. Since

$$\nabla_\xi\hat{\phi}_1(\xi)\hat{\phi}_2(\xi_1) = \int_{\mathbb{R}^3}\int_{\mathbb{R}^3} \phi_1(v)\phi_2(v_1)e^{-i\xi\cdot v - i\xi_1\cdot v_1}(-iv)dvdv_1,$$

and

$$\nabla_{\xi_1}\hat{\phi}_1(\xi)\hat{\phi}_2(\xi_1) = \int_{\mathbb{R}^3}\int_{\mathbb{R}^3} \phi_1(v)\phi_2(v_1)e^{-i\xi\cdot v - i\xi_1\cdot v_1}(-iv_1)dvdv_1,$$

for any ω_* orthogonal to $m_2\xi - m_1\xi_1$ and therefore

$$(\nabla_\xi - \nabla_{\xi_1})\hat{\phi}_1(\xi)\hat{\phi}_2(\xi_1) = \int_{\mathbb{R}^3}\int_{\mathbb{R}^3}\phi_1(v)\phi_2(v_1)e^{-i\xi\cdot v - i\xi_1\cdot v_1}i(v_1 - v)dvdv_1, \quad (1.38)$$

for any ω_* orthogonal to $m_2\xi - m_1\xi_1$. Now, we use Grassmann's identity (see appendix A.1) to deduce

$$\omega_* \times \left[(m_2\xi - m_1\xi_1) \times \left((\nabla_\xi - \nabla_{\xi_1})\hat{\phi}_1(\xi)\hat{\phi}_2(\xi_1)\right)\right]$$
$$= \omega_* \cdot \left((\nabla_\xi - \nabla_{\xi_1})\hat{\phi}_1(\xi)\hat{\phi}_2(\xi_1)\right)(m_2\xi - m_1\xi_1)$$
$$- \omega_* \cdot (m_2\xi - m_1\xi_1)(\nabla_\xi - \nabla_{\xi_1})\hat{\phi}_1(\xi)\hat{\phi}_2(\xi_1) = 0.$$

The term $\omega_* \cdot (m_2\xi - m_1\xi_1)(\nabla_\xi - \nabla_{\xi_1})\hat{\phi}_1(\xi)\hat{\phi}_2(\xi_1)$ is equal to zero since ω_* is orthogonal to $m_1\xi - m_1\xi_1$. The term $\omega_* \cdot \left((\nabla_\xi - \nabla_{\xi_1})\hat{\phi}_1(\xi)\hat{\phi}_2(\xi_1)\right)(m_2\xi - m_1\xi_1)$ is equal to zero since it is equal to $\omega_* \cdot (1.38)$ in the direction of $(m_2\xi - m_1\xi_1)$ which is equal to ω_* times the right-hand side of (1.37). According to the left-hand side of (1.37) this is equal to zero. Now either ω_* is parallel to $(m_1\xi - m_2\xi_1) \times ((\nabla_\xi - \nabla_{\xi_1})\hat{\phi}_1(\xi)\hat{\phi}_2(\xi_1))$ or $m_2\xi - m_1\xi_1$ is parallel to $(\nabla_\xi - \nabla_{\xi_1})\hat{\phi}_1(\xi)\hat{\phi}_2(\xi_1)$. But since ω_* is an arbitrary orthogonal vector to $m_2\xi - m_1\xi_1$, it can also be non-orthogonal to $(m_1\xi - m_2\xi_1) \times ((\nabla_\xi - \nabla_{\xi_1})\hat{\phi}_1(\xi)\hat{\phi}_2(\xi_1))$. So

$$(\nabla_\xi - \nabla_{\xi_1})\hat{\phi}_1(\xi)\hat{\phi}_2(\xi_1) \text{ is linearly dependent to } m_2\xi - m_1\xi_1 \text{ for all } \xi, \xi_1 \in \mathbb{R}^3.$$
$$(1.39)$$

Choose $\xi \neq 0$ and $\xi_1 = 0$. Then according to the definition of the Fourier transform, we have

$$\nabla_\xi\hat{\phi}_1(\xi) \cdot \hat{\phi}_2(0) - \hat{\phi}_1(\xi)\nabla_{\xi_1}\hat{\phi}_2(0) = \nabla_\xi\hat{\phi}_1(\xi)\int_{\mathbb{R}^3}\phi_2(v)dv - \hat{\phi}_1(\xi) \cdot \int_{\mathbb{R}^3}(-iv)\phi_2(v)dv.$$

With the normalization (1.35), we get

$$\nabla_\xi\hat{\phi}_1(\xi) \cdot \hat{\phi}_2(0) - \hat{\phi}_1(\xi)\nabla_{\xi_1}\hat{\phi}_2(0) = \nabla_\xi\hat{\phi}_1(\xi).$$

Together with (1.39), we conclude that $\nabla_\xi\hat{\phi}_1(\xi)$ is linearly dependent to ξ.

This means that $\hat{\phi}_1$ is of the form

$$\hat{\phi}_1(\xi) = \psi_1(|\xi|^2).$$

We replace $\hat{\phi}_1$ by ψ_1 in (1.39) and observe that

$$\psi_1'(|\xi|^2)\hat{\phi}_2(\xi_1)\xi - \psi_1(|\xi|^2)\nabla_{\xi_1}\hat{\phi}_2(\xi_1) \text{ is linearly dependent to } m_2\xi - m_1\xi_1.$$

Choosing $\xi = 0$ and using $\psi_1(|\xi|^2) = \phi_1(\xi) > 0$, we get $\nabla_{\xi_1}\hat{\phi}_2(\xi_1)$ is linearly dependent to ξ_1, and therefore $\hat{\phi}_2$ is of the form

$$\hat{\phi}_2(\xi_1) = \psi_2(|\xi_1|^2).$$

So again we get

$\psi_1'(|\xi|^2)\psi_2(|\xi_1|^2)\xi - \psi_1(|\xi|^2)\psi_2'(|\xi_1|^2)\xi_1$ is linearly dependent to $m_2\xi - m_1\xi_1$.

This is equivalent to

$\frac{\psi_1'(|\xi|^2)\psi_2(|\xi_1|^2)}{m_2}m_2\xi - \frac{\psi_1(|\xi|^2)\psi_2'(|\xi_1|^2)}{m_1}m_1\xi_1$ is linearly dependent to $m_2\xi - m_1\xi_1$.

Whenever ξ and ξ_1 are not linearly dependent, i.e. for a dense subset of all $\xi, \xi_1 \in \mathbb{R}^3$, we then have

$$m_1\psi_1'(|\xi|^2)\psi_2(|\xi_1|^2) = m_2\psi_1(|\xi|^2)\psi_2'(|\xi_1|^2).$$

This is equivalent to the equality

$$m_1\frac{\psi_1'(|\xi|^2)}{\psi_1(|\xi|^2)} = m_2\frac{\psi_2'(|\xi_1|^2)}{\psi_2(|\xi_1|^2)}.$$

Since the left-hand side is independent of ξ_1 and the right-hand side is independent of ξ, both sides are equal to a constant and we get that ψ_1 and ψ_2 are of the form

$$\psi_1(r) = e^{-\alpha r},$$

$$\psi_2(r) = e^{-\beta r},$$

for $r \in \mathbb{R}_0^+$ and $\alpha = \frac{m_2}{m_1}\beta$. This means that both ϕ_1 and ϕ_2 are Maxwell distributions. If we do a re-translation in ϕ_2 to get $\int_{\mathbb{R}^3} v\phi_2(v)dv = u_2$, the function ϕ_1 is then transformed in the same way, so they have the same velocity. The factors α and β are inverted by doing a Fourier transformation, so for ϕ_2 and ϕ_1 we have the relation $\alpha = \frac{m_1}{m_2}\beta$, this means $\frac{m_1}{T_1} = \frac{m_1}{m_2}\frac{m_2}{T_2}$ for $\alpha = \frac{m_1}{T_1}$ from which we can deduce $T_1 = T_2$. □

Lemma 1.3.7. *Assume $y, z \in \mathbb{R}^n$. Then we have the following inequality*

$$(z - y)\ln\frac{y}{z} \le 0,$$

with equality if and only if $z = y$.

Proof. We consider the term

$$(z - y)\ln\frac{y}{z}.$$

For $z > y$, the first factor is positive, but the logarithm is negative. For $y > z$, the logarithm is positive but the first factor is negative. In both cases the whole term is negative. We have equality if and only if $z = y$. □

With these three lemmas we can prove the following inequalities which will result in an inequality for an entropy.

Theorem 1.3.8 (H-theorem of the single collision operator). *Let $f \in L^\infty(dv)$ be a function decaying fast enough at infinity, $f \geq 0$, and $\int(1+|v|^2)f(v)dv < \infty$, then*

$$\int \ln f \, Q_{kk}(f, f)dv \leq 0 \quad for \quad k = 1, 2,$$

with equality if and only if f is a Maxwell distribution.

The proof is given in [46, 47, 83] and is a special case of the proof shown next for the collision operators describing collisions of one species with the other one.

Theorem 1.3.9 (H-theorem of the mixture collision operators). *Let $f_1, f_2 \in L^\infty(dv)$ be two functions decaying fast enough at infinity, $f_1, f_2 \geq 0$, and $\int(1+|v|^2)f_1(v)dv < \infty$ and $\int(1+|v|^2)f_2(v)dv < \infty$, then*

$$\int \left(\ln f_1 \, Q_{12}(f_1, f_2) + \ln f_2 \, Q_{21}(f_2, f_1) \right) dv \leq 0,$$

with equality if and only if f_1 and f_2 are Maxwell distributions with equal mean velocity and temperature.

Proof. We add equations (1.26) and (1.27), insert the definitions of Q_{12} and Q_{21} and choose $h(v) = \ln f_1(v)$ and $g(v) = \ln f_2(v)$ from the proof of properties 1.15, 1.16 and 1.17 and get

$$\int_{\mathbb{R}^3} (h(v)Q_{12}(f_1, f_2) + g(v)Q_{21}(f_2, f_1)) \, dv = \frac{1}{2} \int_{\mathbb{R}^3 \times \mathbb{R}^3 \times S^2} B_{12}(|v - v_1|, \omega)$$
$$\cdot \left[\frac{1}{m_2} (f_2(v')f_1(v_1') - f_2(v)f_1(v_1)) \ln \frac{f_1(v_1)f_2(v)}{f_1(v_1')f_2(v')} \right. \qquad (1.40)$$
$$\left. + \frac{1}{m_1} (f_1(v')f_2(v_1') - f_1(v)f_2(v_1)) \ln \frac{f_1(v)f_2(v_1)}{f_1(v')f_2(v_1')} \right] dv dv_1 d\omega.$$

Now, we apply lemma 1.3.7 choosing $y = f_1(v_1)f_2(v)$ and $z = f_2(v')f_1(v_1')$ in the first term of the sum and $y = f_1(v)f_2(v_1)$ and $z = f_2(v_1')f_1(v')$ in the second term of the sum. Then, we can deduce $\int_{\mathbb{R}^3} (h(v)Q_{12}(f_1, f_2) + g(v)Q_{21}(f_2, f_1)(v)) \, dv \leq 0$, because B_{12} is non-negative, with equality if and only if

$$f_1(v_1)f_2(v) = f_1(v_1')f_2(v').$$

The equality from the second term of the sum is the same just exchanging the notation. In the case of equality, choose $\phi_1 = f_1$ and $\phi_2 = f_2$ in lemma 1.3.6. Then we obtain that f_1 and f_2 are Maxwell distributions with equal mean velocity and temperature. □

In the following we will see three things. We will consider the physical entropy and see that the previous theorem leads to an inequality for this entropy. We will study the equality in this entropy and are able to characterize it with the help of lemma 1.3.6 as Maxwell distributions with equal mean velocity and temperature. In the next part, we want to define equilibrium distributions and finally, we will observe that they correspond to the Maxwell distributions found in the H-theorem. Let's start with the entropy inequality.

Definition 1.3.9 (Physical entropy). The quantity

$$H(f_1, f_2) := \int (f_1 \ln f_1 + f_2 \ln f_2) dv$$

defines the negative of the physical entropy in statistical mechanics.

For a motivation of choosing this expression as the entropy see section 3.4.2 in [81]. It is linked to the number of possibilities to distribute N particles into the cells of phase space. In the following we deduce an inequality for this entropy.

Corollary 1.3.10. *Let $f_1, f_2 \in L^\infty(dv)$ be a solution to (1.14) in the sense of distributions decaying fast enough at infinity in v, $f_1, f_2 \geq 0$, and $\int (1 + |v|^2) f_1(v) dv < \infty$ and $\int (1 + |v|^2) f_2(v) dv < \infty$, then*

$$\partial_t \left(\int f_1 \ln f_1 dv + \int f_2 \ln f_2 dv \right) + \nabla_x \cdot \left(\int v f_1 \ln f_1 dv + \int v f_2 \ln f_2 dv \right) \leq 0,$$

with equality if and only if f_1 and f_2 are Maxwell distributions with equal mean velocity and temperature.

Proof. We multiply the Boltzmann equation (1.14) for species 1 by $\ln f_1$, the Boltzmann equation (1.14) for species 2 by $\ln f_2$, add the result and integrate it with respect to v, use integration by parts and conservation of the number of particles (equation (1.15) and theorem 1.3.1) on the left-hand side and the theorems 1.3.8 and 1.3.9 on the right-hand side. □

The physical meaning of this is the following. In the space-homogeneous case, meaning f_1, f_2 do not depend on x, the entropy is a non-increasing function. This corresponds to the second law of thermodynamics.

The fact that H is a decreasing function unless f_1, f_2 are Maxwell distributions with a common mean velocity and temperature, indicates that f_1, f_2 possibly tend to such a function when $t \to \infty$. The final state will presumably be an equilibrium state.

Definition 1.3.10 (Equilibrium). The system is in equilibrium if and only if f does not depend on x and t.

This means that the left-hand side of the Boltzmann equation vanishes, so we have to consider the equations $Q_{11}(f_1, f_1) + Q_{12}(f_1, f_2) = 0$ and $Q_{22}(f_2, f_2) + Q_{21}(f_2, f_1) = 0$.

Corollary 1.3.11 (Structure of the equilibrium). *Let $f_1, f_2 \in L^\infty(dv)$ be a solution to (1.14) in the sense of distributions and assume $f_1, f_2 > 0$. From the equations $Q_{11}(f_1, f_1) + Q_{12}(f_1, f_2) = 0$ and $Q_{22}(f_2, f_2) + Q_{21}(f_2, f_1) = 0$ we can deduce that f_1 and f_2 are Maxwell distributions with equal velocity and temperature, that means $u_1 = u_2$ and $T_1 = T_2$.*

Proof. If $Q_{11}(f_1, f_1) + Q_{12}(f_1, f_2) = 0$ and $Q_{22}(f_2, f_2) + Q_{21}(f_2, f_1) = 0$, then $Q_{11}(f_1, f_1) + Q_{12}(f_1, f_2) + Q_{22}(f_2, f_2) + Q_{21}(f_2, f_1) = 0$ and so we have equality in the H-theorems 1.3.8 and 1.3.9. □

This shows that the equality in the H-theorem coincides with the distribution in equilibrium. So we can formulate the H-theorem in the space homogeneous case in the following equivalent version: We expect that the entropy H decreases in time until it reaches its equilibrium distribution which is a Maxwell distribution.

The last two lemmas will lead to a comparison of the entropy of a species with the entropy obtained in equilibrium.

Lemma 1.3.12. *Let z, y be arbitrary positive real numbers. Consider the function $h : \mathbb{R}^+ \to \mathbb{R}, h(x) := x \ln x - x$. Then h is strictly convex and the following inequality is satisfied*

$$(z - y) \ln y \leq z \ln z - y \ln y + y - z,$$

or equivalently

$$z \ln y \leq z \ln z + y - z.$$

Proof. If we compute the derivatives of h, we obtain $h'(x) = \ln x$ and $h''(x) = \frac{1}{x}$. So h is strictly convex for $x > 0$ and we have $h'(y)(z - y) \leq h(z) - h(y)$ with equality if and only if $z = y$. □

Lemma 1.3.13 (Estimate of the entropy). *Let $f_1, f_2 \in L^\infty(dv)$ be two functions decaying fast enough at infinity, $f_1, f_2 \geq 0$, and $\int (1 + |v|^2) f_1(v) dv < \infty$ and $\int (1 + |v|^2) f_2(v) dv < \infty$, then*

$$\int f_k \ln f_k dv \geq \int M_k \ln M_k dv, \quad \text{for} \quad k = 1, 2,$$

where M_k is the local Maxwell distribution with the same density, mean velocity and temperature as f_k.

Proof. If we compute $\ln M_k$, we obtain $\ln M_k = \ln \dfrac{n_k}{\sqrt{2\pi \frac{T_k}{m_k}}^3} - \dfrac{|v - u_k|^2}{2\frac{T_k}{m_k}}$. We observe that $\ln M_k$ is a linear combination of $1, v$ and $|v|^2$. Therefore $\int (f_k - M_k) \ln M_k dv = 0$

since f_k and M_k have the same density, mean velocity and temperature. Using this and lemma 1.3.12, we get

$$\int f_k \ln f_k dv - \int M_k \ln M_k dv$$

$$= \int f_k \ln f_k dv - \int M_k \ln M_k dv - \int (f_k - M_k) \ln M_k dv$$

$$= \int f_k \ln f_k dv - \int f_k \ln M_k dv$$

$$\geq \int (f_k - M_k) dv = 0.$$

$\square$

1.4 The BGK model for one species

For one species of particles the Boltzmann equation reduces to

$$\partial_t f + v \cdot \nabla_x f = Q(f, f),$$

for the unknown $f(x, v, t)$. The collision operator corresponds to an operator describing the interactions of the gas with itself as Q_{11} and Q_{22} in the previous section. The collision operator in the Boltzmann equation is very complex. So Bathnagar, Gross and Krook [16] invented a simplification of the collision operator which has still the same main properties mentioned in the previous section as the original collision term Q. It is called BGK model. It satisfies the conservation of the number of particles, momentum and energy. The BGK model has an H-theorem and the same structure in equilibrium as the Boltzmann equation. It is given by

$$\partial_t f + v \cdot \nabla_x f = \nu n(M - f), \tag{1.41}$$

where $\nu(x, t)n(x, t)$ is the collision frequency and M the Maxwell distribution which has the same density, mean velocity and temperatures as f given by (1.34). Moreover BGK models give rise to efficient numerical computations [72, 41, 35, 12, 34, 13, 28]. In [64], the linearized BGK equation is considered and the uncertainties in the case of modelling errors are quantified by Klingenberg, Li and Pirner.

1.4.1 Motivation of the BGK model

The aim of this section is to motivate the choice and structure of the BGK model and the choice of the collision frequency ν. First, we will see under which assumptions and simplifications we can derive the BGK-model from the Boltzmann equation. This is given in [81]. Then, we will see the meaning of the equation by considering a minimization problem of the entropy which also leads to the BGK model. This is given in [2].

Derivation from the Boltzmann equation

The first motivation of the BGK model is to simplify the collision operator in the Boltzmann equation making some simplifying approximations. The collision term in the Boltzmann equation for one species is given by

$$Q(f,f) = \frac{1}{m} \int_{\mathbb{R}^3} \int_{S^2} B(|v - v_1|, \omega)(f(x, v', t)f(x, v_1', t) - f(x, v, t)f(x, v_1, t))d\omega dv_1.$$

In section 1.3.3 we observed that the H-theorem indicates that due to interactions the distribution function f tend to relax towards a Maxwell distribution. Now, we assume that even after one interaction the distribution function becomes a Maxwell distribution. So, we replace the distribution functions with velocities after the collision $f(x, v', t)$ and $f(x, v_1', t)$ in the collision operator by its local Maxwell distribution M. So we get

$$\hat{Q}_M(f,f) = \frac{1}{m} \int_{\mathbb{R}^3} \int_{S^2} B(|v - v_1|, \omega)(M(x, v', t)M(x, v_1', t) - f(x, v, t)f(x, v_1, t))d\omega dv_1.$$

Because of conservation of momentum and kinetic energy (1.10), we have

$$M(x, v', t)M(x, v_1', t) = M(x, v, t)M(x, v_1, t),$$

and so we get

$$\tilde{Q}_M(f,f) = \frac{1}{m} \int_{\mathbb{R}^3} \int_{S^2} B(|v - v_1|, \omega)(M(x, v, t)M(x, v_1, t) - f(x, v, t)f(x, v_1, t))d\omega dv_1$$

$$= \frac{1}{m}(M(x, v, t) \int_{\mathbb{R}^3} \int_{S^2} B(|v - v_1|, \omega)M(x, v_1, t)d\omega dv$$

$$- f(x, v, t) \int_{\mathbb{R}^3} \int_{S^2} B(|v - v_1|, \omega)f(x, v_1, t)d\omega dv_1).$$

Now we assume that the difference between the two integrals can be neglected and get

$$Q_{BGK}(f,f) = \nu n(M - f),$$

where ν is given by

$$\nu n = \frac{1}{m} \int \int M(x, v_1, t)B(|v - v_1|, \omega)d\omega dv_1.$$

With this choice of the collision frequency ν, the conservation properties are only fulfilled when ν does not depend on the velocity v, so one either has to consider Maxwellian molecules, where B does not depend on $|v - v_1|$ or replace ν by a mean collision frequency $\bar{\nu} = \frac{1}{n} \int \nu f dv$.

Minimization of the entropy

The other possible way to motivate the BGK model is due to a minimization problem of the entropy presented in [2]. Suppose we have a model of the form

$$\partial_t f + v \cdot \nabla_x f = \nu n(G(f) - f), \tag{1.42}$$

for a function $G(f)$ which we will determine later. The meaning of the term on the right-hand side is the following. Suppose we are in the space homogeneous case. Then the meaning of the term on the right-hand side is a relaxation of the function f towards the function $G(f)$ in time. If f is smaller than $G(f)$, in the space-homogeneous case the time derivative of f is positive. So f will increase. If f is larger than $G(f)$, the time derivative is negative so f will decrease. This corresponds to a relaxation of f towards the function $G(f)$. See also figure 1.4.

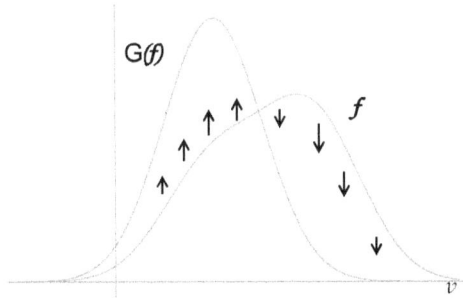

Figure 1.4: Relaxation of f towards the function $G(f)$. The arrows illustrate the time derivative of f under the time evolution of equation (1.42) in the space-homogeneous case.

Now, the idea is to determine the function $G(f)$ as a solution to a minimization problem of the entropy. We consider the following problem

$$S(n, u, T) = \min_{g \in \chi} \int \tilde{H}(g) dv, \tag{1.43}$$

where $\tilde{H}(g)$ is given by $\tilde{H}(g) = g \ln g$ and χ is the following set

$$\chi = \{g \geq 0, (1 + |v|^2)g \in L^1(dv), \int g dv = n, \int v g dv = nu, \int v \otimes v g dv = nu \otimes u + n\frac{T}{m}\mathbf{1}\}.$$

In the case of one species we omit the index in the density, the mean velocity and the temperature given by the definitions 1.3.2, 1.3.3 and 1.3.5, respectively. We want to find the solution to (1.43) and choose $G(f)$ as the minimizer to this problem. The problem is motivated by the H-theorem for the Boltzmann equation, since we expect from it that f relaxes to a function $G(f)$. The set χ can be motivated in the following way. We expect that during the relaxation process the density n, the momentum nu and the energy $\frac{1}{2}n(m|u|^2 + 3T)$ should be conserved, so the solution to (1.43) should have the same density, momentum and energy as f. We observe that we assume more to the function $g \in \chi$ than to have just the same energy as f. We also determine the off-diagonal terms of $\int v \otimes vg\,dv$. The physical meaning of this is the following. If we compute the integral $m \int v \otimes vf\,dv$, we get $mnu \otimes u + \mathbb{P}$. Therefore the restriction on $g \in \chi$ means that we expect that in equilibrium the tensor $\mathbb{P}$ becomes diagonal. We will observe that the solution to this problem is a Maxwell distribution.

Theorem 1.4.1. *The unique minimizer to (1.43) is the Maxwell distribution M given by (1.34).*

Proof. The proof is given in [2]. For the convenience of the reader we will repeat it here. The aim is to show that

$$\int \tilde{H}(M)dv < \int \tilde{H}(g)dv \quad \text{for all} \quad g \in \chi \quad \text{with} \quad g \neq M.$$

Since $\tilde{H}$ is a strictly convex function, we get

$$\tilde{H}(g) > \tilde{H}(M) + \tilde{H}'(M)(g - M) \quad \text{for all} \quad g \in \chi \quad \text{with} \quad g \neq M.$$

Here $'$ denotes the Fréchet derivative of $\tilde{H}$ with respect to g. We are done when we show that $\int \tilde{H}'(M)(g - M)dv = 0$. If we compute $\tilde{H}'(M)$, we get

$$\tilde{H}'(M) = \ln M + 1 = \frac{n}{\sqrt{2\pi \frac{T}{m}}^3} - \frac{m|v - u|^2}{T} + 1.$$

So we observe that $\tilde{H}'(M)$ is a linear combination of $1, v$ and $|v|^2$. Therefore, the integral $\int \tilde{H}'(M)(g - M)dv$ vanishes since g and M have the same density, mean velocity and energy. $\qquad \square$

1.4.2 Fundamental properties of the BGK equation

The reason for this model was to find a simplification of the complicated Boltzmann collision operator but to keep the main features of the Boltzmann collision operator. In this section we prove that the BGK operator for one species satisfies the conservation properties and the H-theorem.

Theorem 1.4.2 (BGK-model). *Assume $f \in L^\infty(dv)$ to be a solution of (1.41) decaying fast enough to zero for $|v| \to \infty$ and $\nu, f > 0$. Then the right-hand side of the equation (1.41) has the same main properties (conservation properties, H-theorem and structure of equilibrium) as the Boltzmann collision operator.*

Proof. The proof is given in section 3.6 in [81]. For the convenience of the reader we want to repeat it here. First, we will discuss the conservation properties, then the H-theorem. Since ν does not depend on the velocity v, the conservation properties as in theorem 1.3.1 are satisfied because f and M have the same density, mean velocity and temperature.

For the same reason, we have

$$\nu n \int \ln M \ (M - f)dv = 0,$$

because $\ln M$ is a linear combination of 1, v and $|v|^2$, so

$$\int Q_{BGK}(f,f)(v)dv = \nu n \int \ln f \, (M-f)dv$$

$$= \nu n \int \ln f \, (M-f)dv - \nu n \int \ln M \, (M-f)dv$$

$$= \nu n \int \ln \frac{f}{M} \, (M-f)dv.$$

With lemma 1.3.7 we can conclude

$$\int Q_{BGK}(f,f)(v)dv = \nu n \int \ln \frac{f}{M} \, (M-f)dv \leq 0,$$

which ensures the property of the H-theorem.

The structure of the equilibrium is clear, because from $\nu(M-f)=0$, we can deduce $f=M$. $\qquad\square$

Chapter 2

A BGK model for mixtures

In applications one often has to deal with gas mixtures instead of a single gas. For example, the air is a gas mixture. Therefore, in this chapter, our attempt is to extend the BGK model for one species described in the previous chapter to gas mixtures. For simplification we consider only two species. This model is also presented in a paper by Klingenberg, Pirner, Puppo in [61].

In the literature one can find two types of BGK models for gas mixtures. Just like the Boltzmann equation for gas mixtures contains a sum of collision terms on the right-hand side, one type of BKG models also has a sum of BGK-type interaction terms in the relaxation operator. Examples are the models of Gross and Krook [49], Hamel [51], Asinari [6], Garzó, Santos, Brey [43], Sofena [79], see also Cercignani [24]. The other type of models contains only one collision term on the right-hand side. Examples for this are Andries, Aoki and Perthame [1] and the models in [22, 48].

For the second type of model which was first presented by Andries, Aoki and Perthame [1], they proved consistency of this type of model. They proved the conservation properties, the positivity of the temperatures, the H-theorem and specified the structure of equilibrium. Whereas for the first type of model this was not done before in the literature, as far as we know. But the first type of BGK model for mixtures is the extension to gas mixtures suggested by physicists and engineers and is still used in numerical applications, see for example [67, 71]. In [54], the second type of model is even seen as a simplification of the first model from a physical point of view. The advantage of the first type of BGK model for mixtures from the physical point of view is that the two different types of interactions, interactions of a species with itself and interactions of a species with the other one, are still kept separated. Therefore, we can still see how these different types of interactions influence the macroscopic equations, the H-theorem and the trend to equilibrium. Especially, when the particles of the two species are very different, then it is desirable to maintain their contribution separately. But, as far as we know, there was no proof of the H-theorem and the structure of equilibrium before for this first type of extension. In order to close this gap, we now present a model which contains all the models of the first type mentioned above as special cases. For this generalized model we then prove the conservation properties, the positivity of all temperatures, the H-theorem and specify the structure of equilibrium.

The outline of the chapter is as follows: in section 2.1 we will present the model for two species and prove the conservation properties and the H-theorem and we show the positivity of all temperatures. In section 2.2, we compare our model with other models presented in the literature. First, we consider special cases of our model and next we compare our model with the model of Andries, Aoki and Perthame [1].

2.1 A two species kinetic BGK model

In this section we will present the model for two species and prove the conservation properties and the H-theorem. Further, we analyse the positivity of the temperatures and characterize the structure of equilibrium.

2.1.1 The general form of the model

We will repeat the most important definitions such that it is possible to read this chapter independently. For simplicity in the following we consider a mixture composed of two different species. Thus, our kinetic model has two distribution functions $f_1(x, v, t) > 0$ and $f_2(x, v, t) > 0$, where $x \in \mathbb{R}^3$ and $v \in \mathbb{R}^3$ are the phase space variables and $t \geq 0$ the time. We expect that they are determined by two equations to describe their time evolution. Since we consider binary interactions, the particles of one species can interact with either themselves or with particles of the other species. In the model this is accounted for by introducing two interaction terms in both equations. These considerations allow us to write formally the system of equations for the evolution of the mixture. The following structure containing a sum of the collision operator is also given in [25, 24]. We describe the time evolution of the number distribution functions f_1 and f_2 by the Boltzmann equation with binary interactions for two species of particles as in [25], chapter 6.2

$$\partial_t f_1 + v \cdot \nabla_x f_1 = Q_{11}(f_1, f_1) + Q_{12}(f_1, f_2),$$
$$\partial_t f_2 + v \cdot \nabla_x f_2 = Q_{22}(f_2, f_2) + Q_{21}(f_2, f_1),$$

where Q_{kl}, $k, l = 1, 2$ are the collision operators for interactions of species k with species l.

Furthermore, we relate the distribution functions to macroscopic quantities by mean-values of f_k

$$\int f_k(v) \begin{pmatrix} 1 \\ v \\ m_k |v - u_k|^2 \end{pmatrix} dv =: \begin{pmatrix} n_k \\ n_k u_k \\ 3 n_k T_k \end{pmatrix}, \tag{2.1}$$

where n_k is the number density, u_k the mean velocity and T_k the temperature which is related to the pressure p_k by $p_k = n_k T_k$. Note that in this chapter we shall write T_k instead of $k_B T_k$, where k_B is Boltzmann's constant.

2.1.2 Requirements on the collision operators

A model for the evolution of a mixture should satisfy the following conservation properties:

Conservation of mass, momentum and energy of the individual species in interaction with the species itself:

1. $\int Q_{kk}(f_k, f_k) dv = 0$ for $k = 1, 2$,

2. $\int m_k v Q_{kk}(f_k, f_k) dv = 0$ for $k = 1, 2$,

3. $\int m_k |v|^2 Q_{kk}(f_k, f_k) dv = 0$ for $k = 1, 2$.

Conservation of total mass, momentum and energy:

1. $\int Q_{kl}(f_k, f_l) dv = 0$ for $k, l = 1, 2$,

2. $\int (m_1 v Q_{12}(f_1, f_2) + m_2 v Q_{21}(f_2, f_1)) dv = 0$,

3. $\int (m_1 |v|^2 Q_{12}(f_1, f_2) + m_2 |v|^2 Q_{21}(f_2, f_1)) dv = 0$.

2.1.3 The BGK approximation

We are interested in a BGK approximation of the interaction terms. This leads us to define equilibrium distributions not only for each species itself but also for the two interspecies equilibrium distributions. Choose the collision terms Q_{11}, Q_{12}, Q_{21} and Q_{22} in section 2.1.1 as BGK operators. Then the model can be written as:

$$\partial_t f_1 + v \cdot \nabla_x f_1 = \nu_{11} n_1 (M_1 - f_1) + \nu_{12} n_2 (M_{12} - f_1),$$
$$\partial_t f_2 + v \cdot \nabla_x f_2 = \nu_{22} n_2 (M_2 - f_2) + \nu_{21} n_1 (M_{21} - f_2),$$

(2.2)

with the Maxwell distributions

$$M_1(x, v, t) = \frac{n_1}{\sqrt{2\pi \frac{T_1}{m_1}}^3} \exp\left(-\frac{|v - u_1|^2}{2\frac{T_1}{m_1}}\right),$$

$$M_2(x, v, t) = \frac{n_2}{\sqrt{2\pi \frac{T_2}{m_2}}^3} \exp\left(-\frac{|v - u_2|^2}{2\frac{T_2}{m_2}}\right),$$

$$M_{12}(x, v, t) = \frac{n_{12}}{\sqrt{2\pi \frac{T_{12}}{m_1}}^3} \exp\left(-\frac{|v - u_{12}|^2}{2\frac{T_{12}}{m_1}}\right),$$

$$M_{21}(x, v, t) = \frac{n_{21}}{\sqrt{2\pi \frac{T_{21}}{m_2}}^3} \exp\left(-\frac{|v - u_{21}|^2}{2\frac{T_{21}}{m_2}}\right),$$

(2.3)

where $\nu_{11} n_1$ and $\nu_{22} n_2$ are the collision frequencies of the particles of each species with itself, while $\nu_{12} n_2$ and $\nu_{21} n_1$ are related to interspecies collisions. The structure of the collision terms ensures that if one collision frequency ν_{kl} tends to ∞, the corresponding distribution function becomes a Maxwell distribution. In addition at global equilibrium, the distribution functions become Maxwell distributions with the same mean velocity and temperature (see later in section 2.1.8). The Maxwell distributions M_1 and M_2 in (2.3) have the same macroscopic quantities (2.1) as f_1 and f_2, respectively. With this choice, we guarantee the conservation of mass, momentum and energy in interactions of one species with itself (see section 2.1.2). The remaining parameters u_{12}, u_{21}, T_{12} and T_{21} will be determined using conservation of total momentum and total energy, together with some symmetry considerations.

2.1.4 Relationship between the collision frequencies

Note, that in our model we have four collision frequencies $\nu_{11}n_1$, $\nu_{22}n_2$, $\nu_{12}n_2$ and $\nu_{21}n_1$. In physical applications the functions ν_{11}, ν_{22}, ν_{12} and ν_{21} are often linked by a constant. We will illustrate this in the case of a plasma in section 5.2.1. To be flexible in choosing the relationship between the collision frequencies, we now assume the relationship

$$\nu_{12} = \varepsilon\nu_{21}, \quad 0 < \varepsilon \leq 1. \tag{2.4}$$

For example, in the case of a plasma, we will see in section 5.2.1 that ε is given by $\varepsilon = \frac{m_1}{m_2}$. The restriction $\varepsilon \leq 1$ is without loss of generality. If $\varepsilon > 1$, exchange the index 1 and 2 and choose $\frac{1}{\varepsilon}$ as new ε.

2.1.5 Conservation properties

This section shows how the macroscopic quantities $n_{12}, n_{21}, u_{12}, u_{21}, T_{12}$ and T_{21} in the interspecies Maxwell distributions M_{12} and M_{21} have to be chosen in order to ensure the macroscopic conservation properties.

Theorem 2.1.1 (Conservation of the number of particles of each species). *Assume that*

$$n_{12} = n_1 \quad and \quad n_{21} = n_2, \tag{2.5}$$

then

$$\int Q_{11}(f_1, f_1)dv = \int Q_{12}(f_1, f_2)dv = \int Q_{22}(f_2, f_2)dv = \int Q_{21}(f_2, f_1)dv = 0.$$

Proof. Since M_k and f_k for $k = 1, 2$ have the same density according to the definition in equation (2.1), we get

$$\int Q_{kk}(f_k, f_k)dv = \int \nu_{kk}n_k(M_k - f_k)dv = 0.$$

The equality

$$\int Q_{12}(f_1, f_2)dv = \nu_{12}n_2 \int (M_{12} - f_1)dv = 0,$$

holds provided that $n_{12} = n_1$. Similarly for the second equation, if $n_{21} = n_2$. $\square$

Theorem 2.1.2 (Conservation of total momentum). *Assume the relationships* (2.4) *and* (2.5) *hold and assume further that u_{12} is a linear combination of u_1 and u_2*

$$u_{12} = \delta u_1 + (1 - \delta)u_2, \quad \delta \in \mathbb{R}. \tag{2.6}$$

Then we have conservation of total momentum

$$\int m_1 v[Q_{11}(f_1, f_1) + Q_{12}(f_1, f_2)]dv + \int m_2 v[Q_{22}(f_2, f_2) + Q_{21}(f_2, f_1)]dv = 0,$$

provided that

$$u_{21} = u_2 - \frac{m_1}{m_2}\varepsilon(1-\delta)(u_2 - u_1). \tag{2.7}$$

Proof. The exchange of momentum of species 1 is given by

$$f_{m_{1,2}} := m_1 \int v\nu_{11}n_1(M_1 - f_1)dv + m_1 \int v\nu_{12}n_2(M_{12} - f_1)dv \tag{2.8}$$

$$= 0 + m_1\nu_{12}n_1n_2(u_{12} - u_1) = m_1\nu_{12}n_1n_2(1-\delta)(u_2 - u_1).$$

The exchange of momentum of species 2 is given by

$$f_{m_{2,1}} := m_2\nu_{21}n_2n_1(u_{21} - u_2). \tag{2.9}$$

In order to get conservation of momentum we therefore need

$$m_1\nu_{12}n_1n_2(1-\delta)(u_2 - u_1) + m_2\nu_{21}n_2n_1(u_{21} - u_2) = 0,$$

which holds provided u_{21} satisfies (2.7) under the assumption that ν_{12} and ν_{21} satisfy (2.4) and u_{12} satisfies (2.6). $\square$

Remark 2.1.1. If we write $\tilde{\varepsilon} = \frac{m_1}{m_2}\varepsilon$ and $\tilde{\delta} = 1 - \tilde{\varepsilon}(1-\delta)$ we obtain a similar structure for u_{21} as for u_{12}

$$u_{21} = \tilde{\delta}u_2 + (1 - \tilde{\delta})u_1.$$

Theorem 2.1.3 (Conservation of total energy). *Assume* (2.4), *conditions* (2.5), (2.6) *and* (2.7) *and assume that* T_{12} *is of the following form*

$$T_{12} = \alpha T_1 + (1-\alpha)T_2 + \gamma|u_1 - u_2|^2, \quad 0 \le \alpha \le 1, \gamma \ge 0. \tag{2.10}$$

Then we have conservation of total energy

$$\int \frac{m_1}{2}|v|^2(Q_{11}(f_1, f_1) + Q_{12}(f_1, f_2))dv + \int \frac{m_2}{2}|v|^2(Q_{22}(f_2, f_2) + Q_{21}(f_2, f_1))dv = 0,$$

provided that

$$T_{21} = \left[\frac{1}{3}\varepsilon m_1(1-\delta)\left(\frac{m_1}{m_2}\varepsilon(\delta-1) + \delta + 1\right) - \varepsilon\gamma\right]|u_1 - u_2|^2 \tag{2.11}$$

$$+\varepsilon(1-\alpha)T_1 + (1 - \varepsilon(1-\alpha))T_2.$$

Proof. Using the energy exchange of species 1

$$F_{E_{1,2}} := \int \frac{m_1}{2}|v|^2\nu_{11}n_1(M_1 - f_1)dv + \int \frac{m_1}{2}|v|^2\nu_{12}n_2(M_{12} - f_1)dv$$

$$= \varepsilon\nu_{21}\frac{1}{2}n_2n_1m_1(|u_{12}|^2 - |u_1|^2) + \frac{3}{2}\varepsilon\nu_{21}n_1n_2(T_{12} - T_1),$$

where we use (2.6) and (2.10). Analogously the energy exchange of species 2 towards 1 is
$$F_{E_{2,1}} = \frac{1}{2}\nu_{21}m_2n_1n_2(|u_{21}|^2 - |u_2|^2) + \frac{3}{2}\nu_{21}n_1n_2(T_{21} - T_2).$$

Substitute u_{21} with (2.7) and T_{21} from (2.11). This permits to rewrite the energy exchange terms as

$$F_{E_{1,2}} = \varepsilon\nu_{21}\frac{1}{2}n_2n_1m_1\left[(\delta^2 - 1)|u_1|^2 + (1 - \delta)^2|u_2|^2 + 2\delta(1 - \delta)u_1 \cdot u_2\right]$$
$$+ \frac{3}{2}\varepsilon\nu_{21}n_1n_2\left[(1 - \alpha)(T_2 - T_1) + \gamma|u_1 - u_2|^2\right], \tag{2.12}$$

$$F_{E_{2,1}} = \frac{1}{2}\nu_{21}m_2n_1n_2\Bigg[\left(\left(1 - \frac{m_1}{m_2}\varepsilon(1 - \delta)\right)^2 - 1\right)|u_2|^2 + \left(\frac{m_1}{m_2}\varepsilon(\delta - 1)\right)^2|u_1|^2$$
$$+ 2(1 - \frac{m_1}{m_2}\varepsilon(1 - \delta))\frac{m_1}{m_2}\varepsilon(1 - \delta)u_1 \cdot u_2\Bigg] + \frac{3}{2}\nu_{21}n_1n_2\Big[\varepsilon(1 - \alpha)(T_1 - T_2)$$
$$+ \left(\frac{1}{3}\varepsilon m_1(1 - \delta)\left(\frac{m_1}{m_2}\varepsilon(\delta - 1) + \delta + 1\right) - \varepsilon\gamma\right)|u_1 - u_2|^2\Big]. \tag{2.13}$$

Adding these two terms, we see that the total energy is conserved. $\qquad\square$

Remark 2.1.2. We have $0 \leq 1 - \varepsilon(1 - \alpha) \leq 1$ and $0 \leq \varepsilon(1 - \alpha) \leq 1$, so that in (2.11) the two terms with the temperatures are also a convex combination of T_1 and T_2.

Remark 2.1.3. The remaining free parameters can be fixed for specific situations. For example, if we see the parameters α, δ, γ and ε from the model presented in this chapter as functions of the masses m_1 and m_2, we can get more restrictions on these parameters by physical considerations.

- In the limit $\frac{m_1}{m_1+m_2} \to 0$, we expect that $u_{12} = u_2$ and $T_{12} = T_2$, since we expect that light particles are driven by the flow of the heavy particles, so they adapt the velocity and the fluctuations to the mean velocity of the heavy particles. If we look at (2.6), (2.7), (2.10) and (2.11), the definitions of u_{12}, u_{21}, T_{12} and T_{21}, we see in order to realize this, we need $\delta \to 0, \alpha \to 0$ and $\gamma \to 0$.

- In the limit $\frac{m_1}{m_1+m_2} \to \frac{1}{2}$, when the mass of the particles become indistinguishable, we expect $T_{12} = T_{21}$ and $u_{12} = u_{21}$. From $u_{12} = u_{21}$ we obtain $\delta = \frac{\varepsilon m_2}{\varepsilon + m_1}$ by using (2.6) and (2.7). From $T_{12} = T_{21}$ we obtain $\alpha = \frac{\varepsilon}{1+\varepsilon}$ and $\gamma = \frac{1}{3}\frac{\varepsilon}{1+\varepsilon}m_1(1 - \delta)(\frac{m_1}{m_2}\varepsilon(\delta - 1) + \delta + 1)$ by using (2.10) and (2.11). So for getting equality in the limit when the masses of the two particles become indistinguishable ($m := m_1 = m_2$), we need we need $\delta \to \frac{\varepsilon}{1+\varepsilon}, \alpha \to \frac{\varepsilon}{1+\varepsilon}$ and $\gamma \to \frac{1}{3}m\frac{\varepsilon}{(1+\varepsilon)^2}$.

- In the limit $\frac{m_1}{m_1+m_2} \to 1$, the heavy particles do not feel the other particles, so we expect that we have no change in the mean velocity and in the temperature, e.g $u_{12} = u_1$ and $T_{12} = T_1$. Here we need $\delta \to 1, \alpha \to 1$ and $\gamma \to 0$.

2.1.6 Positivity of the temperatures

Theorem 2.1.4. *Assume that $f_1(x, v, t), f_2(x, v, t) > 0$. Then all temperatures T_1, T_2, T_{12} given by (2.10) and T_{21} given by (2.11) are positive provided that*

$$0 \le \gamma \le \frac{m_1}{3}(1 - \delta)\left[(1 + \frac{m_1}{m_2}\varepsilon)\delta + 1 - \frac{m_1}{m_2}\varepsilon\right]. \tag{2.14}$$

Proof. T_1 and T_2 are positive as integrals of positive functions. T_{12} is positive because by construction it is a convex combination of T_1 and T_2. For T_{21} we consider the coefficients in front of $|u_1 - u_2|^2$, T_1 and T_2. The term in front of T_1 is positive by definition. The positivity of the term in front of T_2 is equivalent to the condition $\alpha \ge 1 - \frac{1}{\varepsilon}$, which is satisfied since $\varepsilon \le 1$, the positivity of the term in front of $|u_1 - u_2|^2$ is equivalent to the condition (2.14). $\qquad\square$

Remark 2.1.4. According to the definition of γ, it is a non-negative number, so the right-hand side of the inequality in (2.14) must be non-negative. This condition is equivalent to

$$\frac{\frac{m_1}{m_2}\varepsilon - 1}{1 + \frac{m_1}{m_2}\varepsilon} \le \delta \le 1. \tag{2.15}$$

If the collision frequencies are linked as in (2.4) with $\varepsilon = \frac{m_2}{m_1}$ in the case of a plasma, then the right-hand side of (2.14) is positive for $0 \le \delta \le 1$.

2.1.7 H-theorem for mixtures

Now, we want to show the H-theorem for our BGK model for mixtures.

Lemma 2.1.5. *Assuming (2.10) and (2.11) and the positivity of the temperatures (2.14), we have the following inequality*

$$\varepsilon \ln T_{12} + \ln T_{21} \ge \varepsilon \ln T_1 + \ln T_2. \tag{2.16}$$

Proof. We start with the left-hand side of (2.16). First we insert the definition of T_{12} and T_{21} from (2.10) and (2.11). Since γ and the term in front of $|u_1 - u_2|^2$ in (2.11) are positive, we can use the monotonicity of the logarithm and get

$$\varepsilon \ln T_{12} + \ln T_{21}$$
$$= \varepsilon \ln \left[\alpha T_1 + (1 - \alpha)T_2 + \gamma |u_1 - u_2|^2\right]$$
$$+ \ln \left[\left(\frac{1}{3}\varepsilon m_1(1 - \delta)\left(\frac{m_1}{m_2}\varepsilon(\delta - 1) + \delta + 1\right) - \varepsilon\gamma\right)|u_1 - u_2|^2\right.$$
$$\left. + \varepsilon(1 - \alpha)T_1 + (1 - \varepsilon(1 - \alpha))T_2\right]$$
$$\ge \varepsilon \ln\left(\alpha T_1 + (1 - \alpha)T_2\right) + \ln\left(\varepsilon(1 - \alpha)T_1 + (1 - \varepsilon(1 - \alpha))T_2\right).$$

If we now use the concavity of the logarithm and the assumptions $0 \leq \alpha \leq 1, \varepsilon \leq 1$, the expression above can be bounded from below by

$$\varepsilon \alpha \ln T_1 + \varepsilon(1 - \alpha) \ln T_2 + (1 - \varepsilon(1 - \alpha)) \ln T_2 + \varepsilon(1 - \alpha) \ln T_1,$$

which gives the inequality stated in lemma 2.1.5. $\qquad\square$

Theorem 2.1.6 (H-theorem for the mixture). *Assume $f_1, f_2 > 0$. Assume the relationship between the collision frequencies (2.4), the conditions for the interspecies Maxwell distributions (2.5), (2.6), (2.7), (2.10) and (2.11) with $\alpha, \delta \neq 1$ and the positivity of the temperatures (2.14), then*

$$\int (\ln f_1 \, Q_{11}(f_1, f_1) + \ln f_1 \, Q_{12}(f_1, f_2))dv$$

$$+ \int (\ln f_2 \, Q_{22}(f_2, f_2) + \ln f_2 \, Q_{21}(f_2, f_1))dv \leq 0,$$

with equality if and only if f_1 and f_2 are Maxwell distributions with equal mean velocity and temperature.

Proof. The fact that $\int \ln f_k \, Q(f_k, f_k) \leq 0$, $k = 1, 2$ for the single BGK-model was shown in theorem 1.4.2. Note that in the proof it was not necessary that f_k is a solution to the BGK model for one species. It is only a property of the structure of the BGK operator. So here, it is still true. We have equality if and only if $f_1 = M_1$ and $f_2 = M_2$.
Let us define

$$S(f_1, f_2) := \nu_{12} n_2 \int \ln f_1 \, (M_{12} - f_1)dv + \nu_{21} n_1 \int \ln f_2 \, (M_{21} - f_2)dv.$$

The task is to prove that $S(f_1, f_2) \leq 0$. Consider now $S(f_1, f_2)$ and apply the inequality in lemma 1.3.12 to each of the two terms in S.

$$S \leq \nu_{12} n_2 \left[\int M_{12} \ln M_{12} dv - \int f_1 \ln f_1 dv - \int M_{12} dv + \int f_1 dv \right]$$

$$+ \nu_{21} n_1 \left[\int M_{21} \ln M_{21} dv - \int f_2 \ln f_2 dv - \int M_{21} dv + \int f_2 dv \right],$$

with equality if and only if $f_1 = M_{12}$ and $f_2 = M_{21}$. Then $u_{12} = \delta u_1 + (1 - \delta)u_2 = u_1$ from which we can deduce $u_1 = u_2 = u_{21} = u_{12}$ and $T_1 = T_2 = T_{12} = T_{21}$. That means f_1 and f_2 are Maxwell distributions with equal mean velocity and temperature. Since M_{12} and f_1 have the same density and M_{21} and f_2 have the same density, too, the right-hand side reduces to

$$\nu_{12} n_2 \left(\int M_{12} \ln M_{12} dv - \int f_1 \ln f_1 dv \right) + \nu_{21} n_1 \left(\int M_{21} \ln M_{21} dv - \int f_2 \ln f_2 dv \right).$$

Since $\int M \ln M dv = n \ln \left(\frac{n}{\sqrt{\frac{2\pi T}{m}}^3} \right) - \frac{3}{2}n$ for $M = \frac{n}{\sqrt{\frac{2\pi T}{m}}^3} \exp \left(-\frac{|v-u|^2}{\frac{2T}{m}} \right)$, we have

that

$$\nu_{12}n_2 \int M_{12} \ln M_{12} dv + \nu_{21}n_1 \int M_{21} \ln M_{21} dv$$

$$\leq \nu_{21}n_1 \int M_2 \ln M_2 dv + \nu_{12}n_2 \int M_1 \ln M_1 dv,$$

(2.17)

provided that

$$\nu_{12}n_2 n_1 \ln \frac{n_1}{\sqrt{2\pi \frac{T_{12}}{m_1}}^3} + \nu_{21}n_2 n_1 \ln \frac{n_2}{\sqrt{2\pi \frac{T_{21}}{m_2}}^3}$$

$$\leq \nu_{12}n_2 n_1 \ln \frac{n_1}{\sqrt{2\pi \frac{T_1}{m_1}}^3} + \nu_{21}n_2 n_1 \ln \frac{n_2}{\sqrt{2\pi \frac{T_2}{m_2}}^3},$$

which is equivalent to the condition (2.16) proven in lemma 2.1.5. With this inequality we get

$$S(f_1, f_2) \leq \nu_{12}n_2 \left[\int M_1 \ln M_1 dv - \int f_1 \ln f_1 dv \right]$$

$$+ \nu_{21}n_1 \left[M_2 \ln M_2 dv - \int f_2 \ln f_2 dv \right] \leq 0.$$

The last inequality follows from lemma 1.3.13. Here we also have equality if and only if $f_1 = M_1$ and $f_2 = M_2$, but since we already noticed that equality also implies $f_1 = M_{12}$ and $f_2 = M_{21}$, we also have $T_{21} = T_2 = T_1 = T_{12}$ and $u_1 = u_2 = u_{12} = u_{21}$. $\square$

Now, consider the total entropy $H(f_1, f_2)$ from definition 1.3.9. We can compute

$$\partial_t H(f_1, f_2) + \nabla_x \cdot \int (f_1 \ln f_1 + f_2 \ln f_2) v dv = S(f_1, f_2),$$

by multiplying the BGK equation (2.2) for species 1 by $\ln f_1$, the BGK equation (2.2) for the species 2 by $\ln f_2$ and integrating the sum with respect to v.

Corollary 2.1.7 (Entropy inequality for mixtures). *Assume $f_1, f_2 > 0$ to be a solution to (2.2). Assume a fast enough decay of f to zero for $v \to \infty$. Assume relationship (2.4), the conditions (2.5), (2.6), (2.7), (2.10) and (2.11) with $\alpha, \delta \neq 1$ and the positivity of the temperatures (2.14), then we have the following entropy inequality*

$$\partial_t \left(\int f_1 \ln f_1 dv + \int f_2 \ln f_2 dv \right) + \nabla_x \cdot \left(\int v f_1 \ln f_1 dv + \int v f_2 \ln f_2 dv \right) \leq 0,$$

with equality if and only if f_1 and f_2 are Maxwell distributions with equal mean velocity and temperature. Moreover at equilibrium the interspecies Maxwell distributions M_{12} and M_{21} satisfy $u_{12} = u_2 = u_1 = u_{21}$ and $T_{12} = T_2 = T_1 = T_{21}$.

We now explicitly specify the global equilibrium.

2.1.8 The structure of the equilibrium

Theorem 2.1.8 (Equilibrium). *Assume $f_1, f_2 > 0$ and the relationship (2.4), the conditions (2.5), (2.6), (2.7), (2.10) and (2.11) and the positivity of the temperatures (2.14). Then $Q_{11}(f_1, f_1) + Q_{12}(f_1, f_2) = 0$ and $Q_{22}(f_2, f_2) + Q_{21}(f_2, f_1) = 0$ if and only if f_1 and f_2 are Maxwell distributions with equal mean velocity and temperature.*

Proof. If $Q_{11}(f_1, f_1) + Q_{12}(f_1, f_2) = 0$ and $Q_{22}(f_2, f_2) + Q_{21}(f_2, f_1) = 0$, then $\int (\ln f_1 \ Q_{11}(f_1, f_1) + \ln f_1 \ Q_{12}(f_1, f_2) + \ln f_2 \ Q_{22}(f_2, f_2) + \ln f_2 \ Q_{21}(f_2, f_1)) dv = 0$ and so we have equality in the H-theorem. $\qquad\square$

2.1.9 Macroscopic equations

In the macroscopic equations in theorem 1.3.3, we did not compute the integrals on the right-hand side. In the case of Maxwellian molecules, when the collision kernel B in the Boltzmann equation does not depend on the relative velocity $|v - v_1|$, it is possible to compute these integrals. This is done in [1]. In general, this is to complicated or not possible. But, when we use the BGK model as approximations the integrals in the equations in 1.3.3 are just integrals of Maxwell distributions which we can easily compute. In this case we can specify the exchange terms of momentum and energy. This is done in the following theorem.

Theorem 2.1.9 (Macroscopic equations for the BGK equation for mixtures). *If $f_1, f_2 \in L^\infty(\mathbb{R}^3 \times \mathbb{R}^3 \times \mathbb{R}_0^+)$ decay fast enough to zero in the v variable and are a solution to (2.2) in the sense of distributions, they satisfy the following local macroscopic conservation laws.*

$$\partial_t n_1 + \nabla_x \cdot (n_1 u_1) = 0,$$
$$\partial_t n_2 + \nabla_x \cdot (n_2 u_2) = 0,$$
$$\partial_t (m_1 n_1 u_1) + \nabla_x \cdot \mathbb{P}_1 + \nabla_x \cdot (m_1 n_1 u_1 \otimes u_1) = f_{m_{12}},$$
$$\partial_t (m_2 n_2 u_2) + \nabla_x \cdot \mathbb{P}_2 + \nabla_x \cdot (m_2 n_2 u_2 \otimes u_2) = f_{m_{2,1}},$$
$$\partial_t \left(\frac{m_1}{2} n_1 |u_1|^2 + \frac{3}{2} n_1 T_1 \right) + \nabla_x \cdot Q_1 = F_{E_{1,2}},$$
$$\partial_t \left(\frac{m_2}{2} n_2 |u_2|^2 + \frac{3}{2} n_2 T_2 \right) + \nabla_x \cdot Q_2 = F_{E_{2,1}},$$

with $f_{m_{1,2}}, f_{m_{2,1}}, F_{E_{1,2}}$ and $F_{E_{2,1}}$ given by

$$f_{m_{1,2}} = -f_{m_{2,1}} = m_1 \nu_{12} n_1 n_2 (1 - \delta)(u_2 - u_1),$$
$$F_{m_{1,2}} = -F_{m_{2,1}} = \varepsilon \nu_{21} \frac{1}{2} n_2 n_1 m_1 \left((\delta^2 - 1)|u_1|^2 + (1 - \delta)^2 |u_2|^2 + 2\delta(1 - \delta) u_1 \cdot u_2 \right)$$
$$+ \frac{3}{2} \varepsilon \nu_{21} n_1 n_2 \left((1 - \alpha)(T_2 - T_1) + \gamma |u_1 - u_2|^2 \right).$$

Proof. The derivation of the left-hand side of the equations is exactly the same as in the proof of theorem 1.3.3. But for the BGK operators in (2.2) we can compute the integrals on the right-hand side. This is done in the proofs of the theorems 2.1.2 and 2.1.3 using the expressions for the mixture velocities (2.6) and (2.7), and the mixture temperatures (2.10) and (2.11). □

Remark 2.1.5. From theorem 2.1.9 we can observe a physical meaning of α and δ. We see that α and δ show up in the exchange terms of momentum and energy as a parameter in front of the relaxation of u_1 towards u_2 and T_1 towards T_2. So it determines together with the collision frequencies the speed of relaxation of the velocities and the temperatures to a common value.

Remark 2.1.6. The exchange terms of energy can be written in the following equivalent form

$$
\begin{aligned}
F_{E_{1,2}} &= -F_{E_{2,1}} \\
&= \left[\nu_{12} \frac{1}{2} n_1 n_2 m_1 (\delta - 1)(u_1 + u_2 + \delta(u_1 - u_2)) + \frac{1}{2} \nu_{12} n_1 n_2 \gamma (u_1 - u_2) \right] \cdot (u_1 - u_2) \\
&\quad + \frac{3}{2} \varepsilon \nu_{21} n_1 n_2 (1 - \alpha)(T_2 - T_1).
\end{aligned}
$$

2.2 Correlation to other models in the literature

In this section, we review models that have been previously introduced [1], [49] and [51]. [49] and [51] can be considered as special cases of the class described here. Thanks to this, all of them enjoy an H-theorem, conservation properties and positivity of the interspecies temperatures.

2.2.1 Special cases of this model in the literature

In this section we see that the two well-known models [49] and [51] are special cases of our model. In particular, this means that we can proof the H-theorem for these two often used models.

Model of Gross and Krook

In [49], Gross and Krook describe a plasma with two species of particles, species 1 and species 2 .

$$
\partial_t f_k + v \cdot \nabla_x f_k = \frac{n_k}{\sigma_{kk}} (M_k - f_k) + \frac{n_k}{\sigma_{kj}} (M_{kj} - f_k), \quad k, j = 1, 2, \ k \neq j.
$$

The parameters σ_{kj} are collision parameters, the terms $\frac{n_k}{\sigma_{kj}}$ are the collision frequencies from particles of species k with particles of species j. The Maxwell distributions are given by

$$M_k = \frac{n_k}{\sqrt{2\pi \frac{T_k}{m_k}}^3} \exp\left(-\frac{|v - u_k|^2}{2\frac{T_k}{m_k}}\right), \quad k = 1, 2,$$

$$M_{kj} = \frac{n_k}{\sqrt{2\pi \frac{T_{kj}}{m_k}}^3} \exp\left(-\frac{|v - u_{kj}|^2}{2\frac{T_{kj}}{m_k}}\right), \quad k, j = 1, 2, \; k \neq j,$$

with the mixed velocities

$$u_{12} = au_1 + (1 - a)u_2,$$

$$u_{21} = \left(1 + \frac{m_2}{m_1}a\right)u_1 - \frac{m_2}{m_1}au_2,$$

where $a \in \mathbb{R}$ is an undetermined coefficient, and the mixed temperatures

$$T_{21} = bT_2 + (1 - b)T_1 + A|u_2|^2 + Bu_1u_2 + C|u_1|^2,$$

$$T_{12} = bT_1 + (1 - b)T_2 + D|u_2|^2 + Eu_1u_2 + F|u_1|^2.$$

$b \in \mathbb{R}$ is an undetermined coefficient. Five out of the six coefficients A, B, C, D, E and F are determined by the conservation properties, assuming that $\sigma_{12} = \sigma_{21}$, and the assumption that for $t \to \infty$, u_1 and u_2 tend to an equal value and $T_{21}(\infty) = T_1(\infty)$, $T_{12}(\infty) = T_2(\infty)$. The model of Gross and Krook [49] is obtained from our BGK model by choosing $\varepsilon = 1$, while δ, α and γ are free parameters. In the case of a plasma they suggest $\delta = a = \frac{m_1}{m_1 + m_2}$. They also assume (2.5) for conservation of mass. They assume one of the mixture velocities to be a linear combination of u_1 and u_2, similar to (2.6) and deduce (2.7) from conservation of momentum. They further choose T_{12} of the form

$$T_{12} = \alpha T_1 + (1 - \alpha)T_2 + A|u_1|^2 + B|u_2|^2 + Cu_1u_2, \quad \alpha, A, B, C \in \mathbb{R},$$

and deduce with conservation of energy that T_{21} is given by

$$T_{21} = (1 - \alpha)T_1 + \alpha T_2 + D|u_1|^2 + E|u_2|^2 + Fu_1u_2,$$

where five of the variables $A, B, C, D, E \in \mathbb{R}$ are determined in order to get conservation of energy. From the present work the constants must be chosen in order to satisfy (2.14). In this case the model satisfies the H-theorem.

Model of Hamel

In [51], Hamel describes a kinetic model for a gas mixture where the two species interact via a molecular force $F_{kj} = m_k m_j p_{kj} \frac{1}{r^5}$ for $k, j = 1, 2$ where the parameters $p_{kj} \in \mathbb{R}$ describe the strength of the interaction. The model is given by

$$\partial_t f_k + v_k \cdot \nabla_x f_k = n_k \kappa_{kk} \left(\frac{n_k}{\sqrt{2\pi T_k / m_k}^3} \exp\left(-\frac{m_k |v_k - u_k|^2}{2T_k} \right) - f_k \right)$$

$$+ n_j \kappa_{kj} \left(\frac{n_{kj}}{\sqrt{2\pi T_{kj} / m_k}^3} \exp\left(-\frac{m_k |v_k - u_{kj}|^2}{2T_{kj}} \right) - f_k \right),$$

for $k, j = 1, 2, \ k \neq j$, where the functions κ_{kj}, u_{kj} and T_{kj} are given by

$$\kappa_{kj} = 2.66 (p_{ij}(m_i + m_j))^{\frac{1}{2}} \quad \text{for} \quad k, j = 1, 2, \ k \neq j,$$

$$\kappa_{kk} = 2.906 (p_{kk} m_k)^{\frac{1}{2}} \qquad \text{for} \quad k = 1, 2,$$

$$u_{ij} = u_i + \frac{m_j}{m_i + m_j}(u_j - u_i),$$

$$T_{ij} = T_i + 2\frac{m_i m_j}{(m_i + m_j)^2}(T_j - T_i) + 2\frac{m_i m_j}{(m_i + m_j)^2}|u_i - u_j|^2 \frac{m_j}{3}$$

$$= \frac{m_i^2 + m_j^2}{(m_i + m_j)^2} T_i + \frac{2m_i m_j}{(m_i + m_j)^2} T_j + 2\frac{m_i m_j}{(m_i + m_j)^2}|u_i - u_j|^2 \frac{m_j}{3} \text{ for } k, j = 1, 2, \ k \neq j.$$

Hamel's model [51] is obtained from our BGK model (2.2) by choosing $\varepsilon = 1$, $\delta = \frac{m_1}{m_1 + m_2}$, $\alpha = \frac{m_1^2 + m_2^2}{(m_1 + m_2)^2}$ and $\gamma = \frac{m_1 m_2}{(m_1 + m_2)^2} \frac{m_2}{3}$. The parameters are chosen in order to reproduce the fluxes of momentum and energy of Maxwellian molecules. His model also takes into account the physical considerations described in remark 2.1.3. The model satisfies condition (2.5) for conservation of mass. u_{12} and u_{21} satisfy condition (2.6) respectively (2.7) with this chosen δ and ε, so we have conservation of total momentum. T_{12} and T_{21} are of the form (2.10) and (2.11), respectively, so we have conservation of energy. The requirements for positivity of the temperature are satisfied, since

$$1 \geq \alpha = \frac{m_1^2 + m_2^2}{(m_1 + m_2)^2} \geq 0,$$

and conditions (2.14) and (2.15) then reduce to $m_1 \geq 0$ and $m_2 \geq 0$, so Hamel's model has positive temperatures and an H-theorem.

2.2.2 Comparison with the model of Andries, Aoki and Perthame

The next model also describes a gas mixture of Maxwellian molecules, but it contains only one term on the right-hand side [1].

$$\partial_t f_1 + v \cdot \nabla_x f_1 = (\nu_{11} n_1 + \nu_{12} n_2)(M^{(1)} - f_1),$$

$$\partial_t f_2 + v \cdot \nabla_x f_2 = (\nu_{22} n_2 + \nu_{21} n_1)(M^{(2)} - f_2).$$

The Maxwell distributions are given by

$$M^{(k)} = \frac{n_k}{\sqrt{2\pi \frac{T^{(k)}}{m_k}}^3} \exp\left(-\frac{m_k |v - u^{(k)}|^2}{2T^{(k)}}\right), \quad k = 1, 2,$$

with the interspecies velocities

$$u^{(1)} = u_1 + 2\frac{m_2}{m_1 + m_2} \frac{\chi_{12}}{\nu_{11}n_1 + \nu_{12}n_2} n_2(u_2 - u_1),$$

$$u^{(2)} = u_2 + 2\frac{m_1}{m_1 + m_2} \frac{\chi_{21}}{\nu_{22}n_2 + \nu_{21}n_1} n_1(u_1 - u_2),$$

and the interspecies temperatures

$$T^{(1)} = T_1 - \frac{m_1}{3}|u^{(1)} - u_1|^2$$
$$+ \frac{2}{3} \frac{m_1 m_2}{(m_1 + m_2)^2} \frac{4\chi_{12}}{\nu_{11}n_1 + \nu_{12}n_2} n_2 \left(\frac{3}{2}(T_2 - T_1) + m_2 \frac{|u_2 - u_1|^2}{2}\right),$$

$$T^{(2)} = T_2 - \frac{m_2}{3}|u^{(2)} - u_2|^2$$
$$+ \frac{2}{3} \frac{m_1 m_2}{(m_1 + m_2)^2} \frac{4\chi_{21}}{\nu_{22}n_2 + \nu_{21}n_1} n_1 \left(\frac{3}{2}(T_1 - T_2) + m_1 \frac{|u_1 - u_2|^2}{2}\right),$$

where $\chi_{12}, \chi_{21}, \nu_{12}$ and ν_{21} are parameters which are related to the differential cross section. For the detailed expressions see [1].

For $\chi_{12} = \chi_{21}$, the model also satisfies the conservation properties and the H-theorem with equality if and only if the distribution functions are Maxwell distributions with equal mean velocity and temperature.

Remark 2.2.1. The exchange of momentum of the species 1 in this model is given by

$$m_1(\nu_{11}n_1 + \nu_{12}n_2) \int v(M^{(1)} - f_1)dv = 2\frac{m_2 m_1}{m_2 + m_1} \chi_{12} n_1 n_2 (u_2 - u_1).$$

The exchange of the energy of species 1 is given by

$$\int \frac{m_1}{2}|v|^2 (\nu_{11}n_1 + \nu_{12}n_2)(M^{(1)} - f_1)dv$$
$$= n_1 n_2 \frac{2m_2 m_1 \chi_{12}}{(m_1 + m_2)} \left(-\frac{m_1}{m_1 + m_2}|u_1|^2 + \frac{m_2}{m_1 + m_2}|u_2|^2\right.$$
$$\left. + \frac{m_1 - m_2}{m_1 + m_2} u_1 u_2 + \frac{2}{m_1 + m_2}\frac{3}{2}(T_2 - T_1)\right).$$

So the model discussed here reproduces the same momentum and energy fluxes between the species (2.8), (2.9), (2.12) and (2.13) choosing the parameters δ, α and γ as:

$$\delta = -2\frac{m_2}{m_1 + m_2}\frac{\chi_{12}}{\nu_{12}} + 1,$$

$$\alpha = -4\frac{m_1 m_2}{(m_1 + m_2)^2}\frac{\chi_{12}}{\nu_{12}} + 1,$$

$$\gamma = \frac{4}{3}\frac{m_1 m_2^2}{(m_1 + m_2)^2}\frac{\chi_{12}}{\nu_{12}}\left(1 - \frac{\chi_{12}}{\nu_{12}}\right).$$

For $\chi_{12} \leq \nu_{12}$ the parameter γ is non-negative.

The model of Andries, Aoki and Perthame has another property, proposition 3.2 in [1], which the models described above do not have. It is called the indifferentiability principle. It denotes the following property:

Remark 2.2.2 (Indifferentiability principle). When the masses m_1 and m_2 and the collision frequencies $\nu_{11}, \nu_{12}, \nu_{21}$ and ν_{22} are identical, the total distribution function $f = f_1 + f_2$ obeys a single species BGK equation.

See also [21] for another model which also has the indifferentiability principle. The model in this chapter does not satisfy the indifferentiability principle. The indifferentiability principle in our model holds only in the global equilibrium. On physical grounds it is reasonable to assume that two species of identical particles become really indifferentiable when they have the same macroscopic speeds and temperatures.

Chapter 3

Determination of an unknown function in a macroscopic model of Dellacherie

This chapter will show the usefulness of our kinetic description in a macroscopic model by Dellacherie [32]. We want to use the model described in chapter 2 in order to determine an unknown function in the energy exchange in the macroscopic model of Dellacherie [32]. In section 3.1 we introduce the macroscopic model of Dellacherie and compare the moment equations of our kinetic model in section 3.2 with the model of Dellacherie in order to determine his unknown function in the energy exchange. This idea is also presented by Klingenberg and Pirner in [58].

3.1 Macroscopic model of Dellacherie

We consider the macroscopic model for a two component gas mixture from the literature [32]. Each gas consisting of particles of the mass m_k is characterized by a density n_k, a mean velocity u_k and an energy E_k, $k = 1, 2$. Dellacherie in [32] proposes a macroscopic model for gas mixtures given by

$$
\partial_t \begin{pmatrix} m_1 n_1 \\ m_2 n_2 \\ m_1 n_1 u_1 \\ m_2 n_2 u_2 \\ m_1 n_1 E_1 \\ m_2 n_2 E_2 \end{pmatrix} + \nabla_x \cdot \begin{pmatrix} m_1 n_1 u_1 \\ m_2 n_2 u_2 \\ m_1 n_1 u_1 \otimes u_1 + p_1 \mathbf{1} \\ m_2 n_2 u_2 \otimes u_2 + p_2 \mathbf{1} \\ u_1 (m_1 n_1 E_1 + p_1) \\ u_2 (m_2 n_2 E_2 + p_2) \end{pmatrix} = \begin{pmatrix} 0 \\ 0 \\ \lambda_u (u_2 - u_1) \\ \lambda_u (u_1 - u_2) \\ \lambda_T (T_2 - T_1) + \lambda_u U(u_1, u_2) \cdot (u_2 - u_1) \\ \lambda_T (T_1 - T_2) + \lambda_u U(u_1, u_2) \cdot (u_1 - u_2) \end{pmatrix},
$$

$$(3.1)$$

where $U(u_1, u_2)$ is an unknown function of the velocities u_1, u_2 and λ_u, λ_T are relaxation parameters determined by physical experiments. The temperature T_k and the pressure p_k are related by the equation of an ideal gas given by $p_k = n_k T_k$. The unknown function U is inside the relaxation term in the energy equations. Dellacherie [32] has the following restriction on U in the one dimensional case. He can show that his macroscopic model for gas mixtures satisfies an H-theorem as soon as U verifies the condition

$$
\min(u_1, u_2) \leq U(u_1, u_2) \leq \max(u_1, u_2). \tag{3.2}
$$

With this restriction on U in (3.2) Dellacherie is able to prove that for $\lambda_u, \lambda_T \to 0$ the model converges formally to a macroscopic model for the densities, the total momentum and the total energy. For details of the proof see [32].

3.2 Comparison of the energy exchanges

Now, our aim is to derive a macroscopic equation for the energy from the kinetic BGK model (2.2) with zero forces and to determine the parameter γ in the definition of the mixture temperature T_{12} in (2.10).

Lemma 3.2.1. *Assume (2.4), the conditions (2.5), (2.6) and (2.10). Then the momentum and energy exchange term of species 1 of the model (2.2) are given by*

$$F_{m_{1,2}} = m_1 \nu_{12} n_1 n_2 (1 - \delta)(u_2 - u_1),$$

(3.3)

$$F_{E_{1,2}} = \left[\frac{1}{2} m_1 \nu_{12} n_1 n_2 (\delta - 1)(u_1 + u_2 + \delta(u_1 - u_2)) \right.$$
$$\left. + \frac{3}{2} \nu_{12} n_1 n_2 \gamma (u_1 - u_2) \right] \cdot (u_1 - u_2) + \frac{3}{2} \nu_{12} n_1 n_2 (1 - \alpha)(T_1 - T_2).$$

(3.4)

The momentum and energy exchange terms of species 1 are obtained by multiplying the right-hand side of the first equation of (2.2) by v and $|v|^2$, respectively and integrating the result with respect to v, for more details see the proof of theorem 2.1.9 and remark 2.1.6. We will get the following relationship between the energy exchange of the two models (2.2) and (3.1).

Theorem 3.2.2. *Assume $\delta < 1$. The two energy exchange terms (3.4) and the one in (3.1) coincide if U is of the form*

$$U(u_1, u_2) = \frac{1}{2} \frac{(u_1 + u_2) \cdot (u_1 - u_2)}{|u_1 - u_2|^2} (u_1 - u_2) + c(u_1 - u_2) + V_\perp(u_1, u_2), \quad c \in \mathbb{R},$$

where $V_\perp$ is a function orthogonal to $u_1 - u_2$.

Proof. In order to have equality with the exchange term from Dellacherie, we want that

$$F_{E_{1,2}}^{vel} := \left[\frac{1}{2} m_1 \nu_{12} n_1 n_2 (\delta - 1)(u_1 + u_2 + \delta(u_1 - u_2)) + \frac{3}{2} \nu_{12} n_1 n_2 \gamma (u_1 - u_2) \right] \cdot (u_1 - u_2)$$
$$\overset{!}{=} -\lambda_u U(u_1, u_2) \cdot (u_1 - u_2),$$

which is equivalent to

$$\left[\frac{1}{2} m_1 \nu_{12} n_1 n_2 (\delta - 1)(u_1 + u_2 + \delta(u_1 - u_2)) + \frac{3}{2} \nu_{12} n_1 n_2 \gamma (u_1 - u_2) + \lambda_u U(u_1, u_2) \right]$$
$$\cdot (u_1 - u_2) = 0.$$

(3.5)

This means that

$$\left[\frac{1}{2} m_1 \nu_{12} n_1 n_2 (\delta - 1)(u_1 + u_2 + \delta(u_1 - u_2)) + \frac{3}{2} \nu_{12} n_1 n_2 \gamma (u_1 - u_2) + \lambda_u U(u_1, u_2) \right],$$

has to be orthogonal to $u_1 - u_2$.

We split all terms in a term parallel and a term orthogonal to $u_1 - u_2$:

$$U(u_1, u_2) = v(u_1, u_2)(u_1 - u_2) + V_\perp(u_1, u_2),$$

$$u_1 + u_2 = \left[(u_1 + u_2) \cdot \frac{(u_1 - u_2)}{|u_1 - u_2|} \right] \frac{u_1 - u_2}{|u_1 - u_2|} + u_\perp(u_1, u_2).$$

Now the fact that the whole expression has to be orthogonal to $u_1 - u_2$ means that the sum of coefficients in front of $u_1 - u_2$ in (3.5) has to vanish. This leads to

$$\left[\frac{1}{2} m_1 \nu_{12} n_1 n_2 (\delta - 1) \left(\frac{(u_1 + u_2) \cdot (u_1 - u_2)}{|u_1 - u_2|^2} + \delta \right) + \frac{3}{2} \nu_{12} n_1 n_2 \gamma + \lambda_u v(u_1, u_2) \right] = 0. \tag{3.6}$$

In order to get equality in the exchange terms of momentum (3.1) and (3.3), we have to choose

$$\delta = 1 - \frac{\lambda_u}{m_1 \nu_{12} n_1 n_2}. \tag{3.7}$$

If we use this expression for δ given by (3.7) and solve (3.6) for γ, we obtain

$$\begin{aligned}
\gamma &= \frac{1}{3} m_1 (1 - \delta) \frac{(u_1 + u_2) \cdot (u_1 - u_2)}{|u_1 - u_2|^2} + \frac{1}{3} m_1 (1 - \delta) \delta - \frac{2}{3} \lambda_u v(u_1, u_2) \frac{1}{n_1 n_2 \nu_{12}} \\
&= \frac{1}{3} m_1 (1 - \delta) \frac{(u_1 + u_2) \cdot (u_1 - u_2)}{|u_1 - u_2|^2} + \frac{1}{3} m_1 (1 - \delta) \delta - \frac{2}{3} m_1 (1 - \delta) v(u_1, u_2).
\end{aligned} \tag{3.8}$$

Since we assumed γ to be a parameter independent of the velocities, we deduce

$$v(u_1, u_2) = \frac{1}{2} \frac{(u_1 + u_2) \cdot (u_1 - u_2)}{|u_1 - u_2|^2} - c, \quad c \in \mathbb{R},$$

for $\delta < 1$. $\qquad\square$

For γ, this leads to

$$\gamma = \frac{1}{3} m_1 (1 - \delta) \delta + \frac{2}{3} m_1 (1 - \delta) c. \tag{3.9}$$

We also get a restriction on U like Dellacherie.

Lemma 3.2.3 (Restriction on the constant c). *If we assume that all temperatures are positive, we get the following restriction on the constant c given by*

$$-\frac{1}{2} \delta \leq c \leq -\frac{1}{2} \left(\frac{m_1}{m_2} \varepsilon (1 - \delta) \right) + \frac{1}{2}. \tag{3.10}$$

Proof. In order to have positive temperatures in the two species BGK model, we need that γ satisfies the condition (2.14).

We see from (3.7) that $\delta \leq 1$, since $\lambda_u, m_1, \nu_{12}, n_1, n_2$ are assumed to be positive. This leads to the restriction on the constant c given by (3.10). $\qquad\square$

γ is a non-negative number, so the right-hand side of the inequality in (2.14) must be non-negative. This condition is equivalent to (2.15). With this restriction on δ we can deduce from (3.10) the estimate

$$-\frac{1}{2} \le c \le \frac{1}{2}.$$

This corresponds to the estimate (3.2) on U in the one-dimensional case from [32]. With (3.10) we have a more restrictive estimate on the function U and with (3.8) an explicit expression of the parallel part of U. The orthogonal part does not matter because it does not enter in the exchange term.

3.3 Determine the constant c by symmetry arguments

In the kinetic model in chapter 2 the mixture temperature T_{12} of species 1 is given by (2.10) and the one of species 2 by (2.11). Due to symmetry arguments we choose the term in front of $|u_1 - u_2|^2$ in the temperature T_{21} such that it is equal to $\varepsilon\gamma = \frac{1}{3}\varepsilon m_1(1-\delta)\delta + \varepsilon\frac{2}{3}m_1(1-\delta)c$ using γ given by (3.9). Comparing the coefficient in front of $|u_1 - u_2|^2$ with this expression for $\varepsilon\gamma$ leads to a value for the constant c given by

$$c = \frac{1}{4}(1-\delta)\left(1 - \frac{m_1}{m_2}\varepsilon\right).$$

It remains to show that this specific c satisfies the estimates (3.10). First, the estimate from below. If we use (2.15) we obtain

$$c = -\frac{1}{4}\left(\frac{m_1}{m_2}\varepsilon - 1\right)(1-\delta) \ge -\frac{1}{4}\delta\left(1 + \frac{m_1}{m_2}\varepsilon\right)(1-\delta).$$

We rearrange (2.15) to

$$1 - \delta \le \frac{2}{1 + \frac{m_1}{m_2}\varepsilon}. \tag{3.11}$$

This leads to

$$c \ge -\frac{1}{2}\delta.$$

The estimate on this specific c from above is equivalent to

$$\frac{1}{4}\frac{m_1}{m_2}\varepsilon(1-\delta) + \frac{1}{4}(1-\delta) \le \frac{1}{2}.$$

By using (3.11) we get

$$\frac{1}{4}\left(\frac{m_1}{m_2}\varepsilon + 1\right)(1-\delta) \le \frac{1}{4}\frac{2}{1-\delta}(1-\delta) = \frac{1}{2}.$$

In summary we are able to determine more accurately the energy exchange in a model by Dellacherie.

Chapter 4

Existence, uniqueness and positivity of solutions for BGK models for mixtures

In chapter 2, we developed a model which we want to use in applications. Therefore, we are interested in the fact if there exists a solution. If yes, is the solution unique and in which function space lives the solution. Furthermore, the equations describe a time evolution of distribution functions. Therefore, we want to ensure that in our model the distribution functions remain positive in time if we start with positive initial data. We want to prove these properties in this chapter. Moreover, it turns out that the strategy of proving these properties can easily be applied to the model of Andries, Aoki and Perthame in [1] presented in section 2.2.2. So in this chapter, our aim is to prove existence, uniqueness and positivity of solutions to the BGK model for mixtures developed in chapter 2 and the model of Andries, Aoki and Perthame in [1]. This chapter is largely motivated by the paper of Perthame and Pulvirenti [70] where the global existence of mild solutions of the BGK equation for one species was established, and [87] where global existence of mild solutions of the ES-BGK model for one species is shown. The ES-BGK model is an extension of the BGK model and is described later in chapter 7. There is also a result concerning the Boltzmann equation for mixtures in a similar fashion in [50]. These properties in this chapter are also presented in [58] by Klingenberg and Pirner.

The outline of the chapter is as follows: In section 4.1 we want to introduce the notion of solutions which we will consider in this chapter. In section 4.2.1 we repeat the BGK model for two species developed in chapter 2, and in section 4.2.2 we repeat the model of Andries, Aoki and Perthame such that it is possible to read this chapter independently of the others and add further assumptions in order to prove existence and uniqueness. In section 4.3.1, we prove bounds on the macroscopic quantities which we need in order to show existence and uniqueness of non-negative solutions in section 4.3.2. In section 4.4 we will deduce that all classical solutions with positive initial data remain positive for all later times.

4.1 Notion of solutions

In this section we want to present the notion of solutions we will consider in this chapter and compare it with other notions used in the literature. In this section we want to illustrate the notion of solutions with the help of the single inhomogeneous transport equation. The aim of introducing mild solutions is to allow solutions with a lower regularity than in the classical sense.

4.1.1 Motivation of mild solutions

We now consider the problem

$$\partial_t f^{inhom} + v \cdot \nabla_x f^{inhom} + d\, f^{inhom} = \mathcal{F} \quad \text{in} \quad \mathbb{R}^3 \times \mathbb{R}^3 \times (0, T), \qquad (4.1)$$

where $d : \mathbb{R}^3 \times [0, T] \to \mathbb{R}^3$ and $\mathcal{F} : \mathbb{R}^3 \times \mathbb{R}^3 \times [0, T] \times \mathbb{R} \to \mathbb{R}$ are given integrable functions, $d = d(x, t)$, $\mathcal{F} = \mathcal{F}(x, v, t, f^{inhom}(x, v, t))$, and $f^{inhom} : \mathbb{R}^3 \times \mathbb{R}^3 \times \mathbb{R}^+ \to \mathbb{R}$ is the unknown, $f^{inhom} = f^{inhom}(x, v, t)$. Here $(x, v) \in \mathbb{R}^3 \times \mathbb{R}^3$ denotes a point in the position-velocity space called phase space and $t \geq 0$ denotes the time.

First we assume that $\mathcal{F}$ is independent of f^{inhom} and $d = 0$. We require an initial condition given by $f^{inhom}(x, v, 0) = f^{inhom,0}(x, v)$. Assume it exists a classical solution to (4.1). Then we can find the explicit expression of the solution with the following method also described in Evans [40]. We fix $(x, v, t) \in \mathbb{R}^3 \times \mathbb{R}^3 \times \mathbb{R}^+$ and consider the function $z(s) := f^{inhom}(x + sv, v, t + s)$ for $s \in \mathbb{R}^+$. Note that the arguments of f^{inhom} in the definition of z coincide with the characteristic curves of the homogeneous transport equation computed in example 1.1.1 in section 1.1. In this case we obtain for the derivative of z with respect to s

$$\frac{d}{ds} z(s) = \partial_t f^{inhom}(x + sv, v, t + s) + v \cdot \nabla_x f^{inhom}(x + sv, v, t + s)$$
$$= \mathcal{F}(x + sv, v, t + s). \qquad (4.2)$$

Consequently

$$f^{inhom}(x, v, t) - f^{inhom,0}(x - tv, v) = z(0) - z(-t) = \int_{-t}^{0} \frac{d}{ds} z(s) ds$$
$$= \int_{-t}^{0} \mathcal{F}(x + sv, v, t + s) = \int_{0}^{t} \mathcal{F}(x + (s - t)v, v, s) ds.$$

So we obtain a solution of f^{inhom} given by

$$f^{inhom}(x, v, t) = f^{inhom,0}(x - tv, v) + \int_{0}^{t} \mathcal{F}(x + (s - t)v, v, s) ds.$$

If $d(x, t) \neq 0$, we can define $g(x, v, t) = f^{inhom}(x, v, t) e^{\alpha(x, v, t)}$ with some differentiable function $\alpha(x, v, t)$ determined later. Then

$$\partial_t g(x, v, t) = \partial_t f^{inhom}(x, v, t) e^{\alpha(x, v, t)} + f^{inhom}(x, v, t) e^{\alpha(x, v, t)} \partial_t \alpha(x, v, t)$$
$$= \partial_t f^{inhom}(x, v, t) e^{\alpha(x, v, t)} + g(x, v, t) \partial_t \alpha(x, v, t),$$

and

$$\nabla_x g(x, v, t) = \nabla_x f^{inhom}(x, v, t) e^{\alpha(x, v, t)} + g(x, v, t) \nabla_x \alpha(x, v, t).$$

By using (4.1) we get

$$\partial_t g(x, v, t) + v \cdot \nabla_x g(x, v, t)$$
$$= \left(-d(x, t) f^{inhom}(x, v, t) + \mathcal{F}(x, v, t) \right) e^{\alpha(x, v, t)} + g(x, v, t) \left(\partial_t \alpha(x, v, t) + v \cdot \nabla_x \alpha(x, v, t) \right)$$
$$= -d(x, t) g(x, v, t) + \mathcal{F}(x, v, t) e^{\alpha(x, v, t)} + g(x, v, t) \left(\partial_t \alpha(x, v, t) + v \cdot \nabla_x \alpha(x, v, t) \right).$$

Now choose $\alpha(x, v, t)$ such that

$$\partial_t \alpha(x, v, t) + v \cdot \nabla_x \alpha(x, v, t) = d(x, t) \quad \text{and} \quad \alpha(x, v, 0) = 0,$$

so we choose $\alpha(x, v, t)$ as a solution of the inhomogeneous transport equation

$$\alpha(x, v, t) = \int_0^t d(x + (s - t)v, s)ds.$$

The initial value of α is chosen such that f^{inhom} and g have the same initial values. Then g solves

$$\partial_t g(x, v, t) + v \cdot \nabla_x g(x, v, t) = \mathcal{F}(x, v, t)e^{\alpha(x,v,t)},$$

or in integral form

$$g(x, v, t) = g_0(x - tv, v) + \int_0^t \mathcal{F}(x + (s - t)v, v, s)e^{\alpha(x+(s-t)v,v,s)}ds,$$

so

$$f^{inhom}(x, v, t) = e^{-\alpha(x,v,t)} f^{inhom,0}(x - tv, v, t)$$

$$+ e^{-\alpha(x,v,t)} \int_0^t \mathcal{F}(x + (s - t)v, v, s)e^{\alpha(x+(s-t)v,v,s)}ds$$

also for $d \neq 0$. So we obtained a solution f^{inhom} even for $d \neq 0$. If in addition $\mathcal{F}$ depends on f^{inhom}, we do not obtain a solution to f^{inhom} with this strategy but an integral equation for f^{inhom} given by

$$f^{inhom}(x, v, t) = e^{-\alpha(x,v,t)} f^{inhom,0}(x - vt, v, t)$$

$$+ e^{-\alpha(x,v,t)} \int_0^t \mathcal{F}(x + (s - t)v, v, s, f^{inhom}(x, v, t))e^{\alpha(x+(s-t)v,v,s)}ds.$$

$$(4.3)$$

This integral equation is equivalent to equation 4.1 for classical solutions and smooth functions $\mathcal{F}$, $f^{inhom,0}$ and d. But (4.3) can also have solutions with lower regularity. We call this solutions mild solutions.

Definition 4.1.1. Let $f^{inhom,0}$ be measurable, then we call f^{inhom} a mild solution to (4.1) if and only if

- For almost all x, v we have $\mathcal{F}(x - vs, v, s, f^{inhom}(x - vs, v, s)), d(x - vs, s) \in L^1(0, T)$.

- For all $t \in [0, T)$ and for almost all x, f^{inhom} satisfies

$$f^{inhom}(x, v, t) = f^{inhom,0}(x - vt, v, t)$$

$$+ e^{-\alpha(x,v,t)} \int_0^t \mathcal{F}(x + (s - t)v, v, s, f^{inhom}(x + (s - t)v, v, t))e^{\alpha(x+(s-t)v,v,s)}ds.$$

4.1.2 Connection to weak solutions

Another commonly used extension to a solution with lower regularity is given by weak solutions, especially in the study of the macroscopic equations. In this section we want to compare the notion of mild solutions with weak solutions in the sense of distributions.

Definition 4.1.2 (Test functions). We define $\mathcal{D} = \mathcal{D}([0,T] \times \mathbb{R}^3 \times \mathbb{R}^3)$ as the space of all $C^\infty([0,T] \times \mathbb{R}^3 \times \mathbb{R}^3)$ functions with compact support in $[0,T) \times \mathbb{R}^3 \times \mathbb{R}^3$, where by saying that the support is compact in $[0,T)$, we mean that the support can include 0, but must not contain T.

Definition 4.1.3 (Weak solution). By saying that f^{inhom} solves (4.1) in $\mathcal{D}'$ we mean that for all $\phi \in \mathcal{D}$ we have

$$\int_0^T \int_{\mathbb{R}^3} \int_{\mathbb{R}^3} f^{inhom}(-\partial_t \phi - v \cdot \nabla_x \phi + d\,\phi)dxdvdt - \int_{\mathbb{R}^3} \int_{\mathbb{R}^3} f^{inhom,0}\phi(0,x,v)dxdv$$

$$= \int_0^T \int_{\mathbb{R}^3} \int_{\mathbb{R}^3} \mathcal{F}(x,v,t,f^{inhom}(x,v,t))\phi dxdvdt.$$

This makes sense for $f^{inhom} \in L^1$ if we have the conditions on the coefficients

$$d \in L_{loc}^1, \; \mathcal{F} \in L_{loc}^1.$$

We call such a solution weak solution or solution in the sense of distributions.

We have the following equivalence between mild and weak solutions.

Theorem 4.1.1. *If $f^{inhom} \in L_{loc}^1(\mathbb{R}^3 \times \mathbb{R}^3 \times [0,T))$ and $\mathcal{F} \in L_{loc}^1(\mathbb{R}^3 \times \mathbb{R}^3 \times [0,T) \times \mathbb{R})$, then mild and distributional solutions are equivalent.*

Proof. The proof is given in [9] but for $\mathcal{F}$ independent of f^{inhom}. We want to show the proof here but for $\mathcal{F}$ dependent of f^{inhom}. Without loss of generality we assume $d = 0$. Otherwise we can merge it into $\mathcal{F}$. Let f^{inhom} be a distributional solution in the sense of definition 4.1.3. Fix a test function $\phi \in \mathcal{D}(\mathbb{R}^3 \times \mathbb{R}^3)$. Let $\rho_n \in \mathcal{D}([0,T))$, $n \in \mathbb{N}$ with $\rho_n(0) = 0$ and $\rho_n \to \mathbf{1}_{[t_1,t_2]}$ in $\mathcal{D}'([0,T))$. The definition of convergence in the sense of distributions can be found for example in definition 16.2 in volume 2 of [33]. Then we obtain

$$0 = \langle \partial_t f^{inhom} + v \cdot \nabla_x f^{inhom} - \mathcal{F}, \phi(x - vt, v)\rho_n(t)\rangle$$
$$= -\langle f^{inhom}, (\partial_t + v \cdot \nabla_x)(\phi(x - vt, v)\rho_n(t))\rangle - \langle \mathcal{F}, \phi(x - vt)\rho_n(t)\rangle$$
$$= -\langle f^{inhom}, \rho_n(t)(\partial_t + v \cdot \nabla_x)\phi(x - vt, v) + \phi(x - vt)\rho_n'(t)\rangle - \langle \mathcal{F}, \phi(x - vt)\rho_n(t)\rangle$$
$$= -\langle f^{inhom}(x + vt, v, t), 0 + \phi\rho_n'(t)\rangle - \langle \mathcal{F}(x + vt, v, t, f^{inhom}(x + vt, v, t)), \phi\rho_n(t)\rangle.$$
$$\tag{4.4}$$

Here, the notation $\langle \cdot \rangle$ is the action of distributions in $\mathcal{D}'([0,T) \times \mathbb{R}^3 \times \mathbb{R}^3)$. The equality $(\partial_t + v \cdot \nabla_x)\phi(x - tv, v) = 0$ follows from the chain rule applied to the smooth function

ϕ. The last equality follows from a change of variables in the integrations. As ϕ was arbitrary, we have

$$\int_0^T (\rho_n(t)\mathcal{F}(x+vt,v,t,f^{inhom}(x+vt,v,t)) + \rho_n'(t)f^{inhom}(x+vt,v,t))dt = 0,$$

$$(4.5)$$

for almost all (x,v). We then let $n \to \infty$ to obtain that for almost all x

$$f^{inhom}(x+vt_2,v,t_2) - f^{inhom}(x+vt_1,v,t_1)$$

$$= \int_{t_1}^{t_2} \mathcal{F}(x+vs,v,s,f^{inhom}(x+vs,v,s))ds,$$

$$(4.6)$$

which is equivalent to

$$f^{inhom}(x,v,t) - f^{inhom}(x-vt,v,0) = \int_0^t \mathcal{F}(x+v(s-t),v,s,f^{inhom}(x+v(s-t)))ds,$$

for $t = t_2$ and $t_1 = 0$ and a change in the notation $x+vt$ to x. The converse comes from reversing this argument. We start with equation (4.6). We take an arbitrary test function $\phi \in \mathcal{D}(\mathbb{R}^3 \times \mathbb{R}^3)$. We obtain

$$\int_{\mathbb{R}^3}\int_{\mathbb{R}^3} f^{inhom}(x+vt_2,v,t_2)\phi(x,v)dxdy = \int_{\mathbb{R}^3}\int_{\mathbb{R}^3} f^{inhom}(x+vt_1,v,t_1)\phi(x,v)dxdv$$

$$+ \int_{t_1}^{t_2}\int_{\mathbb{R}^3}\int_{\mathbb{R}^3} \mathcal{F}(x+vs,v,s,f^{inhom}(x+vs,v,s))\phi(x,v)dxdvds.$$

Take any $\psi \in \mathcal{D}([0,T))$ and fix $t_1 = 0$, $t_2 = \tau$. Multiply the obtained equation by $\psi'(\tau)$ and integrate with respect to τ.

$$\langle f^{inhom}(x+v\tau,v,\tau)\phi(x,v)\psi'(\tau)\rangle - \int\int \mathcal{F}(x,v,0,f^{inhom}(x,v,t))\phi(x,v)\psi(0)dxdv$$

$$= \int_0^T \int_0^\tau \int_{\mathbb{R}^3}\int_{\mathbb{R}^3} \mathcal{F}(x+vs,v,s,f^{inhom}(x+vs,v,s))\phi(x,v)\psi'(\tau)dxdvdsd\tau$$

$$= \int_0^T \int_s^T \int_{\mathbb{R}^3}\int_{\mathbb{R}^3} \mathcal{F}(x+vs,v,s,f^{inhom}(x+vs,v,s))\phi(x,v)\psi'(\tau)dxdvdsd\tau$$

$$= \int_0^T \int_{\mathbb{R}^3}\int_{\mathbb{R}^3} \mathcal{F}(x+vs,v,s,f^{inhom}(x+vs,v,s))\phi(x,v)\int_s^T \psi'(\tau)d\tau dxdvds$$

$$= -\int_0^T \int_{\mathbb{R}^3}\int_{\mathbb{R}^3} \mathcal{F}(x+vs,v,s,f^{inhom}(x+vs,v,s))\phi(x,v)\psi(s)dxdvds$$

$$= -\langle \mathcal{F}(x+vs,v,s,f^{inhom}(x+vs,v,s))\phi(x,v)\psi(s)\rangle.$$

So we obtain by changing the notation of the variable s into τ in the last integral

$$\langle f^{inhom}(x+v\tau,v,\tau)\phi(x,v)\psi'(\tau)\rangle - \int\int \mathcal{F}(x,v,0,f^{inhom}(x,v,t))\phi(x,v)\psi(0)dxdv$$

$$= -\langle \mathcal{F}(x+v\tau,v,\tau,f^{inhom}(x+v\tau,v,\tau))\phi(x,v)\psi(\tau)\rangle.$$

The corresponds to the last line in (4.4) equal to zero for arbitrary test functions plus a term with additional initial values since the general test function has not be have zero initial data and can be transformed to the first line first line of (4.4) with the same calculation done in (4.4) but in the reversed direction. All in all, we obtain that f^{inhom} is a weak solution according to the definition 4.1.3. □

Remark 4.1.1. This theorem can be easily extended to a system of two equations

$$\partial_t f_k^{inhom} + v \cdot \nabla_x f_k^{inhom} + d_k \, f_k^{inhom} = \mathcal{F}_k \quad \text{in} \quad \mathbb{R}^3 \times \mathbb{R}^3 \times (0, T), \quad k = 1, 2,$$

where $d_k : \mathbb{R}^3 \times [0, T] \to \mathbb{R}^3$ and $\mathcal{F}_k : \mathbb{R}^3 \times \mathbb{R}^3 \times [0, T] \times \mathbb{R} \times \mathbb{R} \to \mathbb{R}$ are given functions, $d_k = d_k(x, t)$, $\mathcal{F}_k = \mathcal{F}_k(x, v, t, f_1^{inhom}(x, v, t), f_2^{inhom}(x, v, t))$, and $f_k^{inhom} : \mathbb{R}^3 \times \mathbb{R}^3 \times \mathbb{R}^+ \to \mathbb{R}$ are the unknown, $f_k^{inhom} = f_k^{inhom}(x, v, t)$, $k = 1, 2$. In the next sections we will see that the BGK model for gas mixtures presented in chapter 2 satisfies the requirements of theorem 4.1.1. Therefore the existence and uniqueness of mild solutions of the model proposed in chapter 2 is equivalent to the existence and uniqueness of weak solutions.

4.2 BGK models for mixtures

In this section we will repeat the two types of BGK models for gas mixtures, the one from chapter 2 developed by Klingenberg, Pirner and Puppo and the one in [1] by Andries, Aoki and Perthame such that this chapter can be read independently of the other chapters.

4.2.1 BGK approximation for mixtures with two collision terms

We will repeat the definitions which are needed to follow this chapter such that it is possible to read this chapter independently. Since we consider a mixture composed of two different species, our kinetic model has two distribution functions $f_1(x, v, t) > 0$ and $f_2(x, v, t) > 0$ where $x \in \mathbb{R}^N$ and $v \in \mathbb{R}^N$, $N \in \mathbb{N}$ are the phase space variables and $t \geq 0$ the time. Note, that the proof of existence, uniqueness and positivity is independent of the dimension. So we replace the dimension 3 from the previous chapters by the variable dimension N.

Furthermore, for any $f_1, f_2 : \Lambda \times \mathbb{R}^N \times \mathbb{R}_0^+ \to \mathbb{R}$, $\Lambda \subset \mathbb{R}^N$ with $(1 + |v|^2) f_1, (1 + |v|^2) f_2 \in L^1(dv)$, $f_1, f_2 \geq 0$ we relate the distribution functions to macroscopic quantities by mean-values of f_k, $k = 1, 2$

$$\int f_k(v) \begin{pmatrix} 1 \\ v \\ m_k |v - u_k|^2 \end{pmatrix} dv =: \begin{pmatrix} n_k \\ n_k u_k \\ N n_k T_k \end{pmatrix}, \quad k = 1, 2, \tag{4.7}$$

where n_k is the number density, u_k the mean velocity and T_k the mean temperature of species k, $k = 1, 2$. Note that we shall write T_k instead of $k_B T_k$, where k_B is Boltzmann's constant.

We consider the model presented in chapter 2 given by

$$\partial_t f_1 + \nabla_x \cdot (v f_1) = \nu_{11} n_1 (M_1 - f_1) + \nu_{12} n_2 (M_{12} - f_1),$$
$$\partial_t f_2 + \nabla_x \cdot (v f_2) = \nu_{22} n_2 (M_2 - f_2) + \nu_{21} n_1 (M_{21} - f_2),$$
$$f_1(t = 0) = f_1^0,$$
$$f_2(t = 0) = f_2^0,$$

(4.8)

with the Maxwell distributions

$$M_k(x, v, t) = \frac{n_k}{\sqrt{2\pi \frac{T_k}{m_k}}^N} \exp(-\frac{|v - u_k|^2}{2\frac{T_k}{m_k}}), \quad k = 1, 2,$$

$$M_{12}(x, v, t) = \frac{n_{12}}{\sqrt{2\pi \frac{T_{12}}{m_1}}^N} \exp(-\frac{|v - u_{12}|^2}{2\frac{T_{12}}{m_1}}),$$

(4.9)

$$M_{21}(x, v, t) = \frac{n_{21}}{\sqrt{2\pi \frac{T_{21}}{m_2}}^N} \exp(-\frac{|v - u_{21}|^2}{2\frac{T_{21}}{m_2}}).$$

Within the next page the unknown variables will be explained. $\nu_{11} n_1$ and $\nu_{22} n_2$ are the collision frequencies of the particles of each species with itself, while $\nu_{12} n_2$ and $\nu_{21} n_1$ are related to interspecies collisions. To be flexible in choosing the relationship between the collision frequencies, we now assume the relationship

$$\nu_{12} = \varepsilon \nu_{21}, \quad 0 < \varepsilon \leq 1.$$

(4.10)

The restriction on ε is without loss of generality. If $\varepsilon > 1$, exchange the notation 1 and 2 and choose $\frac{1}{\varepsilon}$. In addition, we assume that all collision frequencies are positive. For the existence and uniqueness proof we assume the following restrictions on our collision frequencies

$$\nu_{jk}(x, t) n_k(x, t) = \tilde{\nu}_{jk} \frac{n_k(x, t)}{n_1(x, t) + n_2(x, t)}, \quad j, k = 1, 2,$$

(4.11)

with constants $\tilde{\nu}_{11}, \tilde{\nu}_{12}, \tilde{\nu}_{21}, \tilde{\nu}_{22} > 0$. This means that the collision frequencies are given by a constant times the relative density. This makes also sense from the physical point of view. The collision frequencies are related to the speed of relaxation to equilibrium. The speed of the different types of relaxations due to different types of interactions depends on the proportion of the densities compared to the total densities and not from the total values of the densities.

The Maxwell distributions M_1 and M_2 in (4.9) have the same moments as f_1 and f_2, respectively. With this choice, we guarantee the conservation of mass, momentum and energy in interactions of one species with itself (see section 2.1.5). The remaining parameters $n_{12}, n_{21}, u_{12}, u_{21}, T_{12}$ and T_{21} will be determined using conservation of the number of particles, of total momentum and total energy, together with some symmetry considerations. If we assume that

$$n_{12} = n_1 \quad \text{and} \quad n_{21} = n_2,$$

(4.12)

we have conservation of the number of particles, see theorem 2.1.1. If we further assume that u_{12} is a linear combination of u_1 and u_2

$$u_{12} = \delta u_1 + (1 - \delta)u_2, \quad \delta \in \mathbb{R}, \tag{4.13}$$

then we have conservation of total momentum provided that

$$u_{21} = u_2 - \frac{m_1}{m_2}\varepsilon(1 - \delta)(u_2 - u_1), \tag{4.14}$$

see theorem 2.1.2. If we further assume that T_{12} is of the following form

$$T_{12} = \alpha T_1 + (1 - \alpha)T_2 + \gamma|u_1 - u_2|^2, \quad 0 \leq \alpha \leq 1, \gamma \geq 0, \tag{4.15}$$

then we have conservation of total energy provided that

$$T_{21} = \left[\frac{1}{N}\varepsilon m_1(1 - \delta)\left(\frac{m_1}{m_2}\varepsilon(\delta - 1) + \delta + 1 \right) - \varepsilon\gamma \right]|u_1 - u_2|^2 \\ + \varepsilon(1 - \alpha)T_1 + (1 - \varepsilon(1 - \alpha))T_2, \tag{4.16}$$

see theorem 2.1.3. In order to ensure the positivity of all temperatures, we need to restrict δ and γ to

$$0 \leq \gamma \leq \frac{m_1}{N}(1 - \delta)\left[\left(1 + \frac{m_1}{m_2}\varepsilon\right)\delta + 1 - \frac{m_1}{m_2}\varepsilon \right], \tag{4.17}$$

and

$$\frac{\frac{m_1}{m_2}\varepsilon - 1}{1 + \frac{m_1}{m_2}\varepsilon} \leq \delta \leq 1, \tag{4.18}$$

see theorem 2.1.4 in chapter 2 for $N = 3$.

In the following, we want to study mild solutions of (4.8).

Definition 4.2.1. We call (f_1, f_2) with $(1 + |v|^2)f_k \in L^1(dv)$, $f_1, f_2 \geq 0$ a mild solution to (4.8) under the conditions of the collision frequencies (4.11) if and only if f_1, f_2 satisfy

$$f_k(x, v, t) = e^{-\alpha_k(x,v,t)}f_k^0(x - tv, v)$$

$$+ e^{-\alpha_k(x,v,t)} \int_0^t [\tilde{\nu}_{kk} \frac{n_k(x + (s - t)v, s)}{n_k(x + (s - t)v, s) + n_j(x + (s - t)v, s)} M_k(x + (s - t)v, v, s)$$

$$+ \tilde{\nu}_{kj} \frac{n_j(x + (s - t)v, s)}{n_k(x + (s - t)v, s) + n_j(x + (s - t)v, s)} M_{kj}(x + (s - t)v, v, s)]e^{\alpha_k(x + (s - t)v, v, s)}ds,$$

where α_k is given by

$$\alpha_k(x, v, t) = \int_0^t [\tilde{\nu}_{kk} \frac{n_k(x + (s - t)v, s)}{n_k(x + (s - t)v, s) + n_j(x + (s - t)v, s)}$$

$$+ \tilde{\nu}_{kj} \frac{n_j(x + (s - t)v, s)}{n_k(x + (s - t)v, s) + n_j(x + (s - t)v, s)}]ds,$$

for $k, j = 1, 2$, $k \neq j$.

By construction, a classical solution is always a mild solution. But in order to also allow solutions with a lower regularity, in the following, we want to study existence, uniqueness and positivity of mild solutions.

4.2.2 BGK approximation for mixtures with one collision term

The next model describes a gas mixture of Maxwellian molecules, but it contains only one term on the right-hand side [1].

$$\partial_t f_1 + v \cdot \nabla_x f_1 = (\nu_{11} n_1 + \nu_{12} n_2)(M^{(1)} - f_1),$$
$$\partial_t f_2 + v \cdot \nabla_x f_2 = (\nu_{22} n_1 + \nu_{21} n_1)(M^{(2)} - f_2).$$

The Maxwell distributions are given by

$$M^{(k)} = \frac{n_k}{\sqrt{2\pi \frac{T^{(k)}}{m_k}}^3} \exp\left(-\frac{m_k |v - u^{(k)}|^2}{2T^{(k)}}\right), \quad k = 1, 2,$$

with the interspecies velocities

$$u^{(k)} = u_k + 2\frac{m_j}{m_k + m_j}\frac{\chi_{kj}}{\nu_{kk} n_k + \nu_{kj} n_j} n_j (u_k - u_j), \quad k, j = 1, 2, \ k \neq j,$$

and the interspecies temperatures

$$T^{(k)} = T_k - \frac{m_k}{3}|u^{(k)} - u_k|^2$$
$$+ \frac{2}{3}\frac{m_k m_j}{(m_k + m_j)^2}\frac{4\chi_{kj}}{\nu_{kk} n_k + \nu_{kj} n_j} n_j \left(\frac{3}{2}(T_k - T_j) + m_k \frac{|u_j - u_k|^2}{2}\right), \quad (4.19)$$
$$\text{for} \quad k, j = 1, 2, \ k \neq j,$$

where $\chi_{12}, \chi_{21}, \nu_{12}$ and ν_{21} are parameters which are related to the differential cross section. For the detailed expressions see [1]. We still assume for the existence proof that the collision frequencies have the shape given in (4.11).

Definition 4.2.2. We call (f_1, f_2) with $(1+|v|^2) f_k \in L^1(dv)$, $f_1, f_2 \geq 0$ a mild solution to (8.2) under the conditions of the collision frequencies (4.11) if and only if f_1, f_2 satisfy

$$f_k(x, v, t) = e^{-\alpha_k(x,v,t)} f_k^0(x - tv, v)$$
$$+ e^{-\alpha_k(x,v,t)} \int_0^t [\tilde{\nu}_{kk} \frac{n_k(x + (s - t)v, s)}{n_k(x + (s - t)v, s) + n_j(x + (s - t)v, s)}$$
$$+ \tilde{\nu}_{kj} \frac{n_j(x + (s - t)v, s)}{n_k(x + (s - t)v, s) + n_j(x + (s - t)v, s)}] M^{(k)}(x + (s - t)v, v, s)] e^{\alpha_k(x+(s-t)v,v,s)} ds,$$

where α_k is given given as in definition 4.2.1, $k, j = 1, 2, \ k \neq j$.

4.3 Existence and uniqueness of mild solutions for the two species BGK model

In section 4.3.1, we start considering several estimates on the macroscopic quantities which we will use in section 4.3.2 for the existence and uniqueness of mild solutions. This will be done for the model described in section 4.2.1. The proof for the model presented in section 4.2.2 is very similar. So we just illustrate this in remarks.

4.3.1 Estimates on the macroscopic quantities

First, we present some estimates on macroscopic quantities which we need later for the existence and uniqueness proof.

Theorem 4.3.1. *For any pair of functions (f_1, f_2) with $(1+|v|^2)f_k \in L^1(dv)$, $f_1, f_2 \geq 0$, we define the moments and macroscopic parameters as in (4.7), (4.13), (4.14), (4.15) and (4.16) and set*

$$N_q(f_k) = \sup_v |v|^q f_k(v), \quad q \geq 0, k = 1, 2. \tag{4.20}$$

Then the following estimates hold

(i.1) $\frac{n_k}{T_k^{N/2}} \leq C N_0(f_k)$ *for* $k = 1, 2$,

(i.2) $\frac{n_1}{T_{12}^{N/2}} \leq C N_0(f_1)$,

(i.3) $\frac{n_2}{T_{21}^{N/2}} \leq C N_0(f_2)$.

Proof. The proof of $(i.1)$ is exactly the same as the proof of the inequality (2.2) in [70]. We want to repeat it here for the convenience of the reader. We consider $n_k = \int f_k(v)dv$ for $k = 1, 2$. We split the integration with respect to the velocity v into $|u_k - v| > R_k$ and $|u_k - v| \leq R_k$ for some R_k determined later. Then in the first integral we have $1 < \frac{|u_k - v|^2}{R_k^2}$ and obtain

$$n_k \leq \frac{1}{R_k^2} \int_{|u_k - v| > R_k} |v - u_k|^2 f_k(v)dv + \int_{|u_k - v| \leq R_k} f_k(v)dv.$$

The first integral is linked to the temperature T_k defined by (4.7) and the second integral can be estimated by the supremum of f_k. So

$$n_k \leq \frac{n_k N T_k}{m_k R_k^2} + C R_k^N N_0(f_k).$$

Now we choose R_k as $R_k = \left(\frac{n_k T_k}{N_0(f_k)}\right)^{\frac{1}{N+2}}$ and obtain

$$n_k \leq C(n_k N T_k)^{\frac{N}{N+2}} (N_0(f_k))^{\frac{2}{N+2}},$$

which is equivalent to condition $(i.1)$.
We deduce the estimate $(i.2)$ and $(i.3)$ from $(i.1)$. Furthermore, since we assumed that $f_1, f_2 \geq 0, \gamma \geq 0, 0 \leq \alpha \leq 1, \varepsilon \leq 1$ and condition (4.17) both the temperatures

T_1 and T_2 and all coefficients in T_{12} and T_{21} are positive. All in all, with $(i.1)$ this leads to the estimates

$$\frac{n_1}{T_{12}^{N/2}} = \frac{n_1}{(\alpha T_1 + (1-\alpha)T_2 + \gamma|u_1 - u_2|^2)^{N/2}} \leq \frac{n_1}{\alpha^{N/2}T_1^{N/2}} \leq CN_0(f_1),$$

$$\frac{n_2}{T_{21}^{N/2}}$$

$$= \frac{n_2}{\left(\varepsilon(1-\alpha)T_1 + (1-\varepsilon(1-\alpha))T_2 + \left[\frac{1}{N}\varepsilon m_1(1-\delta)\left(\frac{m_1}{m_2}\varepsilon(\delta-1)+\delta+1\right) - \varepsilon\gamma\right]|u_1 - u_2|^2\right)^{N/2}}$$

$$\leq \frac{n_2}{(1-\varepsilon(1-\alpha))^{N/2}T_2^{N/2}} \leq CN_0(f_2).$$

$\square$

Remark 4.3.1. Similar estimates as $(i.2)$ and $(i.3)$ can also be obtained for $T^{(1)}$, $T^{(2)}$ from (4.19) in the model presented in section 4.2.2 in an analogously way if the coefficient in front of $|u_1 - u_2|^2$ in (4.19) is non-negative meaning $\frac{\chi_{12} n_2}{\nu_{11} n_1 + \nu_{12} n_2} \leq 1$ and $\frac{\chi_{21} n_1}{\nu_{22} n_2 + \nu_{21} n_1} \leq 1$. This is reasonable in order to ensure the positivity of the temperatures $T^{(1)}$ and $T^{(2)}$.

Theorem 4.3.2. *For any pair of functions (f_1, f_2) with $(1+|v|^2)f_k \in L^1(dv)$, $f_1, f_2 \geq 0$, we define the moments as in (4.7), (4.13), (4.14), (4.15) and (4.16), then we have*

$(ii.1)$ $n_k(T_k + |u_k|^2)^{\frac{q-N}{2}} \leq C_q N_q(f_k)$ *for $q > N + 2$, $k = 1, 2$,*

$(ii.2)$ $n_1(T_{12} + |u_{12}|^2)^{\frac{q-N}{2}} \leq C_q(N_q(f_1) + \frac{n_1}{n_2}N_q(f_2))$ *for $q > N + 2$,*

$(ii.3)$ $n_2(T_{21} + |u_{21}|^2)^{\frac{q-N}{2}} \leq C_q(\frac{n_2}{n_1}N_q(f_1) + N_q(f_2))$ *for $q > N + 2$.*

Proof. The proof of $(ii.1)$ is exactly the same as the proof of the inequality (2.3) in [70]. We want to repeat it here for the convenience of the reader. We consider $n_k(NT_k + |u_k|^2) = \int |v|^2 f_k(v)dv$ for $k = 1, 2$. We split the integration with respect to the velocity v into $|v| > R_k$ and $|v| \leq R_k$ for some R_k determined later. We obtain

$$n_k(NT_k + |u_k|^2) \leq \int_{|v|>R_k} \frac{|v|^q}{|v|^{q-2}} f_k(v)dv + \int_{|v|\leq R_k} |v|^2 f_k(v)dv.$$

Since $q > N + 2$, we can estimate the integral $\int_{|v|>R_k} \frac{1}{|v|^{q-2}}dv$ from above by $C_q R_k^{N-q+2}$. In the second integral we use that $|v|^2 \leq R_k^2$. Then we get

$$n_k(NT_k + |u_k|^2) \leq CR_k^{N-q+2}N_q(f_k) + n_k R_k^2.$$

Now we choose R_k as $R_k = \left(\frac{n_k}{N_q(f_k)}\right)^{\frac{1}{N-q}}$ and obtain

$$n_k(NT_k + |u_k|^2) \leq C(n_k)^{1-\frac{2}{q-N}}(N_q(f_k))^{\frac{2}{q-N}}.$$

Since $f_k \geq 0$, we have $n_k(T_k + |u_k|^2) \leq n_k(NT_k + |u_k|^2)$ and we can deduce the required inequality $(ii.1)$.

In order to prove $(ii.2)$, estimate $n_1(T_{12}+|u_{12}|^2)$ using that $f_k \geq 0$, (4.13) and (4.15) by

$$n_1(T_{12}+|u_{12}|^2) \leq n_1(NT_{12}+|u_{12}|^2)$$
$$= n_1(\alpha NT_1 + (1-\alpha)NT_2 + N\gamma|u_1-u_2|^2 + |\delta u_1 + (1-\delta)u_2|^2)$$
$$= n_1(\alpha NT_1 + (1-\alpha)NT_2 + (\delta^2 + N\gamma)|u_1|^2 + ((1-\delta)^2 + N\gamma)|u_2|^2$$
$$+ 2(\delta(1-\delta)-N\gamma)u_1 \cdot u_2.$$

Using that $|u_1+u_2|^2 \geq 0$ and $|u_1-u_2|^2 \geq 0$, we can estimate the term $(\delta(1-\delta)-N\gamma)u_1 \cdot u_2$ from above by $|\delta(1-\delta)-N\gamma|\frac{1}{2}(|u_1|^2+|u_2|^2)$ and obtain

$$n_1(T_{12}+|u_{12}|^2) \leq n_1[\alpha NT_1 + (\delta^2 + N\gamma + |\delta(1-\delta)-N\gamma|)|u_1|^2 + (1-\alpha)NT_2$$
$$+ ((1-\delta)^2 + N\gamma + |\delta(1-\delta)-N\gamma|)|u_2|^2$$
$$\leq n_1[\max\{\alpha, \delta^2 + N\gamma + |\delta(1-\delta)-N\gamma|\}(NT_1+|u_1|^2)]$$
$$+ \max\{1-\alpha, ((1-\delta)^2 + N\gamma + |\delta(1-\delta)-N\gamma|)\}(NT_2+|u_2|^2)].$$

Set $A_1 := \max\{\alpha, \delta^2 + N\gamma + |\delta(1-\delta)-N\gamma|\}$ and $A_2 := \max\{1-\alpha, (1-\delta)^2 + N\gamma + |\delta(1-\delta)-N\gamma|\}$. Then

$$n_1(T_{12}+|u_{12}|^2) \leq n_1[A_1(NT_1+|u_1|^2) + A_2(NT_2+|u_2|^2)]$$
$$= A_1 \int |v|^2 f_1(v)dv + A_2\frac{n_1}{n_2} \int |v|^2 f_2(v)dv.$$

We split the integration with respect to the velocity v into $|v| > R_{12}$ and $|v| \leq R_{12}$ for some R_{12} determined later. We obtain

$$n_1(T_{12}+|u_{12}|^2) \leq \int_{|v|>R_{12}} \frac{|v|^q}{|v|^{q-2}} \left(A_1 f_1(v) + A_2\frac{n_1}{n_2}f_2(v) \right) dv$$
$$+ \int_{|v|\leq R_{12}} |v|^2 \left(A_1 f_1(v) + A_2\frac{n_1}{n_2}f_2(v) \right) dv.$$

Again, since $q > N+2$, we can estimate the integral $\int_{|v|>R_{12}} \frac{1}{|v|^{q-2}}dv$ from above by $C_q R_{12}^{N-q+2}$. In the second integral we use that $|v|^2 \leq R_{12}^2$. Then we get

$$n_1(T_{12}+|u_{12}|^2) \leq CR_{12}^{N-q+2}(A_1 N_q(f_1) + A_2\frac{n_1}{n_2}N_q(f_2)) + Cn_1 R_{12}^2.$$

Now we choose $R_{12} = \left(\frac{n_1}{A_1 N_q(f_1)+A_2\frac{n_1}{n_2}N_q(f_2)} \right)^{\frac{1}{N-q}}$ and obtain

$$n_1(T_{12}+|u_{12}|^2) \leq Cn_1^{1-\frac{2}{q-N}}(A_1 N_q(f_1) + \frac{n_1}{n_2}A_2N_q(f_2))^{\frac{2}{q-N}},$$

which is equivalent to the required estimate $(ii.2)$. The proof of $(ii.3)$ is similar to the proof of $(ii.2)$. $\square$

Lemma 4.3.3. *For any pair of functions* (f_1, f_2) *with* $(1+|v|^2)f_k \in L^1(dv)$, $f_1, f_2 \geq 0$, *we define the moments as in (4.7), (4.13), (4.14), (4.15) and (4.16). Let* $q \in \mathbb{N}$ *or* $q - \frac{1}{2} \in \mathbb{N}$, *then there exists a constant* $A > 0$ *such that*

$$|\delta u_1 + (1-\delta)u_2|^q \leq A|u_1|^q + A|u_2|^q,$$

$$(\alpha T_1 + (1-\alpha)T_2 + \gamma|u_1 - u_2|^2)^q \leq A(T_1^q + T_2^q + |u_1 - u_2|^{2q}).$$

Proof. We prove the statement per induction with respect to q. We just prove the first inequality, the proof of the second one is similar to the first one. For $q = 1$ the statement is true, since it is the triangle inequality. Assume now that it is true for a fixed $q \in \mathbb{N}$. Then we get for $q + 1$

$$|\delta u_1 + (1-\delta)u_2|^{q+1} = |\delta u_1 + (1-\delta)u_2|^q |\delta u_1 + (1-\delta)u_2|.$$

With the assumption that the statement is true for this fixed q and for $q = 1$, we obtain

$$|\delta u_1 + (1-\delta)u_2|^{q+1} \leq (A|u_1|^q + A|u_2|^q)(\tilde{A}|u_1| + \tilde{A}|u_2|)$$
$$= A\tilde{A}|u_1|^{q+1} + A\tilde{A}|u_2|^{q+1} + A\tilde{A}|u_1|^q|u_2| + A\tilde{A}|u_2|^q|u_1|,$$

for some constants $A, \tilde{A} > 0$. The requirement that there exists a constant $\hat{A}$ such that this expression is less or equal to

$$\hat{A}|u_1|^{q+1} + \hat{A}|u_2|^{q+1}$$

is equivalent to

$$(A\tilde{A} - \hat{A})|u_1|^{q+1} + (A\tilde{A} - \hat{A})|u_2|^{q+1} + A\tilde{A}|u_1|^q|u_2| + A\tilde{A}|u_2|^q|u_1| \leq 0.$$

Choose $\hat{A} = 2A\tilde{A}$. Then the inequality is equivalent to

$$-A\tilde{A}|u_1|^{q+1} - A\tilde{A}|u_2|^{q+1} + A\tilde{A}|u_1|^q|u_2| + A\tilde{A}|u_2|^q|u_1| \leq 0,$$

which is equivalent to

$$A\hat{A}(|u_1|^q - |u_2|^q)(|u_2| - |u_1|) \leq 0,$$

which is a true statement since $q > 1$.
The proof for q with $q - \frac{1}{2} \in \mathbb{N}$ is similar with induction starting with $q = \frac{1}{2}$. For $q = \frac{1}{2}$ it is right, since it is the triangle inequality for the Euclidean norm. $\square$

Theorem 4.3.4. *For any pair of functions* (f_1, f_2) *with* $(1+|v|^2)f_k \in L^1(dv)$, $f_1, f_2 \geq 0$, *we define the moments as in (4.7), (4.13), (4.14), (4.15) and (4.16), then we have*

(iii.1) $\dfrac{n_k|u_k|^{N+q}}{[(T_k+|u_k|^2)T_k]^{N/2}} \leq C_q N_q(f_k)$ *for any* $q > 1, k = 1, 2,$

(iii.2) $\frac{n_1|u_{12}|^q}{T_{12}^{N/2}} \leq n_1 C(\frac{|u_1|^q}{(T_1)^{N/2}} + \frac{|u_2|^q}{(T_2)^{N/2}})$ *for any $q > 1$,*

(iii.3) $\frac{n_2|u_{21}|^q}{T_{21}^{N/2}} \leq n_2 C(\frac{|u_1|^q}{(T_1)^{N/2}} + \frac{|u_2|^q}{(T_2)^{N/2}})$ *for any $q > 1$.*

Proof. The proof of (iii.1) is exactly the same as the proof of the inequality (2.3) in [70]. For the convenience of the reader we want to repeat it here. For $q > 1$, we get by Hölder's inequality from appendix A.1 that

$$n_k|u_k| \leq \int |v| f_k(v) dv \leq \int_{|u_k-v| \leq R_k} \frac{n_k}{n_k} |v| f_k(v) dv + \int_{|u_k-v| > R_k} |v| f_k(v) dv$$

$$\leq n_k \left(\int_{|u_k-v| \leq R_k} |v|^q \frac{f_k(v)}{n_k} dv \right)^{\frac{1}{q}} + \frac{1}{R_k} \int |u_k - v||v| f_k(v) dv$$

$$\leq C n_k^{1-\frac{1}{q}} N_q(f_k)^{1/q} R_k^{N/q} + \frac{1}{R_k} \left(\int |v|^2 f_k(v) dv \right)^{1/2} \left(\int |u_k - v|^2 f_k(v) dv \right)^{1/2}$$

$$\leq C \left(n_k^{\frac{q-1}{q}} N_q(f_k)^{\frac{1}{q}} R_k^{N/q} + \frac{n_k}{R_k} (NT_k + |u_k|^2)^{1/2} T_k^{1/2} \right).$$

Now if we choose $R_k = \left(\frac{n_k^{1/q}(NT_k+|u_k|^2)^{1/2}T^{1/2}}{N_q(f_k)^{1/q}} \right)^{\frac{q}{N+q}}$, we obtain the inequality

$$n_k|u_k| \leq C n_k^{\frac{1}{q}(q-1+\frac{1}{N+q})} N_q(f_k)^{\frac{1}{q}(1-\frac{N}{N+q})} ((NT + |u|^2)^{\frac{1}{2}} T^{\frac{1}{2}})^{\frac{N}{N+q}},$$

which is equivalent to the required estimate $(iii.1)$.
Estimate $(iii.2)$ is a consequence of lemma 4.3.3 using that $\gamma \geq 0$, $0 \leq \alpha \leq 1$ and condition (4.17), since we have

$$\frac{n_1|u_{12}|^q}{T_{12}^{N/2}} = \frac{n_1|\delta u_1 + (1-\delta)u_2|^q}{(\alpha T_1 + (1-\alpha)T_2 + \gamma|u_1 - u_2|^2)^{N/2}} \leq n_1 \frac{A(|u_1|^q + |u_2|^q)}{(\alpha T_1 + (1-\alpha)T_2)^{N/2}}$$

$$\leq n_1 \frac{A|u_1|^q}{(\alpha T_1)^{N/2}} + n_1 \frac{A|u_2|^q}{((1-\alpha)T_2)^{N/2}}.$$

The proof of $(iii.3)$ is similar to the proof of $(iii.2)$. $\qquad \square$

Consequence 4.3.5. For any pair of functions (f_1, f_2) with $(1 + |v|^2) f_k \in L^1(dv)$, $f_1, f_2 \geq 0$, we define the moments as in (4.7), (4.13), (4.14), (4.15) and (4.16), then we have

(iv.1) $\sup_v |v|^q M_k[f_k] \leq C_q N_q(f_k)$ for $q > N + 2$ or $q = 0$,

(iv.2) $\sup_v |v|^q M_{12}[f_1, f_2] \leq C_q(N_q(f_1) + \frac{n_1}{n_2} N_q(f_2))$ for $q > N + 2$ or $q = 0$,

(iv.3) $\sup_v |v|^q M_{21}[f_1, f_2] \leq C_q(\frac{n_2}{n_1} N_q(f_1) + N_q(f_2))$ for $q > N + 2$ or $q = 0$.

Note that here and in the following we write $M_k[f_k], M_{12}[f_1, f_2]$ and $M_{21}[f_1, f_2]$ instead of M_k, M_{12} and M_{21} in order to emphasize the dependence of the Maxwell distributions on the distribution functions f_1 and f_2 via the macroscopic quantities as densities, velocities and temperatures.

Proof. The proof of $(iv.1)$ is exactly the same as the proof of the inequality (2.3) in [70]. For the convenience of the reader we want to repeat it here. First, we consider the case $q > N + 2$. According to lemma 4.3.3, we obtain

$$\sup_v |v|^q M_k[f_k] \leq C \sup_v (|v - u_k|^q + |u_k|^q) M_k[f_k]$$

$$= C \sup_v (|v - u_k|^q + |u_k|^q) \frac{n_k}{(2\pi T_k)^{N/2}} e^{-\frac{|v-u_k|^2}{2T_k/m_k}}.$$

By computing derivatives of $|v - u_k|^q M_k[f_k]$ with respect to v, we see that the maximum of $|v - u_k|^q M_k[f_k]$ is given by $n_k T_k^{\frac{q-N}{2}}$. The computation is done in more details in the proof of estimate $(iv.2)$, so we will omit the details here. We obtain

$$\sup_v |v|^q M_k[f_k] \leq C(n_k T_k^{\frac{q-N}{2}} + n_k \frac{|u_k|^q}{T_k^{N/2}}).$$

Define $E_k := C(n_k T_k^{\frac{q-N}{2}} + n_k \frac{|u_k|^q}{T_k^{N/2}})$. Now we consider two cases, if $|u_k| > T_k^{1/2}$ and $|u_k| \leq T_k^{1/2}$. First $|u_k| > T_k^{1/2}$. In this case

$$E_k \leq C n_k \frac{|u_k|^q}{T_k^{N/2}} = C n_k \frac{|u_k|^{q+N}}{|u_k|^N T_k^{N/2}} = C n_k \frac{|u_k|^{q+N}}{(\frac{1}{2}|u_k|^2 + \frac{1}{2}|u_k|^2)^{N/2} T_k^{N/2}}$$

$$\leq C n_k \frac{|u_k|^{q+N}}{(|u_k|^2 + T_k)^{N/2} T_k^{N/2}}.$$

Finally the estimate (iii.1) leads to

$$E_k \leq C_q N_q[f_k].$$

When $|u_k| \leq T_k^{1/2}$, we obtain with (ii.1)

$$E_k \leq C n_k T_k^{\frac{q-N}{2}} \leq C_q N_q[f_k].$$

So all in all, we obtain

$$\sup_v |v|^q M_k[f_k] \leq C_q N_q[f_k] \quad \text{for} \quad q > N + 2.$$

For $q = 0$ we obtain with (i.1)

$$\sup_v M_k[f_k] \leq n_k T_k^{-N/2} \leq C N_0(f_k).$$

Now, the proof of $(iv.2)$. First for $q > N + 2$. First we compute the maximum of $M_{12}[f_1, f_2]$ and $|v - u_{12}|M_{12}[f_1, f_2]$ similar to the case of one species. The maximum of the Maxwell distribution $M_{12}[f_1, f_2]$ in v is reached when $v = u_{12}$. Therefore

$$\max_v M_{12}[f_1, f_2] = \frac{n_1}{(2\pi \frac{T_{12}}{m_1})^{N/2}}.$$

For the maximum of $|v - u_{12}|^q M_{12}[f_1, f_2]$, we compute the gradient in v and obtain by using the product rule

$$\nabla_v(|v - u_{12}|^q M_{12}[f_1, f_2]) = (v - u_{12})q|v - u_{12}|^{q-2}M_{12}[f_1, f_2] - \frac{m_1}{T_{12}}|v - u_{12}|^q(v - u_{12})M_{12}[f_1, f_2].$$

The condition that this expression is equal to zero is equivalent to

$$(q(v - u_{12}) - \frac{m_1}{T_{12}}|v - u_{12}|(v - u_{12})) = 0$$

for $v \neq u_{12}$. We can exclude $v = u_{12}$ since it is a minimum. From this expression, we can deduce

$$|v - u_{12}|^2 = \frac{T_{12}}{m_1}q.$$

If we insert this into $|v - u_{12}|^q M_{12}[f_1, f_2]$, we obtain

$$\max_v(|v - u_{12}|^q M_{12}[f_1, f_2]) = \max_v((\frac{T_{12}}{m_1}q)^{\frac{q}{2}} \frac{n_1}{(2\pi \frac{T_{12}}{m_1})^{N/2}}e^{-q}).$$

For $|v| \to \infty$, the expression $|v - u_{12}|^q M_{12}[f_1, f_2]$ tends to zero, so it is equal to the supremum. All in all, we obtain

$$\sup_v |v|^q M_{12}[f_1, f_2] \leq \sup_v |v - u_{12}|^q M_{12}[f_1, f_2] + \sup_v |u_{12}|^q M_{12}[f_1, f_2]$$

$$\leq C\left(n_1 T_{12}^{\frac{q-N}{2}} + n_1 \frac{|u_{12}|^q}{T_{12}^{N/2}}\right) = C\left(n_1(\alpha T_1 + (1-\alpha)T_2 + \gamma|u_1 - u_2|^2) + n_1 \frac{|u_{12}|^q}{T_{12}^{N/2}}\right).$$

Since $q - N > 0$, we can use lemma 4.3.3 in the first term twice and $(iii.2)$ in the second term on the right-hand side and obtain

$$\sup_v |v|^q M_{12}[f_1, f_2]$$

$$\leq C\left(n_1(T_1 + |u_1|^2)^{\frac{q-N}{2}} + n_1 \frac{|u_1|^q}{T_1^{N/2}} + \frac{n_1}{n_2}n_2\left((T_2 + |u_2|^2)^{\frac{q-N}{2}} + \frac{|u_2|^q}{T_2^{N/2}}\right)\right).$$

The first and the third term on the right-hand side can be estimated using $(ii.1)$ and the other two terms can be estimated in the same way as in the proof of $(iv.1)$ for one species by $CN_q(f_1)$ and $C\frac{n_1}{n_2}N_q(f_2)$, respectively. Combining both, we get

$$\sup_v |v|^q M_{12}[f_1, f_2] \leq C_q(N_q(f_1) + \frac{n_1}{n_2}N_q(f_2)).$$

For $q = 0$ we use

$$\sup_{v} M_{12}[f_1, f_2] \leq \frac{n_1}{T_{12}^{N/2}} \leq \frac{n_1}{T_1^{N/2}} \leq CN_0(f_1),$$

using $(i.1)$. The proof of $(iv.3)$ is similar to the proof of $(iv.2)$. $\qquad\square$

Remark 4.3.2. For the multi-species model of Andries, Aoki and Perthame in section 4.2.2 we can obtain the same estimates

$(i.2^*/\,i.3^*)$ $\qquad \frac{n_k}{(T^{(k)})^{\frac{N}{2}}} \leq CN_0(f_k), \quad k = 1, 2,$

$(ii.2^*/ii.3^*)$ $\qquad n_k(T^{(k)} + |u^{(k)}|^2)^{\frac{q-N}{2}} \leq C_q(N_q(f_j) + \frac{n_j}{n_k}N_q(f_k))$ for $q > N + 2, j \neq k,$

$(iii.2^*/iii3^*)$ $\qquad \frac{n_k|u^{(k)}|^q}{(T^{(k)})^{\frac{N}{2}}} \leq n_k C \left(\frac{|u_1|^2}{T_1^{N/2}} + \frac{|u_2|^2}{T_2^{N/2}} \right),$

$(iv.2^*/iv.3^*)$ $\qquad \sup_{v} |v|^q M^{(k)}[f_1, f_2] \leq C_q(\frac{n_j}{n_k}N_q(f_k) + N_q(f_j))$
$\qquad\qquad$ for $q > N + 2$ or $q = 0, j \neq k,$

analogously to the estimates $(i.2/i.3)/(ii.2/ii.3)/(iii.2/iii.3)/(iv.2/iv.3)$, since $u^{(1)}, u^{(2)}$ are also linear combinations of u_1 and u_2 and $T^{(1)}, T^{(2)}$ are also combinations of $T_1, T_2, |u_1 - u_2|^2$.

4.3.2 Existence and uniqueness proof

In this section we want to show existence and uniqueness of non-negative solutions in a certain function space using the estimates of the previous section. For the existence and uniqueness proof, we make the following assumptions:

Assumptions 4.3.1.

1. We assume periodic boundary conditions. Equivalently we can construct solutions satisfying

$$f_k(t, x_1, ..., x_N, v_1, ..., v_N) = f_k(t, x_1, ..., x_{i-1}, x_i + a_i, x_{i+1}, ...x_N, v_1, ...v_N),$$

   for all $i = 1, ..., N$ and a suitable $\{a_i\} \in \mathbb{R}^N$ with positive components, for $k = 1, 2.$

2. We require that the initial values $f_k^0, k = 1, 2$ satisfy assumption 1.

3. We are on the bounded domain in space $\Lambda = \{x \in \mathbb{R}^N | x_i \in (0, a_i)\}.$

4. Suppose that f_k^0 satisfies $f_k^0 \geq 0, (1 + |v|^2)f_k^0 \in L^1(\Lambda \times \mathbb{R}^N)$ with $\int f_k^0 dx dv = 1, k = 1, 2.$

5. Suppose $N_q(f_k^0) := \sup_{v} f_k^0(x, v)(1 + |v|^q) = \frac{1}{2}A_0 < \infty$ for some $q > N + 2.$

6. Suppose $\gamma_k(x, t) := \int f_k^0(x - vt, v)dv \geq C_0 > 0$ for all $t \in \mathbb{R}.$

7. Assume that the collision frequencies are written as in (4.11) and are positive.

With this assumptions we can show the following theorem.

Theorem 4.3.6. *Under the assumptions 4.3.1 and the definitions* (4.7), (4.13), (4.14), (4.15) *and* (4.16), *there exists a unique non-negative mild solution* $(f_1, f_2) \in C(\mathbb{R}^+; L^1((1 + |v|^2)dvdx))$ *of the initial value problem* (4.8). *Moreover, for all* $t > 0$ *the following bounds hold:*

$$|u_k(t)|, |u_{12}(t)|, |u_{21}(t)|, T_k(t), T_{12}(t), T_{21}(t), N_q(f_k)(t) \le A(t) < \infty,$$

$$n_k(t) \ge C_0 e^{-t} > 0,$$

$$T_k(t), T_{12}(t), T_{21}(t) \ge B(t) > 0,$$

for $k = 1, 2$ *and some constants* $A(t), B(t)$.

Proof. The idea of the proof is to find a Cauchy sequence of functions in a certain space which converges towards a solution to (4.8). The sequence will be constructed in a way such that each member of the sequence satisfies an inhomogeneous transport equation. In this case we know results of existence and uniqueness. In order to show that this sequence is a Cauchy sequence we need to show that the Maxwell distributions on the right-hand side of (4.8) are Lipschitz continuous with respect to f_1, f_2.

The proof is structured as follows: First we proof some estimates on the macroscopic quantities (4.7), (4.13), (4.14), (4.15) and (4.16). From this we can deduce Lipschitz continuity of the Maxwell distributions M_1, M_2, M_{12}, M_{21} with respect to f_1 and f_2 which finally leads to the convergence of this Cauchy sequence to a solution to (4.8).

Step 1: Gronwall estimate on $N_q(f_k(t))$ given by (4.20)

If f_1 is a mild solution according to definition 4.2.1, we have

$$N_q(f_1) = \sup_v |v|^q f_1 \le e^{-\alpha_1(x,v,t)} \sup_v |v|^q f_1^0(x - tv, v)$$

$$+ \sup_v |v|^q [e^{-\alpha_1(x,v,t)} \int_0^t [\tilde{\nu}_{11} \frac{n_1(x + (s-t)v, s)}{n_1(x + (s-t)v, s) + n_2(x + (s-t)v, s)} M_1(x + (s-t)v, v, s)$$

$$+ \tilde{\nu}_{12} \frac{n_2(x + (s-t)v, s)}{n_1(x + (s-t)v, s) + n_2(x + (s-t)v, s)} M_{12}(x + (s-t)v, v, s)] e^{\alpha_1(x+(s-t)v, v, s)} ds].$$

Since α_1 is non-negative, we can estimate $e^{-\alpha_1(x,v,t)}$ in front of the initial data from above by 1. Since we assumed that the collision frequencies have the shape given in (4.11), we can estimate the integrand in the exponential function $e^{-\alpha_1(x,v,t)} e^{\alpha_1(x+(s-t)v,v,s)}$ by a constant and obtain

$$N_q(f_1) = \sup_v |v|^q f_1 \le \sup_v |v|^q f_1^0(x - tv, v)$$

$$+ \sup_v |v|^q [\int_0^t e^{-C(t-s)} [C \frac{n_1(x + (s-t)v, s)}{n_1(x + (s-t)v, s) + n_2(x + (s-t)v, s)} M_1(x + (s-t)v, v, s)$$

$$+ C \frac{n_2(x + (s-t)v, s)}{n_1(x + (s-t)v, s) + n_2(x + (s-t)v, s)} M_{12}(x + (s-t)v, v, s)] ds].$$

Using assumption 5 (in the assumption 4.3.1) and the fact that we can estimate $e^{-C(t-s)}$ from above by 1 since s is between 0 and t, we get

$$N_q(f_1) = \sup_v |v|^q f_1 \leq \frac{1}{2} A_0 + \int_0^t C \sup_x [\frac{n_1(x,s)}{n_1(x,s) + n_2(x,s)} \sup_v |v|^q M_1(x,v,s)$$

$$+ \frac{n_2(x,s)}{n_1(x,s) + n_2(x,s)} \sup_v |v|^q M_{12}(x,v,s)] ds.$$

With $(iv.1)$ and $(iv.2)$, we obtain

$$N_q(f_1) = \sup_{x,v} |v|^q f_1$$

$$\leq \frac{1}{2} A_0 + \int_0^t C_q \sup_x [\frac{n_1(x,s) + n_2(x,t)}{n_1(x,s) + n_1(x,s)} N_q(f_1)(s) + \frac{n_1(x,s)}{n_1(x,s) + n_2(x,s)} N_q(f_2(s))] ds$$

$$\leq \frac{1}{2} A_0 + \int_0^t C_q [\sup_x N_q(f_1)(s) + \sup_x N_q(f_2)(s)] ds.$$

Similarly, we can estimate $N_q(f_2)$ by

$$N_q(f_2) = \sup_v |v|^q f_2 \leq \frac{1}{2} A_0 + \int_0^t C_q [\sup_x N_q(f_1)(s) + \sup_x N_q(f_2)(s)] ds.$$

We add both inequalities and obtain

$$N_q(f_1) + N_q(f_2) \leq A_0 + \int_0^t C_q [\sup_x N_q(f_1)(s) + \sup_x N_q(f_2)(s)] ds.$$

With Gronwall's lemma, we obtain

$$N_q(f_1)(t) + N_q(f_2)(t) \leq A_0 e^{C_q t} \quad \text{for} \quad q > N+2 \quad \text{or} \quad q = 0. \tag{4.21}$$

Step 2: Estimate on the densities

If $f_k \geq 0$ is a solution, it satisfies

$$\partial_t f_k + v \cdot \nabla_x f_k = \tilde{\nu}_{kk} \frac{n_k}{n_k + n_j} (M_k - f_k) + \tilde{\nu}_{kj} \frac{n_j}{n_k + n_j} (M_{kj} - f_k)$$

$$\geq -(\tilde{\nu}_{kk} + \tilde{\nu}_{kj}) f_k.$$

If we write this in the mild formulation, this leads to

$$f_k(x,v,t) \geq e^{-(\tilde{\nu}_{kk} + \tilde{\nu}_{kj})t} f_k^0(x - tv).$$

Integrating this with respect to v leads with assumption 6 (in assumptions 4.3.1) to the estimate of the densities

$$n_k(x,t) \geq e^{-(\tilde{\nu}_{kk} + \tilde{\nu}_{kj})t} \int f_k^0(x - vt, v) dv$$

$$\geq e^{-(\tilde{\nu}_{kk} + \tilde{\nu}_{kj})t} \gamma_k(x,t) \geq e^{-(\tilde{\nu}_{kk} + \tilde{\nu}_{kj})t} C_0 > 0. \tag{4.22}$$

Step 3: Estimate on the temperatures

Now, we estimate the temperatures from below. First, we consider $T_k^{N/2}$. We can estimate it from below using $(i.1)$

$$T_k^{N/2}(t) \geq \frac{Cn_k(t)}{N_0(f_k(t))}.$$

Using (4.21) and (4.22), we obtain

$$T_k^{N/2}(t) \geq \frac{Ce^{-(\tilde{\nu}_{kk}+\tilde{\nu}_{kj})t}C_0}{Ae^{C_q t}} =: B(t) > 0.$$

We obtain the same estimate for $T_{12}^{N/2}$ using $(i.2)$, (4.21) and (4.22), and for $T_{21}^{N/2}$ using $(i.3)$, (4.21) and (4.22).

Step 4: Estimates on the velocities

We estimate $T_k + |u_k|^2$, $T_{12} + |u_{12}|^2$, and $T_{21} + |u_{21}|^2$ first using $(ii.1), (ii.2)$ and $(ii.3)$, respectively and then using (4.21) and (4.22). For example

$$T_{12} + |u_{12}|^2 \leq \frac{C_q(N_q(f_1) + \frac{n_1}{n_2}N_q(f_2))^{\frac{2}{q-N}}}{n_1^{2/(q-N)}} \leq \frac{C_q Ae^{C_q \frac{2}{q-N}t}}{e^{-C\frac{2}{q-N}t}C_0^{\frac{2}{q-N}}} < A(t) < \infty.$$

Step 5: Lipschitz continuity

The next step of the proof is to show Lipschitz continuity of the operators $f_k \mapsto M_k[f_k]$, $(f_1, f_2) \mapsto \frac{n_2}{n_1+n_2}M_{12}[f_1, f_2]$ and $(f_1, f_2) \mapsto \frac{n_1}{n_1+n_2}M_{21}[f_1, f_2]$, when (f_1, f_2) are restricted to

$$\Omega = \{(f_1, f_2) \in L^1(\Lambda \times \mathbb{R}^N; (1+|v|^2)dvdx)|f_k \geq 0, N_q(f_k) < A, \min(n_k, T_k) > C, k = 1, 2\}. \tag{4.23}$$

The proof for $f_k \mapsto M_k[f_k]$ is given in [70]. So it remains to show Lipschitz continuity for $(f_1, f_2) \mapsto \frac{n_2}{n_1+n_2}M_{12}[f_1, f_2]$ and for $(f_1, f_2) \mapsto \frac{n_1}{n_1+n_2}M_{21}[f_1, f_2]$. We only prove the first case since the second one is similar to the first one. For any pair $(f_1^i, f_2^i), i = 1, 2$ in the subset Ω, define $(n_1^i, u_{12}^i, T_{12}^i)$ as their corresponding moments. Set

$$(n_1^\Theta, n_2^\Theta, u_{12}^\Theta, T_{12}^\Theta) = \Theta(n_1^1, n_2^1, u_{12}^1, T_{12}^1) + (1-\Theta)(n_1^2, n_2^2, u_{12}^2, T_{12}^2),$$

and

$$M_{12}(\Theta) = \frac{n_1^\Theta}{(2\pi T_{12}^\Theta/m_1)^{N/2}}e^{-\frac{|v-u_{12}^\Theta|^2}{2T_{12}^\Theta/m_1}}\frac{n_2^\Theta}{n_1^\Theta + n_2^\Theta}.$$

Then we have

$$\int \left| \frac{n_2^1}{n_1^1 + n_2^1}M_{12}[f_1^1, f_2^1] - \frac{n_2^2}{n_1^2 + n_2^2}M_{12}[f_1^2, f_2^2] \right|(1 + |v|^2)dv$$

$$= \int |M_{12}(1) - M_{12}(0)|(1 + |v|^2)dv.$$

Now, we use the Taylor formula with first derivative as remainder and the chain rule and obtain

$$\int \left| \frac{n_2^1}{n_1^1 + n_2^1} M_{12}[f_1^1, f_2^1] - \frac{n_2^2}{n_1^2 + n_2^2} M_{12}[f_1^2, f_2^2] \right| (1 + |v|^2) dv = \int \left| \frac{\partial M_{12}}{\partial \Theta}(\Theta) \right| (1 + |v|^2) dv$$

$$\leq \int_0^1 \int \left(\left| \frac{\partial M_{12}}{\partial n_1^\Theta}(\Theta) \frac{\partial n_1^\Theta}{\partial \Theta} \right| + \left| \frac{\partial M_{12}}{\partial u_{12}^\Theta}(\Theta) \frac{\partial u_{12}^\Theta}{\partial \Theta} \right| + \left| \frac{\partial M_{12}}{\partial T_{12}^\Theta}(\Theta) \frac{\partial T_{12}^\Theta}{\partial \Theta} \right| \right.$$

$$\left. + \left| \frac{\partial M_{12}}{\partial n_2^\Theta}(\Theta) \frac{\partial n_2^\Theta}{\partial \Theta} \right| \right) (1 + |v|^2) dv d\Theta$$

$$= \int_0^1 \int \left(\left| \frac{\partial M_{12}}{\partial n_1^\Theta}(\Theta) \right| |n_1^1 - n_1^2| + \left| \frac{\partial M_{12}}{\partial u_{12}^\Theta}(\Theta) \right| |u_{12}^1 - u_{12}^2| \right.$$

$$\left. + \left| \frac{\partial M_{12}}{\partial T_{12}^\Theta}(\Theta) \right| |T_{12}^1 - T_{12}^2| + \left| \frac{\partial M_{12}}{\partial n_2^\Theta}(\Theta) \right| |n_2^1 - n_2^2| \right) (1 + |v|^2) dv d\Theta.$$

An explicit calculation of the derivatives leads to

$$\int \left| \frac{n_2^1}{n_1^1 + n_2^1} M_{12}[f_1^1, f_2^1] - \frac{n_2^2}{n_1^2 + n_2^2} M_{12}[f_1^2, f_2^2] \right| (1 + |v|^2) dv$$

$$\leq \int_0^1 \left((1 + |u_{12}^\Theta|^2 + NT_{12}^\Theta)|n_1^1 - n_1^2| + C[\frac{n_2^\Theta}{n_1^\Theta + n_2^\Theta} \frac{n_1^\Theta}{(T_{12}^\Theta)^{1/2}} (1 + |u_{12}^\Theta|^2 + T_{12}^\Theta)]|u_{12}^1 - u_{12}^2| \right.$$

$$+ C[\frac{n_2^\Theta}{n_1^\Theta + n_2^\Theta} \frac{n_1^\Theta}{T_{12}^\Theta} (1 + |u_{12}^\Theta|^2 + T_{12}^\Theta)]|T_{12}^1 - T_{12}^2| + (1 + |u_{12}^\Theta|^2 + NT_{12}^\Theta)|n_2^1 - n_2^2|) d\Theta.$$

The main difference to the one species case is the additional term $|\frac{\partial M_{12}}{\partial n_2}(\Theta)|$ and the term $\partial_{n_1^\Theta}(\frac{n_1^\Theta n_2^\Theta}{n_1^\Theta + n_2^\Theta})$. For the second term we compute $\partial_{n_1^\Theta}(\frac{n_1^\Theta n_2^\Theta}{n_1^\Theta + n_2^\Theta}) = \frac{n_2^\Theta}{n_1^\Theta + n_2^\Theta} - \frac{n_1^\Theta n_2^\Theta}{(n_1^\Theta + n_2^\Theta)^2}$ which we can estimate from above by $\frac{n_2^\Theta}{n_1^\Theta + n_2^\Theta} \leq 1$. All terms in front of the norms $|\cdot|$ are bounded by a constant due to the estimate on the temperature $T_{12}^{N/2}$ and the estimate on $T_{12} + |u_{12}|^2$ proven in step 2 and 3. Furthermore, we can estimate

$$|n_1^1 - n_1^2| \leq \int (1 + |v|^2)|f_1^1 - f_1^2| dv,$$

and

$$|n_2^1 - n_2^2| \leq \int (1 + |v|^2)|f_2^1 - f_2^2| dv,$$

$$U := \frac{n_1^\Theta n_2^\Theta}{n_1^\Theta + n_2^\Theta} |u_{12}^1 - u_{12}^2| = \frac{n_1^\Theta n_2^\Theta}{n_1^\Theta + n_2^\Theta} |\alpha u_1^1 + (1 - \alpha) u_2^1 - \alpha u_1^2 - (1 - \alpha) u_2^2|$$

$$\leq \frac{n_2^\Theta (n_1^1 + n_2^1)}{n_1^\Theta + n_2^\Theta} \alpha |u_1^1 - u_1^2| + \frac{n_1^\Theta (n_2^1 + n_2^2)}{n_1^\Theta + n_2^\Theta} (1 - \alpha) |u_2^1 - u_2^2|.$$

Since $\dfrac{n_2^\Theta}{n_1^\Theta + n_2^\Theta}$ and $\dfrac{n_1^\Theta}{n_1^\Theta + n_2^\Theta}$ are smaller or equal 1, we can estimate

$$
\begin{aligned}
U &\leq (n_1^1 + n_1^2)\alpha|u_1^1 - u_1^2| + (n_2^1 + n_2^2)(1-\alpha)|u_2^1 - u_2^2| \\
&\leq \alpha|n_1^1 u_1^1 - n_1^1 u_1^2 + n_1^2 u_1^1 - n_1^2 u_1^2| + (1-\alpha)|n_2^1 u_2^1 - n_2^1 u_2^2 + n_2^2 u_2^1 - n_2^2 u_2^2| \\
&\leq \alpha|n_1^1 u_1^1 - n_1^1 u_1^2| + \alpha|n_1^2 u_1^1 - n_1^2 u_1^2| + (1-\alpha)|n_2^1 u_2^1 - n_2^1 u_2^2| + (1-\alpha)|n_2^2 u_2^1 - n_2^2 u_2^2| \\
&\leq \alpha|n_1^1 u_1^1 - n_1^2 u_1^2 + n_1^2 u_1^2 - n_1^1 u_1^2| + \alpha|n_1^1 u_1^1 - n_1^2 u_1^2 + n_1^2 u_1^1 - n_1^1 u_1^1| \\
&\quad + (1-\alpha)|n_2^1 u_2^1 - n_2^2 u_2^2 + n_2^2 u_2^2 - n_2^1 u_2^2| + (1-\alpha)|n_2^1 u_2^1 - n_2^2 u_2^2 + n_2^2 u_2^1 - n_2^1 u_2^1| \\
&\leq \alpha[|n_1^1 u_1^1 - n_1^2 u_1^2| + |u_1^2||n_1^2 - n_1^1| + |n_1^1 u_1^1 - n_1^2 u_1^2| + |u_1^1||n_1^2 - n_1^1|] \\
&\quad + (1-\alpha)[|n_2^1 u_2^1 - n_2^2 u_2^2| + |u_2^2||n_2^2 - n_2^1| + |n_2^1 u_2^1 - n_2^2 u_2^2| + |u_2^1||n_2^2 - n_2^1|].
\end{aligned}
$$

Due to the previous estimates on the velocities in step 4, the velocities are bounded and therefore

$$
U \leq C\Big[\int (1+|v|^2)|f_1^1 - f_1^2|\, dv + \int (1+|v|^2)|f_2^1 - f_2^2|\, dv\Big].
$$

In an analogous way, we can estimate

$$
\frac{n_1^\Theta n_2^\Theta}{n_1^\Theta + n_2^\Theta}|T_{12}^1 - T_{12}^2| \leq C\Big[\int (1+|v|^2)|f_1^1 - f_1^2|\, dv + \int (1+|v|^2)|f_2^1 - f_2^2|\, dv\Big].
$$

This all combines to the desired Lipschitz estimate.

Step 6: Existence and Uniqueness of non-negative solutions in $\bar{\Omega}$ (see definition of Ω in (4.23))

Now, introduce the sequence $\{(f_1^n, f_2^n)\}$ of mild solutions to

$$
\begin{aligned}
\partial_t f_1^n + v \cdot \nabla_x f_1^n &= \tilde{\nu}_{11}\frac{n_1^{n-1}}{n_1^{n-1} + n_2^{n-1}}(M_1[f_1^{n-1}] - f_1^n) \\
&\quad + \tilde{\nu}_{12}\frac{n_2^{n-1}}{n_1^{n-1} + n_2^{n-1}}(M_{12}[f_1^{n-1}, f_2^{n-1}] - f_1^n), \\
\partial_t f_2^n + v \cdot \nabla_x f_2^n &= \tilde{\nu}_{22}\frac{n_2^{n-1}}{n_1^{n-1} + n_2^{n-1}}(M_2[f_2^{n-1}] - f_2^n) \\
&\quad + \tilde{\nu}_{21}\frac{n_1^{n-1}}{n_1^{n-1} + n_2^{n-1}}(M_{21}[f_1^{n-1}, f_2^{n-1}] - f_2^n), \\
f_1^0 &= f_1(t=0), \\
f_2^0 &= f_2(t=0).
\end{aligned}
$$

Since the zeroth functions are known as the initial values, these are inhomogeneous transport equations for fixed $n \in \mathbb{N}$. For an inhomogeneous transport equation we know the existence of a unique mild solution in the periodic setting

$$f_1^n(x,v,t) = e^{-\alpha_1^{n-1}(x,v,t)} f_1^0(x - tv, v)$$

$$+ e^{-\alpha_1^{n-1}(x,v,t)} \int_0^t [\tilde{\nu}_{11} \frac{n_1^{n-1}(x + (s-t)v, s)}{n_1^{n-1}(x + (s-t)v, s) + n_2^{n-1}(x + (s-t)v, s)} M_1^{n-1}(x + (s-t)v, v, s)$$

$$+ \tilde{\nu}_{12} \frac{n_2^{n-1}(x + (s-t)v, s)}{n_1^{n-1}(x + (s-t)v, s) + n_2^{n-1}(x + (s-t)v, s)} M_{12}^{n-1}(x + (s-t)v, v, s)] e^{\alpha_1^{n-1}(x + (s-t)v, v, s)} ds,$$

$$f_2^n(x,v,t) = e^{-\alpha_2^{n-1}(x,v,t)} f_2^0(x - tv, v)$$

$$+ e^{-\alpha_2^{n-1}(x,v,t)} \int_0^t [\tilde{\nu}_{22} \frac{n_2^{n-1}(x + (s-t)v, s)}{n_1^{n-1}(x + (s-t)v, s) + n_2^{n-1}(x + (s-t)v, s)} M_2^{n-1}(x + (s-t)v, v, s)$$

$$+ \tilde{\nu}_{21} \frac{n_1^{n-1}(x + (s-t)v, s)}{n_1^{n-1}(x + (s-t)v, s) + n_2^{n-1}(x + (s-t)v, s)} M_{21}^{n-1}(x + (s-t)v, v, s)] e^{\alpha_2^{n-1}(x + (s-t)v, v, s)} ds.$$

Now, we show that $\{(f_1^n, f_2^n)\}$ is a Cauchy sequence in Ω. Then, since $\bar{\Omega}$ is complete, we can conclude convergence in $\bar{\Omega}$. First, we show that $\{(f_1^n, f_2^n)\}$ is in Ω.

- f_1^n, f_2^n are in $L^1((1 + |v|^2)dvdx)$ since f_1^0, f_2^0 are in $L^1((1 + |v|^2)dvdx)$.

- $f_1^n, f_2^n \geq 0$ since $f_1^0, f_2^0 \geq 0$.

- $N_q(f_k^n) < A$, $\min(n_k^n, T_k^n) > C$, since all estimates in step $1, 2$ and 4 are independent of n.

Now, $\{(f_1^n, f_2^n)\}$ is a Cauchy sequence in Ω since we have

$$\|f_1^n - f_1^{n-1}\|_{L^1((1+|v|^2)dvdx)}$$

$$\leq \int_\Lambda \int_{\mathbb{R}^n} e^{-\alpha_1^{n-1}(x,v,t)} \int_0^t e^{\alpha_1^{n-1}(x+(s-t)v,v,s)} |\tilde{\nu}_{11}^{n-1} \frac{n_1^{n-1}(x + (s-t)v, s)}{n_1^{n-1}(x + (s-t)v, s) + n_2^{n-1}(x + (s-t)v, s)}$$

$$M_1^{n-1}(x + (s-t)v, v, s) - \tilde{\nu}_{11}^{n-2} \frac{n_1^{n-2}(x + (s-t)v, s)}{n_1^{n-2}(x + (s-t)v, s) + n_2^{n-2}(x + (s-t)v, s)}$$

$$M_1^{n-2}(x + (s-t)v, v, s)|ds(1 + |v|^2)dxdv$$

$$+ \int_\Lambda \int_{\mathbb{R}^n} e^{-\alpha_1^{n-1}(x,v,t)} \int_0^t e^{\alpha_1^{n-1}(x+(s-t)v,v,s)} |\tilde{\nu}_{12}^{n-1} \frac{n_2^{n-1}(x + (s-t)v, s)}{n_1^{n-1}(x + (s-t)v, s) + n_2^{n-1}(x + (s-t)v, s)}$$

$$M_{12}^{n-1}(x + (s-t)v, v, s) - \tilde{\nu}_{12}^{n-2} \frac{n_2^{n-2}(x + (s-t)v, s)}{n_1^{n-2}(x + (s-t)v, s) + n_2^{n-2}(x + (s-t)v, s)}$$

$$M_{12}^{n-2}(x + (s-t)v, v, s)|ds(1 + |v|^2)dxdv.$$

Now we use the Lipschitz continuity of the Maxwell distributions

$$\|f_1^n - f_1^{n-1}\|_{L^1((1+|v|^2)dvdx)}$$

$$\leq C \int_\Lambda \int_{\mathbb{R}^n} e^{-\alpha_1^{n-1}(x,v,t)} \int_0^t e^{\alpha_1^{n-1}(x+(s-t)v,v,s)} |f_1^{n-1}(x + (s-t)v, v, s)$$

$$- f_1^{n-2}(x + (s-t)v, v, s)|ds(1 + |v|^2)dxdv$$

$$+ \int_\Lambda \int_{\mathbb{R}^n} e^{-\alpha_1^{n-1}(x,v,t)} \int_0^t e^{\alpha_1^{n-1}(x+(s-t)v,v,s)} [|f_1^{n-1}(x + (s-t)v, v, s) - f_1^{n-2}(x + (s-t)v, v, s)|$$

$$+ |f_2^{n-1}(x + (s-t)v, v, s) - f_2^{n-2}(x + (s-t)v, v, s)|]ds(1 + |v|^2)dxdv$$

$$\leq e^{-Ct} \int_0^t e^{Cs} [\|f_1^{n-1}(s) - f_1^{n-2}(s)\|_{L^1((1+|v|^2)dvdx)} + \|f_2^{n-1}(s) - f_2^{n-2}(s)\|_{L^1((1+|v|^2)dvdx)}]ds.$$

Similarly, we get for species 2

$$||f_2^n - f_2^{n-1}||_{L^1((1+|v|^2)dvdx)} \leq e^{-Ct} \int_0^t e^{Cs} [||f_1^{n-1}(s) - f_1^{n-2}(s)||_{L^1((1+|v|^2)dvdx)}$$
$$+ ||f_2^{n-1}(s) - f_2^{n-2}(s)||_{L^1((1+|v|^2)dvdx)}]ds.$$

Doing this inductively, we obtain

$$||f_1^n - f_1^{n-1}||_{L^1((1+|v|^2)dvdx)}$$
$$\leq (e^{-Ct})^n \int_0^t \cdots \int_0^t e^{Cs_1} \cdots e^{Cs_n} [||f_1^1(s_n) - f_1^0||_{L^1((1+|v|^2)dvdx)}$$
$$+ ||f_2^1(s_n) - f_2^0||_{L^1((1+|v|^2)dvdx)}]ds_1 \cdots ds_n$$
$$\leq \frac{1}{C^n}(1 - e^{-Ct})^n [\sup_{0 \leq s \leq t} ||f_1^1(s) - f_1^0||_{L^1((1+|v|^2)dvdx)} + \sup_{0 \leq s \leq t} ||f_2^1(s) - f_2^0||_{L^1((1+|v|^2)dvdx)}],$$

with a constant $C > 1$. So, for species 1, we obtain

$$\sup_{0 \leq t \leq T} ||f_1^{n+m} - f_1^n||_{L^1((1+|v|^2)dvdx)}$$
$$\leq \sup_{0 \leq t \leq T} [||f_1^{n+m} - f_1^{n+m-1}||_{L^1((1+|v|^2)dvdx)} + \cdots + ||f_1^{n+1} - f_1^n||_{L^1((1+|v|^2)dvdx)}]$$
$$\leq \sup_{0 \leq t \leq T} ((\frac{1}{C}(1 - e^{-Ct}))^{n+m} + \cdots + (\frac{1}{C}(1 - e^{-t}))^n)[\sup_{0 \leq s \leq t} ||f_1^1(s) - f_1^0||_{L^1((1+|v|^2)dvdx)}$$
$$+ \sup_{0 \leq s \leq t} ||f_2^1(s) - f_2^0||_{L^1((1+|v|^2)dvdx)}]$$
$$\leq ((C(T))^{n+m} + \cdots + C(T)^n)[\sup_{0 \leq s \leq T} ||f_1^1(s) - f_1^0||_{L^1((1+|v|^2)dvdx)}$$
$$+ \sup_{0 \leq s \leq T} ||f_2^1(s) - f_2^0||_{L^1((1+|v|^2)dvdx)}]$$
$$\leq C(T)^n \sum_{j=1}^{\infty} (C(T))^j [\sup_{0 \leq s \leq T} ||f_1^1(s) - f_1^0||_{L^1((1+|v|^2)dvdx)} + \sup_{0 \leq s \leq T} ||f_2^1(s) - f_2^0||_{L^1((1+|v|^2)dvdx)}]$$
$$\leq \frac{C(T)^n}{1 - C(T)}[\sup_{0 \leq s \leq T} ||f_1^1(s) - f_1^0||_{L^1((1+|v|^2)dvdx)} + \sup_{0 \leq s \leq T} ||f_2^1(s) - f_2^0||_{L^1((1+|v|^2)dvdx)}],$$

which converges to zero as $n \to \infty$ since $C(T) = \frac{1-e^{-CT}}{C} < 1$. In order to prove that the limit is a mild solution to (4.8) and the uniqueness of solutions to (4.8) we use standard arguments similar as in the proof of the fix point theorem of Banach and the theorem of Picard-Lindelöf. $\qquad\square$

Remark 4.3.3. $M^{(1)}$ and $M^{(2)}$ in the model of Andries, Aoki and Perthame in [1] and section 4.2.2 have the same structure as M_{12} and M_{21}, respectively, meaning that the velocities $u^{(1)}$ and $u^{(2)}$ and the temperatures $T^{(1)}$ and $T^{(2)}$ of $M^{(1)}$ and $M^{(2)}$, respectively have the same structure as the velocities u_{12} and u_{21} and the temperatures T_{12} and T_{21}. So the proof of theorem 4.3.6 for the model in section 4.2.2 goes through analogously as for the model in section 4.2.1.

4.4 Positivity of solutions of the BGK approximation for two species

4.4.1 Idea of the proof

Our aim is to prove that all classical solutions to (4.8) - (4.16) under the assumptions 4.3.1 with positive initial data are positive for all larger times $t > 0$. The idea of the proof is as follows. In the previous section, we stated our result about existence and uniqueness of non-negative solutions.Then, with a Gronwall estimate on the densities, we deduce that this non-negative solution can be estimated from below by an exponential function. Considering the solution along characteristics we will see that when the densities are positive the solution is also positive. With this and continuity in time, we can conclude that for positive initial data there cannot be a solution which becomes zero or negative at a time $t > 0$. So all classical solutions to (4.8) - (4.16) are positive.

4.4.2 Estimate on the densities

Lemma 4.4.1. *If* $f_k \geq 0$ *is a mild solution to (4.8) - (4.16) and*

$$\gamma_k(x,t) := \int f_k^0(x - vt, v)dv \geq C_0 > 0,$$

for all $t \geq 0$, $k = 1, 2$, *then the densities satisfy the estimate*

$$n_k(x,t) \geq C_0 e^{-(\tilde{\nu}_{kk} + \tilde{\nu}_{kj})t},$$

for all $t \geq 0$ *where* $C_0 > 0$ *is a positive constant.*

Proof. See step 2 in the proof of theorem 4.3.6. $\qquad\square$

4.4.3 Positivity of non-negative solutions

Lemma 4.4.2 (Positivity of non-negative solutions). *Let* (f_1, f_2) *with* $f_1, f_2 \geq 0$ *be a mild solution to (4.8)-(4.16) with positive initial data under the assumptions 4.3.1. Then* f_1, f_2 *are even positive, that means* $f_1, f_2 > 0$ *a.e.*

Proof. We prove the statement for f_1, the proof for f_2 is analogously. Let f_1 be part of the non-negative mild solution to (4.8)-(4.16). Then it satisfies by definition

$$f_1(x, v, t) = e^{-\alpha_1(x,v,t)} f_1^0(x - tv, v)$$
$$+ e^{-\alpha_1(x,v,t)} \int_0^t [\tilde{\nu}_{11} \frac{n_1(x+(s-t)v,s)}{n_1(x+(s-t)v,s)+n_2(x+(s-t)v,s)} M_1(x + (s-t)v, v, s)$$
$$+ \tilde{\nu}_{12} \frac{n_2(x+(s-t)v,s)}{n_1(x+(s-t)v,s)+n_2(x+(s-t)v,s)} M_{12}(x + (s-t)v, v, s)] e^{\alpha_1(x+(s-t)v,v,s)} ds.$$
$$\tag{4.24}$$

We assumed that all collision frequencies are positive and according to lemma 4.4.1 all densities are positive. So the right-hand side of (4.24) is positive, therefore

$$f_1(x, v, t) > 0,$$

for positive initial data. So non-negative solutions to (4.8) - (4.16) are even positive.

□

4.4.4 Positivity of classical solutions

Theorem 4.4.3. *Let (f_1, f_2) be a classical solution to (4.8) - (4.16) with positive initial data under the assumptions 4.3.1. Then the solution is positive meaning $f_1, f_2 > 0$.*

Proof. According to theorem 4.3.6 there exists a non-negative solution to (4.8) - (4.16) and it is the only non-negative solution to (4.8) - (4.16). So there could exist another classical solution which at a certain time becomes zero and negative afterwards. But due to continuity in time, it could only happen if it reaches zero first. According to lemma 4.4.2 this is not possible, because a non-negative solution always stays positive. So the unique solution to (4.8) - (4.16) with positive initial data is positive meaning $f_1, f_2 > 0$.   □

Chapter 5

Extension to a kinetic model for plasmas consisting of electrons and ions

In the introduction of chapter 2 we mentioned the advantages of this type of BGK models for mixtures having a sum of BGK-type interaction terms on the right-hand side. It is that the two different types of interactions, interactions of a species with itself and interactions of a species with the other one, are still kept separated. This allows to model the influence of the contribution to momentum and energy exchanges and the trend to equilibrium of two species which are very different. One example of two species which differ in some physical quantities is a gas mixture consisting of electrons and positively charged ions. In this case, the particles differ in their charge. Electrons have negative charges, the ions a positive one. Moreover, the mass of electrons is very small compared to the mass of ions. Both facts influences the behaviour in a significant way. This is why we want to use the model presented in chapter 2 as a starting point. However, we have to extend it to charged particles. This is done in this chapter. Models for electrons and ions are widely used in applications. One appearance of electrons and ions in applications and in nature is in form of a plasma. When a gas is brought to a very high temperature (10^4 K or more), electrons leave their orbit around the nuclei of the atom to which they belong. This gives an overall neutral mixture of charged particles, ions and electrons, which is called plasma. There are a lot of research areas and applications of a plasma. For example in astrophysics, plasmas play an important role since for example stars are formed from plasma and gain energy from the process of fusion, which may be a future possibility to gain energy in applications. The physical principle behind fusion is to gain energy by removing mass. This is one big research field in physics. It has the goal of developing fusion as a new energy source. It will be motivated later in chapter 6.

The outline of this chapter is as follows: In section 5.1 we introduce the basic physical principles when we deal with charged particles. In section 5.2 we want to present a model for a two-component mixture of charged particles. In section 5.3, we conclude with a formal derivation of the equations of ideal magnetohydrodynamics called MHD equations from the model presented in section 5.2 in order to put the model presented in section 5.2 in the context of ideal magnetohydrodynamics and observe how the typical quantities in the kinetic model will show up and influence the macroscopic equations. This derivation is also presented in [61] by Klingenberg, Pirner and Puppo.

5.1 Physical prerequisites for the following

In this section we want to introduce the basic physical knowledge in order to deal with charged particles. This is presented in a lot of introductory physics books for example in [44].

5.1.1 The Maxwell equations

We consider a system with an electric field $E^0(x) \in \mathbb{R}^3$ and a magnetic field $B^0(x) \in \mathbb{R}^3$ at time $t = 0$. Here $x \in \mathbb{R}^3$ denotes the position in space. The vector E^0 is a vector in $\mathbb{R}^3$ which describes the strength and the orientation of forces on charged particles due to an electric field. Assume we have a small positive test charge q. Then E^0 is defined as force on this test charge over this charge q. Similarly, the magnetic field is a measure of the strength and the orientation of forces on charged particles due to a magnetic field.

The time evolution of the electric field $E(x, t) \in \mathbb{R}^3$ and the magnetic field $B(x, t) \in \mathbb{R}^3$, where $t > 0$ denotes the time, is given by the Maxwell equations. We will present them in the following. The first one is called Gauß's law and describes the relationship between a static electric field and electric charges $\rho_c(x, t)$. It is given by

$$\nabla_x \cdot E = \frac{1}{\varepsilon_0} \rho_c, \tag{5.1}$$

where ε_0 denotes a constant called the vacuum permittivity or electric constant. The meaning of the equation is the following. Electric charges induce an electric field. Positive charges are sources of an electric field, negative charges are sinks of an electric field. This can be seen if we integrate (5.1) over a volume in x and use the theorem of Gauß. Then we see that the electric flux which leaves or enters this volume is proportional to the charge inside. Especially, if we take a particle with a small positive charge, the particle would travel from the positive charge towards the negative charge along the lines determined by the vector field $E(x, t)$.

Physicists suppose that in the case of the magnetic field there are neither sinks nor sources meaning

$$\nabla_x \cdot B = 0. \tag{5.2}$$

This is called Gauß's law for magnetism.

We can also get electric and magnetic fields which have no sinks and sources. In such a field, a particle with a small positive charge would travel along a closed circle. This is called a rotational field. From physical experiments, we expect that a rotational field is created by a time varying magnetic field

$$\nabla_x \times E = -\partial_t B. \tag{5.3}$$

This is called Faraday's law. Physicists also observes the reversed process. A time dependent electric field induces a rotational magnetic field. But a rotational magnetic field can also be generated in another way. It can be induced by a charge current $j(x,t)$ meaning that we have moving charged particles. This is called Ampère's law.

$$\nabla_x \times B = \mu_0 j + \mu_0 \varepsilon_0 \partial_t E. \tag{5.4}$$

μ_0 is the analogous magnetic constant to ε_0 and is called vacuum permeability or magnetic constant.

5.1.2 Lorentz force on charged particles

From physical experiments, we expect that the force F^L acting on a particle of electric charge q with velocity v under the influence of an electric field E and a magnetic field B is given by

$$F^L = q(E + v \times B).$$

This force is called Lorentz force and the effect on a charged particle is the following. We observe that a positively charged particle will be accelerated in the same direction as the electric field E and will travel on a curve which is orthogonal to the magnetic field B. A negatively charged particle will be accelerated in the opposite direction.

We observe that the Lorentz force due to an electric or magnetic field changes the velocity of a particle. If we model the physical system by a distribution function $f(x,v,t)$, the distribution function will change in time when we have electric or magnetic fields since the velocity of the particles change. We take this into account by replacing the left-hand side of the kinetic equation, the transport part,

$$\partial_t f + v \cdot \nabla_x f$$

by

$$\partial_t f + v \cdot \nabla_x f + F^L \cdot \nabla_v f.$$

5.2 A kinetic BGK model for ions and electrons

In this section we will present the Vlasov-BGK model for a mixture of two species and mention its fundamental properties like the conservation properties.

We consider a plasma consisting of electrons denoted by the index e and one species of ions denoted by the index i. Thus, our kinetic model has two distribution functions $f_e(x,v,t) > 0$ and $f_i(x,v,t) > 0$ where $x \in \mathbb{R}^3$ and $v \in \mathbb{R}^3$ are the phase space variables and $t \geq 0$ is the time.

Furthermore, for any $f_i, f_e : \mathbb{R}^3 \times \mathbb{R}^3 \times \mathbb{R}_0^+ \to \mathbb{R}$ with $(1 + |v|^2)f_i$, $(1 + |v|^2)f_e \in L^1(dv), f_i, f_e \geq 0$ we relate the distribution functions to macroscopic quantities by mean-values of f_k, $k = i, e$

$$\int f_k(v) \begin{pmatrix} 1 \\ v \\ m_k|v - u_k|^2 \end{pmatrix} dv =: \begin{pmatrix} n_k \\ n_k u_k \\ 3 n_k T_k \end{pmatrix}, \quad k = i, e, \tag{5.5}$$

where m_k is the mass, n_k the number density, u_k the mean velocity and T_k the temperature of species k, $k = i, e$. Note that in this chapter we shall write T_k instead of $k_B T_k$, where k_B is Boltzmann's constant.

We want to model the time evolution of the distribution functions by a Vlasov-BGK equation. The distribution functions are determined by two equations to describe their time evolution given by

$$\partial_t f_i + v \cdot \nabla_x f_i + \frac{F_i^L}{m_i} \cdot \nabla_v f_i = \nu_{ii} n_i (M_i - f_i) + \nu_{ie} n_e (M_{ie} - f_i),$$
$$\partial_t f_e + v \cdot \nabla_x f_e + \frac{F_e^L}{m_e} \cdot \nabla_v f_e = \nu_{ee} n_e (M_e - f_e) + \nu_{ei} n_i (M_{ei} - f_e),$$

(5.6)

with the mean-field forces F_i^L and F_e^L specified later and the Maxwell distributions

$$M_k(x, v, t) = \frac{n_k}{\sqrt{2\pi \frac{T_k}{m_k}}^3} \exp\left(-\frac{|v - u_k|^2}{2 \frac{T_k}{m_k}}\right), \quad k = i, e,$$

$$M_{kj}(x, v, t) = \frac{n_k}{\sqrt{2\pi \frac{T_{kj}}{m_k}}^3} \exp\left(-\frac{|v - u_{kj}|^2}{2 \frac{T_{kj}}{m_k}}\right), \quad k, j = i, e, k \neq j,$$

(5.7)

where $\nu_{ii} n_i$ and $\nu_{ee} n_e$ are the collision frequencies of the particles of each species with itself, while $\nu_{ie} n_e$ and $\nu_{ei} n_i$ are related to interspecies collisions. In the previous chapters we assumed the general relationship between the collision frequencies

$$\nu_{ie} = \varepsilon \nu_{ei}, \quad 0 < \varepsilon \leq 1. \tag{5.8}$$

In the next section we want to derive a specific value for ε in the case of electrons and ions. We assume that all collision frequencies are positive. In addition, we take into account an acceleration due to interactions using the Lorentz forces F_i^L, F_e^L given by

$$F_i^L(x, t) = e\left(E(x, t) + v \times B(x, t)\right) \quad \text{and} \quad F_e^L(x, t) = -e\left(E(x, t) + v \times B(x, t)\right),$$

where e denotes the elementary charge. In order to determine the time evolution of the electric and magnetic field, we couple the system with the Maxwell equations (5.1), (5.2), (5.3) and (5.4). We assume that there are no external electric and magnetic fields. The only electric and magnetic field we have is a mean electric and magnetic field generated by the particles themselves. We therefore assume that the charge density is given by

$$\rho_c = e(n_i - n_e), \tag{5.9}$$

and the charge current given by

$$j = e(n_i u_i - n_e u_e). \tag{5.10}$$

We note that the choice of the Maxwell distributions M_i, M_e, M_{ie} and M_{ei} is the same as in the model from chapter 2. The quantities u_{12}, u_{21}, T_{12} and T_{21} are still given by the expressions (2.6), (2.7), (2.10) and (2.11). This guarantees that the mass exchange, the sum of the two momentum exchanges and the sum of the two energy exchanges are still zero and we still guarantee that all temperatures are positive under the restrictions

$$0 \leq \gamma \leq m_i(1 - \delta) \left[(1 + \frac{m_i}{m_e}\varepsilon)\delta + 1 - \frac{m_i}{m_e}\varepsilon \right],$$ (5.11)

and

$$\frac{\frac{m_i}{m_e}\varepsilon - 1}{1 + \frac{m_i}{m_e}\varepsilon} \leq \delta \leq 1,$$ (5.12)

see theorems 2.1.1, 2.1.2, 2.1.3 and 2.1.4 in chapter 2.

5.2.1 Relationship between the collision frequencies

The goal of this section is to derive an expression for the ratio of all the relaxation parameters $\nu_{ii}, \nu_{ie}, \nu_{ee}$ and ν_{ei} in the case of a plasma, for example a value for ε in (5.8).

The parameters ν_{ie} and ν_{ei} are linked to the interspecies collision frequency. In plasmas, the mass ratio of the two kinds of particles is $\frac{m_e}{m_i} \ll 1$, where i denotes ions and e denotes electrons. In this case a common relationship found in the literature [11] is

$$\nu_{ie} = \frac{m_e}{m_i}\nu_{ei}.$$ (5.13)

A motivation for this relationship in the case of a plasma can be found in [11], chapter 1.9, which we want to mention here shortly. From physical experiments, we expect that the collision frequency is proportional to the differential cross section and the relative velocity. For the typical velocity of ions and electrons close to equilibrium we take the peak of the Boltzmann distribution called the thermal velocity $v_{T_k} = \left(\frac{2T_k}{m_k}\right)^{\frac{1}{2}}$, $k = i, e$ and assume that the temperatures are of the same order, $T_i \approx T_e$. The cross sections are considered equal, because they depend on the interaction potential, which in this case is the Coulomb force, that is the same for both particles. So the only thing which remains to consider is the relative velocity. Since the mass of the ions m_i is much larger than the mass of the electrons m_e, we get in the case of ν_{ei} for the relative velocity of an ion and an electron

$$\left(\frac{2T_e}{m_e}\right)^{\frac{1}{2}} - \left(\frac{2T_i}{m_i}\right)^{\frac{1}{2}} \approx (2T_e)^{\frac{1}{2}} \left(\left(\frac{1}{m_e}\right)^{\frac{1}{2}} - \left(\frac{1}{m_i}\right)^{\frac{1}{2}} \right)$$

$$= (2T_e)^{\frac{1}{2}} \frac{1 - \left(\frac{m_e}{m_i}\right)^{\frac{1}{2}}}{m_e^{\frac{1}{2}}} \approx (2T_e)^{\frac{1}{2}} \left(\frac{1}{m_e}\right)^{\frac{1}{2}},$$

which is the order of magnitude of the mean velocity of the electrons. We expect the relative velocity of two electrons to have the same order of magnitude as the thermal velocity of an electron. Since ν_{ee} is proportional to the relative velocity of two electrons and we only want to compare the order of magnitudes of ν_{ei} and ν_{ee}, we conclude that ν_{ei} and ν_{ee} are of the same order of magnitude, so we have

$$\nu_{ei} \approx \nu_{ee}.$$

Now consider ν_{ii}. The ion thermal velocity is lower by an amount of $(\frac{m_e}{m_i})^{\frac{1}{2}}$ with respect to the electrons, since

$$\left(\frac{2T_i}{m_i}\right)^{\frac{1}{2}} = \left(\frac{m_e}{m_i}\right)^{\frac{1}{2}} \left(\frac{2T_i}{m_e}\right)^{\frac{1}{2}} \approx \left(\frac{m_e}{m_i}\right)^{\frac{1}{2}} \left(\frac{2T_e}{m_e}\right)^{\frac{1}{2}}.$$

Therefore

$$\nu_{ii} \approx \left(\frac{m_e}{m_i}\right)^{\frac{1}{2}} \nu_{ee}.$$

For an estimate of ν_{ie} and ν_{ei} we consider a collision of an electron head-on with an ion. The velocities after a collision of an ion with an electron are given by

$$v_i' = v_i - \frac{2m_e}{m_i + m_e}[(v_i - v_e) \cdot \omega]\omega,$$

$$v_e' = v_e - \frac{2m_i}{m_i + m_e}[(v_e - v_i) \cdot \omega]\omega,$$

see corollary 1.2.5. The vector ω was defined as the unit vector along the line with the minimal distance of the two particles during the interaction, in the direction of particle 2. Since we consider a head-on collision, ω is parallel to $\frac{v_i - v_i}{|v_i - v_e|}$. So the formulas for the velocities after the interaction simplify to

$$v_i' = v_i - \frac{2m_e}{m_i + m_e}(v_i - v_e),$$

$$v_e' = v_e - \frac{2m_i}{m_i + m_e}(v_e - v_i).$$

Since m_e is small compared to m_i, we get

$$v_i' = v_i + O\left(\frac{m_e}{m_i}\right),$$

$$v_e' = v_e + O(1),$$

which reflects the physical fact that collisions of a heavy particle with a light one have a bigger influence on the lighter one than on the heavy one. Hence $\nu_{ie} = \frac{m_e}{m_i}\nu_{ee}$.

To summarize, in the case of ions and electrons, the collision frequencies can be ordered as follows:

$$\nu_{ei} \approx \nu_{ee} \approx \left(\frac{m_i}{m_e}\right)^{\frac{1}{2}} \nu_{ii} \approx \left(\frac{m_i}{m_e}\right) \nu_{ie}.$$

5.2.2 Macroscopic equations

In section 2.1.9 we derived macroscopic equations for the model (2.2) presented in chapter 2. For the model (5.6) we can also derive macroscopic equations. In this case we will obtain additional terms due to the additional force terms in (5.6).

Theorem 5.2.1 (Macroscopic equations for the BGK equation for mixtures). *If $f_1, f_2 \in L^\infty(dv)$ decay fast enough to zero in the v variable and are a solution to (5.6) in the sense of distributions, they satisfy the following local macroscopic conservation laws.*

$$\partial_t n_i + \nabla_x(n_i u_i) = 0,$$
$$\partial_t n_e + \nabla_x(n_e u_e) = 0,$$
$$\partial_t(m_i n_i u_i) + \nabla_x \cdot \mathbb{P}_i + \nabla_x \cdot (m_i n_i u_i \otimes u_i) - e n_i(E + u_i \times B) = f_{m_{i,e}},$$
$$\partial_t(m_e n_e u_e) + \nabla_x \cdot \mathbb{P}_e + \nabla_x \cdot (m_e n_e u_e \otimes u_e) + e n_e(E + u_e \times B) = f_{m_{e,i}},$$
$$\partial_t \left(\frac{m_i}{2} n_i |u_i|^2 + \frac{3}{2} n_i T_i \right) + \nabla_x \cdot Q_i - e E u_i n_i = F_{E_{i,e}},$$
$$\partial_t \left(\frac{m_e}{2} n_e |u_e|^2 + \frac{3}{2} n_e T_e \right) + \nabla_x \cdot Q_e + e E u_e n_e = F_{E_{e,i}},$$

with the pressure tensor $\mathbb{P}_k$ and the energy flux Q_k given by

$$\mathbb{P}_k(x,t) = \int (v - u_k(x,t)) \otimes (v - u_k(x,t)) f_k(x,v,t) dv,$$
$$Q_k(x,t) = \frac{1}{2} \int |v|^2 v f_k(x,v,t) dv,$$

for $k = 1, 2$ and with $f_{m_{i,e}}$, $f_{m_{e,i}}$, $F_{E_{i,e}}$ and $F_{E_{e,i}}$ given by the expressions in theorem 2.1.9 with $1 = i$ and $2 = e$ and E, B given by the Maxwell equations (5.1), (5.2), (5.3) and (5.4) with charge density and charge current given by (5.9) and (5.10).

Proof. We derive only the additional terms due to the force terms in (5.6). The rest is shown in the proof of theorem 2.1.9. We start with the derivation of conservation of the number of particles. If we integrate the equation for ions in (5.6) with respect to v, we get

$$\int \partial_t f_i(x,v,t) dv + \int v \cdot \nabla_x f_i dv + \int \nabla_v \cdot \left(\frac{e}{m_i}(E + v \times B) f_i \right) dv = 0.$$

We assume that f_i and $v f_i$ are decreasing fast enough to zero for $|v| \to \infty$, so with the Gauß theorem we get, that the third integral on the left-hand side vanishes. So the equation is equivalent to

$$\partial_t n_i + \nabla_x \cdot (n_i u_i) = 0.$$

We can do the same with f_e. In this case we get

$$\partial_t n_e + \nabla_x \cdot (n_e u_e) = 0.$$

Multiplying the equation for ions by $m_i v$ and integrating it with respect to the velocity v, leads to

$$m_i \int v \partial_t f_i dv + m_i \int vv \cdot \nabla_x f_i dv + \int v \nabla_v \cdot (e(E + v \times B) f_i) dv = f_{m_{i,e}}.$$

In the third term integration by parts with the assumption that $v f_i$ is decreasing fast enough to zero for $|v| \to \infty$ such that we have no contribution from the boundary terms leads to

$$- \int e(E + v \times B) f_i dv,$$

which becomes

$$-e n_i (E + u_i \times B).$$

So all in all, we get

$$\partial_t (m_i n_i u_i) + \nabla_x \cdot \mathbb{P}_i + \nabla_x \cdot (m_i n_i u_i \otimes u_i) - e n_i (E + u_i \times B) = f_{m_{i,e}}.$$

We can do the same with f_e.
Multiplying the equation for the ions by $\frac{m_i}{2}|v|^2$ and integrating it with respect to v leads to

$$\frac{m_i}{2} \int |v|^2 \partial_t f_i dv + \frac{m_i}{2} \int |v|^2 v \cdot \nabla_x f_i dv + \frac{1}{2} \int |v|^2 \nabla_v \cdot (e(E + v \times B) f_i) dv = F_{E_{i,e}}.$$

In the third term integration by parts and the assumption that $|v|^2 f_i$ and $v|v|^2 f_i$ are decreasing fast enough to zero for $|v| \to \infty$ leads to

$$\frac{1}{2} \int |v|^2 \nabla_v \cdot (e(E + v \times B) f_i) dv = - \int ev \cdot (E + v \times B) f_i dv = -eE n_i u_i.$$

The term with the magnetic field vanishes since v is orthogonal to $v \times B$.
We obtain the macroscopic equation

$$\partial_t \left(\frac{m_i}{2} n_i |u_i|^2 + \frac{3}{2} n_i T_i \right) + \nabla_x \cdot Q_i - e E u_i n_i = F_{E_{i,e}}.$$

So all in all, we get the system in theorem 5.2.1. $\qquad \square$

5.2.3 Entropy inequality

In this section we want to show that the entropy inequality from theorem 2.1.7 remains true for the model (5.6) with its additional force terms.

Theorem 5.2.2 (Entropy inequality). *Assume $f_i, f_e > 0$. With the assumption f_i, f_e, $v f_i$ and $v f_e$ are decreasing fast enough to zero for $|v| \to \infty$, we have the following entropy inequality*

$$\partial_t \left(\int f_i \ln f_i dv + \int f_e \ln f_e dv \right) + \nabla_x \cdot \left(\int v f_i \ln f_i dv + \int v f_e \ln f_e dv \right) \leq 0,$$

with equality if and only if f_i and f_e are Maxwell distributions with equal mean velocity and temperature.

Proof. We multiply the first equation of (5.6) by $\ln f_i$, the second one by $\ln f_e$, integrate with respect to v and add both. The terms which are new compared to the case of neutral particles are the terms with the magnetic and electric fields. There we use integration by parts and Gauß theorem.

$$\int \ln f_i \nabla_v \cdot \left(\frac{e}{m_i}(E + v \times B) f_i \right) dv = - \int \frac{1}{f_i} \nabla_v f_i \cdot \left(\frac{e}{m_i}(E + v \times B) f_i \right) dv$$
$$= - \int \nabla_v f_i \cdot \left(\frac{e}{m_i}(E + v \times B) \right) dv = - \int \nabla_v \cdot \left(\frac{e}{m_i}(E + v \times B) f_i \right) dv = 0.$$

The last but one equality is due to the fact that the ith component of $v \times B$ is independent of v_i. We can do the same for the term coming from the electrons, so

$$\int \ln f_e \nabla_v \cdot \left(\frac{-e}{m_e}(E + v \times B) f_e \right) dv = 0.$$

So the additional terms turn out to be zero and we obtain the same inequality for the entropy as in the case of neutral particles. $\qquad \square$

5.3 Deriving macroscopic MHD equations

In this section we want to illustrate the model in the case of ions and electrons. We want to show that it is possible to put the proposed kinetic model in the context of the typical macroscopic equations for charged particles. Therefore we want to derive the equations of ideal magnetohydrodynamics, from our model. You can also find a similar derivation in [15] but for an isothermal flow.

5.3.1 The BGK model for ions and electrons

We consider the case of ions and electrons and set $\varepsilon = \frac{m_2}{m_1}$ as it is motivated in section 5.2.1. For simplicity, we take $\delta = 0$, $\alpha = \frac{m_2}{m_1+m_2}$ and $\gamma = 0$ in (2.6) and (2.10), although the MHD equations can also derived from the general model. We use the index i for the ions and e for the electrons. Then the particles are subjected to the Lorentz force $F_i = e(E + v \times B)$ and $F_e = -e(E + v \times B)$, where e is the elementary charge and E and B the mean electric and magnetic fields given by the Maxwell equations (5.1), (5.2), (5.3) and (5.4) with the charge density given by (5.9) and the charge current given by (5.10). In this case the model (5.6) rewrites as

$$\partial_t f_i + v \cdot \nabla_x f_i + \frac{e(E + v \times B)}{m_i} \cdot \nabla_v f_i = \nu_{ii} n_i (M_i - f_i) + \nu_{ie} n_e (M_{ie} - f_i),$$

$$\partial_t f_e + v \cdot \nabla_x f_e - \frac{e(E + v \times B)}{m_e} \cdot \nabla_v f_e = \nu_{ee} n_e (M_e - f_e) + \frac{m_i}{m_e} \nu_{ie} n_i (M_{ei} - f_e).$$

$$\tag{5.14}$$

5.3.2 Macroscopic equations for electrons and ions

In order to derive macroscopic equations, we multiply the first equation of (5.14) with $(1, m_i v, \frac{m_i}{2}|v|^2)$, and the second with $(1, m_e v, \frac{m_e}{2}|v|^2)$. Then we integrate them with respect to the velocity. We obtain the system derived in theorem 5.2.1 with the specific values for α, δ and γ. The obtained macroscopic system is not closed since we obtain terms of the form $\int v \otimes v f_i dv, \int v \otimes v f_e dv, \int |v|^2 v f_i dv$ and $\int |v|^2 v f_e dv$. All the other terms are functions of known quantities given in (5.5). We propose the following closure. There are plasmas where the two species first relax to its own equilibrium and then to a global one. According to chapter 1.9 in [11], we expect that plasmas are typically not in global equilibrium, although the components may be in a partial equilibrium. This means, the electrons are in equilibrium with itself but not with the ions, and the other way round. So in our considerations we assume that each species is in equilibrium with itself, e.g. setting $f_i = M_i$ and $f_e = M_e$ in the equations from theorem 5.2.1. In this way, we obtain a closed system of equations for the conservation of mass, momentum and energy.

$$\partial_t n_k + \nabla_x (n_k u_k) = 0, \quad k = i, e,$$

$$\partial_t (m_i n_i u_i) + \nabla_x (n_i T_i) + \nabla_x \cdot (m_i u_i \otimes u_i n_i) - e n_i (E + u_i \times B)$$
$$= \nu_{ie} m_i n_e n_i (u_e - u_i),$$

$$\partial_t (m_e n_e u_e) + \nabla_x (n_e T_e) + \nabla_x \cdot (m_e u_e \otimes u_e n_e) + e n_e (E + u_e \times B)$$
$$= \nu_{ei} m_e n_e n_i (u_i - u_e),$$

$$\partial_t \left(\frac{m_i}{2} n_i |u_i|^2 + \frac{3}{2} n_i T_i \right) + \nabla_x \cdot \left(\frac{5}{2} n_i T_i u_i \right) + \nabla_x \cdot \left(\frac{m_i}{2} n_i |u_i|^2 u_i \right) - e n_i E u_i$$
$$= \frac{1}{2} \nu_{ie} n_e n_i m_i (|u_e|^2 - |u_i|^2) + \nu_{ie} \frac{3}{2} n_i n_e \frac{m_i}{m_e + m_i} (T_e - T_i),$$

$$\partial_t \left(\frac{m_e}{2} n_e |u_e|^2 + \frac{3}{2} n_e T_e \right) + \nabla_x \cdot \left(\frac{5}{2} n_e T_e u_e \right) + \nabla_x \cdot \left(\frac{m_e}{2} n_e |u_e|^2 u_e \right) + e n_e E u_e$$
$$= \frac{1}{2} \nu_{ei} n_e n_i m_e (|u_i|^2 - |u_e|^2) + \nu_{ei} \frac{3}{2} n_i n_e \frac{m_e}{m_e + m_i} (T_i - T_e).$$

Note that we used the specific values $\delta = 0$, $\alpha = \frac{m_2}{m_1+m_2}$ and $\gamma = 0$ in the exchange terms. In order to determine the time evolution of the electric and magnetic field, we couple the system with the Maxwell equations.

$$\nabla_x \cdot E = \frac{1}{\varepsilon_0} \rho_c,$$
$$\nabla_x \times E + \partial_t B = 0,$$
$$\nabla_x \times B = \mu_0 j + \mu_0 \varepsilon_0 \partial_t E,$$
$$\nabla_x \cdot B = 0,$$
$$\rho_c = e(n_i - n_e),$$
$$j = e(n_i u_i - n_e u_e),$$

where μ_0, ε_0 are the magnetic and electric vacuum permittivity. From physics, we expect that these two constant are linked by the speed of light c via $c^2 = \frac{1}{\mu_0 \varepsilon_0}$.

5.3.3 Dimensionless equations

First we define dimensionless variables of the time t, the length x, the velocities u_e, u_i, the number densities n_e, n_i, the temperatures T_e, T_i, the magnetic field B, the electric field E, the electron-ion collision frequency ν_{ei}, the ion-electron collision frequency ν_{ie} and the current density j, for example $t' = {}^t/_{\bar{t}}$ for a typical time scale $\bar{t}$. In particular, the order of magnitudes of some quantities are assumed to be linked: We assume that both species have densities, mean velocities and temperatures of the same order of magnitude, e.g. $\bar{n}_i = \bar{n}_e = \bar{n}$, $\bar{u}_i = \bar{u}_e = \bar{u} = {}^{\bar{x}}/_{\bar{t}}$ and $\bar{T}_i = \bar{T}_e = \bar{T}$. The last two assumptions allow to assume that we are close to a thermodynamic equilibrium in which the two mean velocities and temperatures would be equal. Further, we assume that $\bar{E} = \bar{B}\bar{u}$. From non-dimensionalizing the first two Maxwell equations we see that this means that the electric field induced by a change of the magnetic field in time dominates over the fields which arise from charges and currents. Furthermore, we assume that $\bar{B} = \mu_0 \bar{x} \bar{j}$, which means that the magnetic field induced by currents dominates over the magnetic field due to changes of the electric field in time.

This leads to the following equations, where now the variables are non-dimensional

$$\partial_t n_k + \nabla_x \cdot (n_k u_k) = 0, \quad k = i, e,$$

$$\partial_t (n_i u_i) + C_1 \nabla_x (n_i T_i) + \nabla_x \cdot (n_i u_i \otimes u_i) - C_2 n_i (E + u_i \times B) = \\ C_3 \nu_{ie} n_e n_i (u_e - u_i),$$

$$C_4 \partial_t (n_e u_e) + C_1 \nabla_x (n_e T_e) + C_4 \nabla_x \cdot (n_e u_e \otimes u_e) + C_2 n_e (E + u_e \times B) = \\ C_3 \nu_{ie} n_e n_i (u_i - u_e),$$

$$C_1 \, \partial_t \left(\frac{3}{2} n_i T_i \right) + \partial_t \left(\frac{1}{2} n_i |u_i|^2 \right) + C_1 \, \nabla_x \cdot \left(\frac{5}{2} n_i T_i u_i \right) + \nabla_x \cdot \left(\frac{1}{2} n_i |u_i|^2 u_i \right)$$

$$= C_2 \, E n_i u_i + C_3 \, \frac{1}{2} \nu_{ie} n_e n_i (|u_e|^2 - |u_i|^2) + \frac{1}{1+C_4} C_3 C_1 \nu_{ie} \frac{3}{2} n_i n_e (T_e - T_i),$$

$$C_1 \, \partial_t \left(\frac{3}{2} n_e T_e \right) + C_4 \, \partial_t \left(\frac{1}{2} n_e |u_e|^2 \right) + C_1 \, \nabla_x \cdot \left(\frac{5}{2} n_e T_e u_e \right) + C_4 \nabla_x \cdot \left(\frac{1}{2} n_e |u_e|^2 u_e \right)$$

$$= -C_2 \, E n_e u_e + C_3 \, \frac{1}{2} \nu_{ei} n_e n_i (|u_i|^2 - |u_e|^2) + \frac{1}{1+C_4} C_3 C_1 \, \nu_{ie} \frac{3}{2} n_i n_e (T_i - T_e),$$

together with the Maxwell equations

$$C_5 M \, \nabla_x \cdot E = \rho_c,$$
$$\nabla_x \times E + \partial_t B = 0,$$
$$\nabla_x \times B = j + M \partial_t E,$$
$$\nabla_x \cdot B = 0,$$
$$\rho_c = (n_i - n_e),$$
$$C_5 \, j = (n_i u_i - n_e u_e).$$

The constants $C_i, i = 1, ..., 5$ and M are dimensionless parameters. In particular,

$$C_1 = \frac{\bar{n}\bar{T}}{m_i \bar{n}\bar{u}^2}, \quad C_2 = \frac{e\bar{B}\bar{t}}{m_i}, \quad C_3 = \bar{\nu}_{ie} \bar{n}\bar{t}, \quad C_4 = \frac{m_e}{m_i}, \quad C_5 = \frac{\bar{j}}{e\bar{n}\bar{u}} \quad \text{and} \quad M = \frac{\bar{u}^2}{c^2},$$

coming from non-dimensionalizing. The physical meaning is the following: C_1 describes the ratio of the typical scale of thermal energy $\bar{n}\bar{T}$ and of the kinetic energy $m_i \bar{n}\bar{u}^2$ of ions. For the meaning of C_2, we consider an ion travelling with a speed perpendicular to a magnetic field at distance r, the force due to the magnetic field $e\bar{B}\bar{u}$ on the particle acts as a centripetal force $\frac{m\bar{u}^2}{r}$, so the norm of the forces is equal

$$\frac{m\bar{u}^2}{r} = e\bar{B}\bar{u},$$

which is equivalent to $\omega := \frac{\bar{u}}{r} = \frac{e\bar{B}\bar{u}}{m}$ which describes a frequency called cyclotron frequency. So C_2 is the product of the typical scale of the cyclotron frequency and the typical time scale. C_3 is the ratio of the macroscopic time scale and the time scale induced by the collisions. C_4 is the mass ratio and M the typical scale of the speed squared and the speed of light squared. Finally, C_5 is the typical scale of the current density induced by electric fields over the typical scale of the current induced by the flow of the particles.

5.3.4 The limits to the MHD equations

Now we consider the formal limit of the mass ratio $C_4 \to 0$ and the non-relativistic limit $M \to 0$.

Theorem 5.3.1. *The formal limit of the mass ratio $C_4 \to 0$ and the non-relativistic limit $M \to 0$ of the non-dimensionalized system with the remaining parameters C_1, C_2, C_3 and C_5 remaining finite is the system*

$$\partial_t n + \nabla_x \cdot (nu) = 0,$$

$$\partial_t (nu) + C_1 \, \nabla_x(nT) + \nabla_x \cdot (nu \otimes u) = C_2 C_5 \, j \times B,$$

$$C_1 \, \partial_t \left(\frac{3}{2} nT \right) + \partial_t \left(\frac{1}{2} n|u|^2 \right) + C_1 \, \nabla_x \cdot \left(\frac{5}{2} nu(T_e - T_i) \right) - C_1 C_5 \nabla_x \cdot \left(\frac{5}{2} Tj \right)$$

$$+ \nabla_x \cdot \left(\frac{1}{2} n|u|^2 u \right) = C_2 C_5 \, Ej,$$

$$\frac{C_3}{C_2} C_1 \, \nabla_x(nT_e) + C_3 \, n(E + u \times B) - C_3 C_5 (j \times B) = \frac{C_3^2}{C_2} C_5 \, \nu_{ei} nj,$$

$$\frac{C_3}{C_2} C_1 \, \partial_t \left(\frac{3}{2} nT_e \right) + \frac{C_3}{C_2} C_1 \, \nabla_x \cdot \left(\frac{5}{2} nT_e u \right) - \frac{C_3}{C_2} C_1 C_5 \nabla_x \cdot \left(\frac{5}{2} T_e j \right) = -C_3 \, En \left(u - C_5 \frac{j}{n} \right)$$

$$+ \frac{C_3^2}{C_2} C_5 \, \nu_{ei} nju - \frac{C_3^2}{C_2} C_5^2 \frac{1}{2} \nu_{ei} |j|^2 + \frac{C_3}{C_2} C_3 C_1 \, \nu_{ei} \frac{3}{2} n^2 (T_i - T_e),$$

$$\nabla_x \times E + \partial_t B = 0,$$

$$\nabla_x \cdot B = 0,$$

$$\nabla_x \times B = j,$$

$$C_5 j = n(u - u_e).$$

Proof. We start with the non-dimensionalized system from section 5.3.3. In the limit $M \to 0$, we get from the first Maxwell equation that n_i and n_e converge formally to the same limit n. The third Maxwell equation simplifies to

$$\nabla_x \times B = j.$$

We denote the limit of u_i by u. Then we get from conservation of the number of ions

$$\partial_t n + \nabla_x \cdot (nu) = 0.$$

In the limit $C_4 \to 0$, the momentum equation of the electrons turns into

$$C_1 \, \nabla_x(nT_e) + C_2 \, n(E + u_e \times B) = C_3 \, \nu_{ie} nn(u - u_e). \tag{5.15}$$

The limit of the sum of the momentum equations with $T := T_i + T_e$ gives

$$\partial_t (nu) + C_1 \, \nabla_x(nT) + \nabla_x \cdot (u \otimes un) + C_2 \, n(u_e - u) \times B = 0. \tag{5.16}$$

The other Maxwell equations turn into

$$\nabla_x \times E + \partial_t B = 0,$$
$$\nabla_x \cdot B = 0,$$
$$C_5 j = n(u - u_e). \tag{5.17}$$

The energy equation of the electrons leads to

$$C_1 \partial_t \left(\frac{3}{2} n T_e \right) + C_1 \nabla_x \cdot \left(\frac{5}{2} n T_e u_e \right)$$
$$= -C_2 \, En_e u_e + C_3 \, \frac{1}{2} \nu_{ei} n^2 (|u|^2 - |u_e|^2) + C_3 C_1 \, \nu_{ei} \frac{3}{2} n^2 (T_i - T_e). \tag{5.18}$$

From the sum of the energy equations we get

$$C_1 \partial_t \left(\frac{3}{2} n T \right) + \partial_t \left(\frac{1}{2} n |u|^2 \right) + C_1 \nabla_x \cdot \left(\frac{5}{2} (n T_e u_e + n T_i u) \right) + \nabla_x \cdot \left(\frac{1}{2} n |u|^2 u \right)$$
$$= C_2 \, En(u - u_e). \tag{5.19}$$

Using $C_5 j = n(u - u_e)$, we get from (5.15), (5.16) and (5.19)

$$C_1 \nabla_x (n T_e) + C_2 \, n(E + u_e \times B) = C_3 C_5 \, \nu_{ei} n j, \tag{5.20}$$

$$\partial_t (nu) + C_1 \nabla_x (n T) + \nabla_x \cdot (nu \otimes u) = C_2 C_5 \, j \times B,$$

$$C_1 \partial_t \left(\frac{3}{2} n T \right) + \partial_t \left(\frac{1}{2} n |u|^2 \right) + C_1 \nabla_x \cdot \left(\frac{5}{2} (n T_e u_e + n T_i u) \right) + \nabla_x \cdot \left(\frac{1}{2} n |u|^2 u \right)$$
$$= C_2 C_5 \, E j. \tag{5.21}$$

Writing $|u|^2 - |u_e|^2$ as $(u - u_e) \cdot (u + u_e)$ and again replacing j by $C_5 j = n(u - u_e)$, we obtain from (5.18)

$$C_1 \partial_t \left(\frac{3}{2} n T_e \right) + C_1 \nabla_x \cdot \left(\frac{5}{2} n T_e u_e \right)$$
$$= -C_2 \, En_e u_e + C_3 C_5 \, \frac{1}{2} \nu_{ei} n j (u + u_e) + C_3 C_1 \, \nu_{ei} \frac{3}{2} nn(T_i - T_e). \tag{5.22}$$

Equations (5.20) and (5.22) are equivalent to

$$C_1 \nabla_x (n T_e) + \frac{C_2}{C_3} C_3 \, n(E + u_e \times B) = \frac{C_3^2}{C_2} \frac{1}{\frac{C_3}{C_2}} C_5 \, \nu_{ei} n j, \tag{5.23}$$

$$C_1 \, \partial_t \left(\frac{3}{2} n T_e \right) + C_1 \, \nabla_x \cdot \left(\frac{5}{2} n T_e u_e \right)$$

$$= -\frac{C_2}{C_3} C_3 \, E n u_e + \frac{C_3^2}{C_2} \frac{1}{\frac{C_3}{C_2}} C_5 \, \frac{1}{2} \nu_{ei} n j (u + u_e) + C_3 C_1 \, \nu_{ei} \frac{3}{2} n^2 (T_i - T_e).$$

$$(5.24)$$

We multiply (5.23) and (5.24) by $\frac{C_3}{C_2}$ and insert $u_e = u - C_5 \frac{j}{n}$ from (5.17), we get from (5.21), (5.23) and (5.24)

$$C_1 \, \partial_t \left(\frac{3}{2} n T \right) + \partial_t \left(\frac{1}{2} n |u|^2 \right) + C_1 \, \nabla_x \cdot \left(\frac{5}{2} n u (T_e - T_i) \right) - C_1 C_5 \nabla_x \cdot \left(\frac{5}{2} T j \right)$$

$$+ \nabla_x \cdot \left(\frac{1}{2} n |u|^2 u \right) = C_2 C_5 \, E j,$$

$$\frac{C_3}{C_2} C_1 \, \nabla_x (n T_e) + C_3 \, n (E + u \times B) - C_3 C_5 (j \times B) = \frac{C_3^2}{C_2} C_5 \, \nu_{ei} n j,$$

$$\frac{C_3}{C_2} C_1 \, \partial_t \left(\frac{3}{2} n T_e \right) + \frac{C_3}{C_2} C_1 \, \nabla_x \cdot \left(\frac{5}{2} n T_e u \right) - \frac{C_3}{C_2} C_1 C_5 \nabla_x \cdot \left(\frac{5}{2} T_e j \right)$$

$$= -C_3 \, E n \left(u - C_5 \frac{j}{n} \right) + \frac{C_3^2}{C_2} C_5 \, \nu_{ei} n j u - \frac{C_3^2}{C_2} C_5^2 \frac{1}{2} \nu_{ei} |j|^2 + \frac{C_3}{C_2} C_3 C_1 \, \nu_{ei} \frac{3}{2} n^2 (T_i - T_e).$$

□

Next, we consider the formal limit $C_5 \to 0$ and $\frac{C_3}{C_2} \to 0$, such that $C_2 C_5$ and $\frac{C_3^2 C_5}{C_2}$ remain bounded away from zero. Physically the first limit means that the current from moving particles $e \bar{n} \bar{u}$ dominates over the current due to electric forces $\bar{j}$. The second limit means that the cyclotron frequency $\frac{e \bar{B}}{m_i}$ dominates over the collision frequency $\bar{\nu}_{ie} \bar{n}$, while the current due to electric fields $\bar{j}$ per cyclotron time $1/\frac{e \bar{B}}{m_i}$ over the current induced by the flow $e \bar{n} \bar{u}$ in a typical time scale $\bar{t}$, remains bounded away from zero. Moreover, the ratio of the collision frequency and the cyclotron frequency is assumed to be of the same order of the electric current per collision time $\frac{1}{\bar{\nu}_{ie} \bar{n}}$ over the current induced by the flow per typical time scale. All in all, we get the following theorem.

Theorem 5.3.2. As $C_5 \to 0$ and $\frac{C_3}{C_2} \to 0$, such that $C_2 C_5$ and $\frac{C_3^2 C_5}{C_2}$ remain bounded away from zero, formally the solution of the system in theorem 5.3.1 tends to the solution of

$$\partial_t n + \nabla_x \cdot (n u) = 0, \qquad (5.25)$$

$$\partial_t (n u) + C_1 \nabla_x (n T) + \nabla_x \cdot (n u \otimes u) = C_2 C_5 \, j \times B, \qquad (5.26)$$

$$C_1 \partial_t \left(\frac{3}{2} n T \right) + \partial_t \left(\frac{1}{2} n |u|^2 \right) + C_1 \nabla_x \cdot \left(\frac{5}{2} n T u \right) + \nabla_x \cdot \left(\frac{1}{2} n |u|^2 u \right) = C_2 C_5 \, E j,$$

$$(5.27)$$

$$(E + u \times B) = \frac{C_3 C_5}{C_2} \nu_{ei} j, \tag{5.28}$$

$$Eu = \frac{C_3^2 C_5}{C_2} \nu_{ei} j u, \tag{5.29}$$

$$\nabla_x \times B = j, \tag{5.30}$$

$$\nabla_x \times E + \partial_t B = 0, \tag{5.31}$$

$$\nabla_x \cdot B = 0. \tag{5.32}$$

This is a direct consequence of theorem 5.3.1. Last we consider the formal limit $C_3 \to 0$ which means that interactions of ions and electrons can be neglected. In addition, we choose the special regime where $C_1 = 1$, that is $\bar{n}\bar{T} = m_i \bar{n}\bar{u}^2$ and $C_2 C_5 = 1$ in order to obtain the well-known conservation form for ideal MHD.

Theorem 5.3.3. *As $C_3 \to 0$ and in the special regime $C_1 = 1$ and $C_2 C_5 = 1$, formally, we obtain the system of ideal MHD equations*

$$\partial_t n + \nabla_x \cdot (nu) = 0,$$

$$\partial_t (nu) + \nabla_x (u \otimes un + (p + \frac{1}{2}|B|^2)\mathbf{1} - B \otimes B) = 0,$$

$$\partial_t \left(\frac{1}{2}n|u|^2 + \frac{3}{2}p + \frac{1}{2}|B|^2\right) + \nabla_x \cdot \left(\frac{1}{2}n|u|^2 u + \frac{5}{2}pu + |B|^2 u - B \cdot (B \otimes u)\right) = 0,$$

$$\partial_t B + \nabla_x \cdot (B \otimes u - u \otimes B) = 0,$$

$$\nabla_x \cdot B = 0.$$

Proof. In the limit $C_3 \to 0$, the equations (5.28) and (5.29) turn into

$$E + u \times B = 0, \tag{5.33}$$

$$Eu = 0. \tag{5.34}$$

We insert $j = \nabla_x \times B$ from (5.30) into (5.26). The l-th component of the term $(\nabla_x \times B) \times B$ can be simplified to $\sum_{n=1}^{3} B_n(\partial_{x_l} B_n - \partial_{x_n} B_l)$, $l = 1, 2, 3$. Since $\nabla_x \cdot B = 0$, we can add $(\nabla_x \cdot B)B_l$, so we get $\sum_{n=1}^{3} B_n(\partial_{x_l} B_n - \partial_{x_n} B_l) + (\nabla_x \cdot B)B_l$ which is the l–th component of $-\nabla_x \cdot (\frac{1}{2}|B|^2\mathbf{1} - B \otimes B)$. Thus, (5.26) turns into

$$\partial_t (nu) + \nabla_x \cdot \left(u \otimes un + \left(nT + \frac{1}{2}|B|^2\right)\mathbf{1} - B \otimes B\right) = 0.$$

Now, we insert $E = -u \times B$ from (5.33) into (5.31). In a similar way again using $\nabla_x \cdot B = 0$, we obtain $-\nabla_x \times (u \times B) = \nabla_x \cdot (B \otimes u - u \otimes B)$, so (5.31) leads to

$$\partial_t B + \nabla_x \cdot (B \otimes u - u \otimes B) = 0. \tag{5.35}$$

Finally, inserting (5.30), (5.33), $\nabla_x \cdot B = 0$ and (5.35) into (5.27), leads to

$$\partial_t \left(\frac{1}{2}n|u|^2 + \frac{3}{2}nT + \frac{1}{2}|B|^2\right) + \nabla_x \cdot \left(\frac{1}{2}n|u|^2 u + \frac{5}{2}nTu + |B|^2 u - B \cdot (B \otimes u)\right) = 0.$$

$\square$

Chapter 6

Application to plasmas

As we mentioned in the previous chapter, there are several applications where a mixture of ions and electrons needs to be modelled and simulated. In this chapter we develop the prerequisites in order to simulate a mixture of ions and electrons in a certain regime. The regime of the mixture is the following. In some regions the mixture is near equilibrium in other regions it is far away from equilibrium. In section 6.1 we extend a method called micro-macro decomposition from one species to a gas mixture. Using this method we can derive a system of coupled kinetic and macroscopic equations from the system of kinetic equations. The advantage of this new system is that it is more appropriate in order to simulate a mixture which is partly close to equilibrium. This is needed for example in the case of a plasma in a Tokamak, a toroidal-shaped chamber where a plasma can be confined for the research on controlled fusion. More details are described in the next section. In the following section we want to consider two test cases which can be used to test the results of a numerical simulation with the theory of the model. The first one considers the space-homogeneous case. In the space-homogeneous case we can prove convergence rates of the velocities and the temperatures to a common value and of the distribution functions to Maxwell distributions. These estimates are derived in section 6.2.1. They can be compared to the results of a numerical simulation. The second test case is an extension of the theory of Landau damping to two species in section 6.2.2. It is a theory which shows a physical behaviour of a mixture of ions and electrons. This theoretical result can also be compared to a result of a numerical simulation.

6.1 Kinetic/Fluid micro-macro decomposition for two component plasmas

We want to model a plasma consisting of two species, electrons and one species of ions. The kinetic description of a plasma is based on the Vlasov equation. In [28], Crestetto, Crouseilles and Lemou developed a numerical simulation of the Vlasov-BGK equation in the fluid limit using particles. They consider a Vlasov-BGK equation for the electrons and treat the ions as a background charge. In [28] a micro-macro decomposition is used as in [12] where asymptotic preserving schemes have been derived in the fluid limit. In [28], the approach in [12] is modified by using a particle approximation for the kinetic part, the fluid part being always discretized by standard Finite Volume schemes. Other approaches where a kinetic description of one species is written in a micro-macro decomposition can be seen in [30, 31].

In this section, we want to model both the electrons and the ions by a Vlasov-BGK equation instead of treating one only as a background charge. Such a two component kinetic description of the gas mixture has for example importance in a Tokamak plasma. In regions next to the wall of the Tokamak, the plasma is close to a fluid, but the kinetic description is mandatory in the core plasma so that a hybrid fluid/kinetic description is adequate. For this, we want to use the approach in [28], since it has the following advantages: the presented scheme has a much less level of noise compared to the standard particle method and the computational cost of the micro-macro model is reduced in the fluid regime since a small number of particles is needed for the micro part.

From the modelling point of view, we want to describe this gas mixture using two distribution functions via the Vlasov equation with interaction terms on the right-hand side. For the interactions we use the BGK approach, since BGK models give rise to efficient numerical computations, see for example [72, 41, 35, 12, 34, 13, 28]. In previous sections we have seen that there are two types of BGK models for gas mixtures, one with one relaxation term on the right-hand side, one with two interaction terms one the right-hand side. In this section we are interested in the second type of models, and use the model developed in chapter 2. In this type of model the two different types of interactions, interactions of a species with itself and interactions of a species with the other one, are kept separated. Therefore we can see how these different types of interactions influence the trend to equilibrium. From the physical point of view, we expect two different types of trends to equilibrium. For example, if the collision frequencies of the particles of each species with itself are larger compared to the collision frequencies related to interspecies collisions, we expect that we first observe that the relaxation of the two distribution functions to its own equilibrium distribution is faster compared to the relaxation towards a common velocity and a common temperature. This effect is clearly seen in the model presented in chapter 2 since the two types of interactions are separated.

The outline of this section is as follows: In section 6.1.1 we illustrate the idea of a micro-macro decomposition in the case of one species. In section 6.1.2 we want to mention the usefulness of this micro-macro decomposition by giving a physical application where it is reasonable to apply it. In section 6.1.3 we present the model for a plasma consisting of electrons and one species of ions and write it in dimensionless form. In section 6.1.4 we derive the micro-macro decomposition of the model presented in section 6.1.3. Sections 6.1.3, 6.1.3 and 6.1.4 are also presented in [29] by Crestetto, Klingenberg and Pirner.

6.1.1 The idea of the micro-macro decomposition for one species

We want to illustrate the method of the micro-macro decomposition in the case of one species for the Vlasov-Poisson-BGK equation using particles done in [28]. We consider a distribution function $f(x, v, t) \geq 0$ and $E(x, t)$ is a self consistent electric field where $x \in [0, L]$ is the position and $v \in \mathbb{R}$ the velocity and $t \geq 0$ the time. The Vlasov-BGK equation with initial data reads

$$\begin{cases} \partial_t f + v \, \partial_x f &= \frac{1}{\varepsilon}(M - f), \\ f(x, v, 0) &= f^0(x, v), \end{cases} \tag{6.1}$$

coupled with the Maxwell equation

$$\begin{cases} \partial_x E(x, t) &= \int f(x, v, t) dv - 1, \\ \int_0^L E(x, t) dx &= 0, \end{cases} \tag{6.2}$$

and f and E are assumed to have periodic boundary conditions in x. The Maxwell distribution M in one dimension is given by

$$M = \frac{n}{\sqrt{2\pi \frac{T}{m}}} \exp\left(-\frac{|v - u|^2}{T/m}\right),$$

with mass m and the moments

$$\int f(v) \begin{pmatrix} 1 \\ v \\ m|v - u|^2 \end{pmatrix} dv =: \begin{pmatrix} n \\ nu \\ nT \end{pmatrix}.$$

The idea of the micro-macro decomposition is the following. We decompose the distribution function f into

$$f = M + g.$$

With this decomposition we can derive the system

$$\partial_t g + (1 - \Pi_M)(v\partial_x g + E\partial_v g) = \frac{1}{\varepsilon}[-g - \varepsilon(1 - \Pi_M)(v\partial_x M + E\partial_v M)],$$

$$\partial_t \begin{pmatrix} n \\ nu \\ \frac{1}{2}nu^2 + \frac{1}{2}n\frac{T}{m} \end{pmatrix} + \partial_x \begin{pmatrix} nu \\ nu^2 + n\frac{T}{m} \\ \frac{1}{2}nu^3 + \frac{3}{2}n\frac{T}{m}u \end{pmatrix} + \partial_x \left(\int \begin{pmatrix} 1 \\ v \\ \frac{v^2}{2} \end{pmatrix} g dv \right) = \begin{pmatrix} 0 \\ nE \\ nEu \end{pmatrix}, \tag{6.3}$$

where Π_M denotes the orthogonal projection in $L^2(\frac{1}{M}dv)$ onto the null space of the BGK operator. All in all, one can show the following theorem.

Theorem 6.1.1. *If (f, E) is a solution of (6.1), (6.2), then $((n, u, T), g, E) = (\int(1, \frac{v}{n}, \frac{m}{n}|v - u|^2)f dv, f - M, E)$ is a solution of (6.3), (6.2), with the associated initial data*

$$(n(0), u(0), T(0)) = \left(\int (1, v, \frac{v^2}{2})f^0(v)dv \right), \quad g(0) = f(0) - M(0). \tag{6.4}$$

Conversely, if (n, u, T, g, E) is a solution of (6.3), (6.2) with initial data (6.4), then $\int(1, \frac{v}{n}, \frac{m}{n}|v - u|^2)g dv = 0$ and $f = M + g$ is a solution to (6.1), (6.2).

The proof is given in [12] without the electric field. In [28] the first part of the theorem is shown with an electric field. We will not repeat it here since we carry it out later in the two species case. The advantage from the numerical point of view is the following. The equation (6.1) is a kinetic equation. The system (6.3) is a system consisting of a kinetic type equation and a macroscopic type equation. The kinetic equation is solved by a particle in cell (PIC) method, the macroscopic equation by Finite Volume schemes. The particle-in-cell method is more costly from the computational point of view. So the advantage of the system (6.3) is the following. If we are close to equilibrium we can take less particles in the particle-in-cell method since the time evolution is mainly described by the macroscopic equation which reduces the computational cost.

6.1.2 Physical application for the micro-macro decomposition

One application with a plasma is the controlled thermonuclear fusion. It is described in [80]. The aim of a controlled nuclear fusion is to develop new energy sources. According to principles of relativistic physics we can produce energy by performing a transformation that removes the mass. There are two possibilities to obtain this, a fission reaction which generates two lighter nuclei from the nucleus of a heavy atom; and a fusion reaction that creates a heavier nucleus from two light atoms. Controlled fusion is still in the research stage. The temperatures required for thermonuclear fusion are larger than one hundred million degrees. At this temperatures the electrons are totally freed from their atoms so that one obtains a gas of electrons and ions which is a totally ionized plasma. One project concerning thermonuclear fusion is the ITER project. There, the plasma is confined in a toroidal shaped chamber called a Tokamak. Inside the Tokamak we have a regime in which it is useful to use the micro-macro decomposition. In regions next to the wall of the Tokamak, the plasma is close to a fluid, but the kinetic description is mandatory in the core plasma so that a hybrid fluid/kinetic description is adequate.

6.1.3 The two species model for electrons and ions

In this section we present in 1D the Vlasov-BGK model for a mixture of two species developed in chapter 5 and mention its fundamental properties like the conservation properties. Then, we present its dimensionless form. For numerical reasons it is useful to use the dimensionless form, since in this case we only have to deal with the mass ratio $\frac{m_e}{m_i}$ and not with the individual masses of electrons and ions which are very small.

1D Vlasov-BGK model for a mixture of two species

In this section we will repeat the Vlasov-BGK model for a mixture of two species without magnetic field in the one species case for the convenience of the reader.

We consider a plasma consisting of electrons denoted by the index e and one species of ions denoted by the index i. Thus, our kinetic model has two distribution functions $f_e(x, v, t) > 0$ and $f_i(x, v, t) > 0$ where $x \in [0, L], L > 0$ and $v \in \mathbb{R}$ are the phase space variables and $t \geq 0$ the time.

Furthermore, for any $f_i, f_e : [0, L] \times \mathbb{R} \times \mathbb{R}_0^+ \to \mathbb{R}$ with $(1 + |v|^2) f_i$, $(1 + |v|^2) f_e \in L^1(dv)$, $f_i, f_e \geq 0$ we relate the distribution functions to macroscopic quantities by mean-values of f_k, $k = i, e$

$$\int f_k(v) \begin{pmatrix} 1 \\ v \\ m_k |v - u_k|^2 \end{pmatrix} dv =: \begin{pmatrix} n_k \\ n_k u_k \\ n_k T_k \end{pmatrix}, \quad k = i, e, \tag{6.5}$$

where m_k is the mass, n_k the number density, u_k the mean velocity and T_k the temperature of species k, $k = i, e$. Note that in this chapter we shall write T_k instead of $k_B T_k$, where k_B is Boltzmann's constant.

We want to model the time evolution of the distribution functions by Vlasov-BGK equations.

$$\partial_t f_i + v \partial_x f_i + \frac{F_i^L}{m_i} \partial_v f_i = \nu_{ii} n_i (M_i - f_i) + \nu_{ie} n_e (M_{ie} - f_i),$$

$$\partial_t f_e + v \partial_x f_e + \frac{F_e^L}{m_e} \partial_v f_e = \nu_{ee} n_e (M_e - f_e) + \nu_{ei} n_i (M_{ei} - f_e), \tag{6.6}$$

with the mean-field forces F_i^L and F_e^L specified later and the Maxwell distributions

$$M_k(x, v, t) = \frac{n_k}{\sqrt{2\pi \frac{T_k}{m_k}}} \exp\left(-\frac{|v - u_k|^2}{2 \frac{T_k}{m_k}}\right), \quad k = i, e,$$

$$M_{kj}(x, v, t) = \frac{n_{kj}}{\sqrt{2\pi \frac{T_{kj}}{m_k}}} \exp\left(-\frac{|v - u_{kj}|^2}{2 \frac{T_{kj}}{m_k}}\right), \quad k, j = i, e, k \neq j, \tag{6.7}$$

where $\nu_{ii} n_i$ and $\nu_{ee} n_e$ are the collision frequencies of the particles of each species with itself, while $\nu_{ie} n_e$ and $\nu_{ei} n_i$ are related to interspecies collisions. To be flexible in choosing the relationship between the collision frequencies, we now assume the relationship

$$\nu_{ie} = \varepsilon \nu_{ei}, \qquad\qquad\qquad 0 < \varepsilon \leq 1,$$

$$\nu_{ii} = \beta_i \nu_{ie}, \quad \nu_{ee} = \beta_e \nu_{ei} = \frac{\beta_e}{\varepsilon} \nu_{ie}, \qquad \beta_i, \beta_e > 0. \tag{6.8}$$

The restriction $\varepsilon \leq 1$ is without loss of generality. If $\varepsilon > 1$, exchange the notation i and e and choose $\frac{1}{\varepsilon}$. We assume that all collision frequencies are positive. In addition, we take into account an acceleration due to interactions using mean-field Lorentz forces F_i^L, F_e^L. We assume that the magnetic field is negligible compared to the electric field. Therefore the Lorentz forces are given by

$$F_i^L(x, t) = e\, E(x, t) \quad \text{and} \quad F_e^L(x, t) = -e\, E(x, t),$$

where e denotes the elementary charge. For simplicity, we assumed that the ions have the charge e. The electric field is given by the Maxwell equation

$$\partial_x E(x,t) = \rho(x,t), \tag{6.9}$$

where

$$\rho(x,t) = e \int_{-\infty}^{\infty} (f_i(x,v,t) - f_e(x,v,t))dv, \tag{6.10}$$

describes the charge density.

The functions f_k and E are submitted to the following periodic boundary conditions

$$f_k(0,v,t) = f_k(L,v,t), \quad \text{for every} \quad v \in \mathbb{R}, t \geq 0,$$
$$E(0,t) = E(L,t), \quad \text{for every} \quad t \geq 0.$$

In order to get a well-posed problem, a zero-mean electrostatic condition has to be added,

$$\int_0^L E(x,t)dx = 0, \quad \text{for every} \quad t \geq 0,$$

together with an initial condition

$$f_k(x,v,0) = f_k^0(x,v), \quad \text{for every} \quad x \in [0,L], v \in \mathbb{R}.$$

From the initial condition on f_k, we can compute an initial condition of the charge density ρ given by (6.10). From this we can compute the initial data of E using (6.9). In the velocity v we either assume periodic boundary conditions on a bounded subset of $\mathbb{R}$ or a fast enough decay of f_i and f_e to zero for $|v| \to \infty$.

The Maxwell distributions M_i and M_e in (6.7) have the same moments as f_i and f_e, respectively. With this choice, we guarantee the conservation of mass, momentum and energy in interactions of one species with itself (see section 2.1.2). The remaining parameters $n_{ie}, n_{ei}, u_{ie}, u_{ei}, T_{ie}$ and T_{ei} will be determined using conservation of the number of particles, conservation of total momentum and conservation of total energy, together with some symmetry considerations.

If we assume that

$$n_{ie} = n_i \quad \text{and} \quad n_{ei} = n_e, \tag{6.11}$$

$$u_{ie} = \delta u_i + (1-\delta)u_e, \quad \delta \in \mathbb{R}, \tag{6.12}$$

and

$$T_{ie} = \alpha T_i + (1-\alpha)T_e + \gamma|u_i - u_e|^2, \quad 0 \leq \alpha \leq 1, \gamma \geq 0, \tag{6.13}$$

we have conservation of the number of particles, of total momentum and total energy provided that

$$u_{ei} = u_e - \frac{m_i}{m_e}\varepsilon(1-\delta)(u_e - u_i), \tag{6.14}$$

$$T_{ei} = \left[\varepsilon m_i(1-\delta)\left(\frac{m_i}{m_e}\varepsilon(\delta-1)+\delta+1\right) - \varepsilon\gamma\right]|u_i - u_e|^2$$
$$+\varepsilon(1-\alpha)T_i + (1-\varepsilon(1-\alpha))T_e, \tag{6.15}$$

see section 2.1.5 and section 5.2. In order to ensure the positivity of all temperatures, we need to impose restrictions on δ and γ given by

$$0 \le \gamma \le m_i(1-\delta)\left[(1+\frac{m_i}{m_e}\varepsilon)\delta + 1 - \frac{m_i}{m_e}\varepsilon\right], \tag{6.16}$$

and

$$\frac{\frac{m_i}{m_e}\varepsilon - 1}{1 + \frac{m_i}{m_e}\varepsilon} \le \delta \le 1, \tag{6.17}$$

see theorem 2.1.4.

Notes on the existence of solutions

In chapter 4 we presented an existence and uniqueness result of the model presented in chapter 2. In this chapter we consider the extended model for ions and electrons given by (6.6). This model has additional force terms due to an acceleration of charged particles. The force is determined by the Maxwell equation (6.9) together with the charge density (6.10). From physics we expect that the electric field $E(x,t)$ can be written as a gradient of a potential $\phi(x,t)$. Then ϕ solves a Poisson equation and we have to consider kinetic equations with force terms coupled to a Poisson equation. So it is not clear if we still have existence and uniqueness of mild solutions. In the case of one species with a force term given by a Poisson equation, there is an existence result of mild solutions given by Rejeb in [75]. The difference compared to the BGK equation for one species without a force term is the following. In the case of the BGK equation for one species without a force term we are able to compute the characteristic curves. This ended in a fixed-point argument as in the proof of the Picard-Lindelöf theorem or the fixed-point theorem of Banach since we are able to show that the right hand side in the definition of a mild solution is a contraction. In the case of the BGK equation for one species with force term in one velocity and one space dimension, we obtain the characteristics

$$\frac{dt(s)}{ds} = 1,$$
$$\frac{dx(s)}{ds} = v(s), \tag{6.18}$$
$$\frac{dv(s)}{ds} = -\partial_x\phi(x(s),s),$$

where ϕ solves a Poisson equation in one velocity and one space dimension with a right-hand side given by $\int f(x, v, t)dv$. We are not able to solve the characteristics explicitly. So at this point it is not possible to show that the right-hand side in the definition of a mild solution is a contraction. So Rejeb in [75] suggested to use the fixed point theorem of Schauder (in Werner, [86]).

Theorem 6.1.2 (Fix point theorem of Schauder). *Let S be a convex, compact and non-empty subset of a Banach space X and let $\Phi : S \to S$ be a continuous map. Then Φ has at least one fix point in S.*

So the two essential things in the existence proof of Rejeb in [75] are to find an appropriate space S which has the properties of theorem 6.1.2 and to show that the right hand side in the definition of a mild solution is a continuous map in f. The latter thing is done in a similar way as the Lipschitz continuity in the case without a force term. Since we can not compute the characteristic curves explicitly this leads to a more careful study of the characteristics. Rejeb uses an estimate which shows how the characteristics changes if one changes the function f on the right hand side of the Poisson equation into a function g. This estimate has been proven by Ukai and Okabe in [82], it can be summarized as the following lemma.

Lemma 6.1.3. *Let $(x^f(s), v^f(s))$ and $(x^g(s), v^g(s))$ be solutions to the characteristic equations (6.18) coupled with the Poisson equation for ϕ where the charge density is computed via (6.10) with the distribution functions f and g, respectively. Then we have the following estimate*

$$\max\{|x^f(s) - x^g(s)|, |v^f(s) - v^g(s)|\} \le C(t)\|E^f - E^g\|_{L^1(\Omega)}.$$

The set S is a subset of Ω where Ω contains the position space, the velocity space and the time interval. Rejeb defined a nonempty, convex, compact subset S of Ω and thus Schauder's theorem could be applied. The estimate is used to prove that the Maxwell distribution on the right-hand side is Lipschitz continuous in f. This estimate is actually proven for two species in [82]. So when we want to extend the existence result from one species with force term from [75], we have to combine the idea of [75] with the analysis of the characteristics for two species from [82] and the estimates of the right-hand side of the BGK model for mixtures proven in chapter 4.

Dimensionless form of the two species model

We want to write the BGK model presented in section 6.1.3 in dimensionless form. The principle of non-dimensionalization can also be found in chapter 2.2.1 in [78] for the Boltzmann equation and in [15] for macroscopic equations. First, we define dimensionless variables for the time $t \in \mathbb{R}_0^+$, the length $x \in [0, L]$, the velocity $v \in \mathbb{R}$, the distribution functions f_i, f_e, the number densities n_i, n_e, the mean velocities u_i, u_e, the temperatures T_i, T_e, the electric field E and of the collision frequency ν_{ie}. Then, dimensionless variables of the other collision frequencies $\nu_{ii}, \nu_{ee}, \nu_{ei}$ can be derived by using the relationships (6.8). We start with choosing typical scales denoted by a bar and then define dimensionless quantities by dividing the previous quantity by its typical scale.

$$t' = t/\bar{t}, \quad x' = x/\bar{x}, \quad v' = v/\bar{v},$$

$$f_i'(x', v', t') = \frac{\bar{x}\bar{v}}{N_i} f_i(x, v, t), \quad f_e'(x', v', t') = \frac{\bar{x}\bar{v}}{N_e} f_e(x, v, t),$$

where N_i is the total number of ions and N_e the total number of electrons in the volume $\bar{x}^3$. We assume $N_i = N_e =: N$. This assumption is in accordance with the typical values in a plasma described in [15]. Further, we choose

$$n_i' = n_i/\bar{n}_i, \quad n_e' = n_e/\bar{n}_e, \quad \bar{n}_i = \bar{n}_e = \frac{N}{\bar{x}},$$

$$E' = E/\bar{E},$$

$$u_i' = u_i/\bar{u}_i, \quad u_e' = u_e/\bar{u}_e, \quad \bar{u}_e = \bar{u}_i = \bar{v},$$

$$T_i' = T_i/\bar{T}_i, T_e' = T_e/\bar{T}_e, \quad \bar{T}_e = \bar{T}_i = m_i\bar{v}^2,$$

$$\nu_{ie}' = \nu_{ie}/\bar{\nu}_{ie}.$$

Now we want to write equations (6.6) in dimensionless variables. We start with the Maxwell distribution in (6.7) and (6.12)-(6.15). We replace the macroscopic quantities n_i, u_i and T_i in M_i by their dimensionless expressions and obtain

$$M_i = \frac{n_i'\bar{n}_i}{\sqrt{2\pi \frac{\bar{T}_i T_i'}{m_i}}} \exp\left(-\frac{|v'\bar{v} - u_i'\bar{u}_i|^2 m_i}{2T_i'\bar{T}_i}\right).$$

If we assume that $\bar{v}^2 = |\bar{u}_i|^2 = \frac{\bar{T}_i}{m_i}$, we obtain

$$M_i = \frac{\bar{n}_i}{\bar{v}} \frac{n_i'}{\sqrt{2\pi T_i'}} \exp\left(-\frac{|v' - u_i'|^2}{2T_i'}\right) =: \frac{\bar{n}_i}{\bar{v}} M_i'. \tag{6.19}$$

The relationship which is used on $\bar{u}_i$ and $\bar{T}_i$ is in accordance with the typical values in a plasma described in [15]. In the Maxwell distribution M_e we assume $\bar{T}_i = \bar{T}_e =: \bar{T}$ and obtain in the same way as for M_i

$$M_e = \frac{\bar{n}_e}{\bar{v}} \left(\frac{m_e}{m_i}\right)^{\frac{1}{2}} \frac{n_e'}{\sqrt{2\pi T_e'}} \exp\left(-\frac{|v' - u_e'|^2}{2T_e'} \frac{m_e}{m_i}\right) =: \frac{\bar{n}_e}{\bar{v}} M_e'.$$

Now, we consider the Maxwell distribution M_{ie} in (6.7) and its velocity u_{ie} in (6.12) and its temperature T_{ie} in (6.13). Again we use $\bar{v} = \bar{u}_i = \bar{u}_e$ and $\bar{v}^2 = \frac{\bar{T}}{m_i} = \frac{\bar{T}_i}{m_i} = \frac{\bar{T}_e}{m_e} \frac{m_e}{m_i}$ and obtain

$$u_{ie} = \delta u_i'\bar{u}_i + (1-\delta)u_e'\bar{u}_e = (\delta u_i' + (1-\delta)u_e')\bar{v} =: \bar{v}u_{ie}',$$

$$T_{ie} = \alpha T_i'\bar{T}_i + (1-\alpha)T_e'\bar{T}_e + \gamma|\bar{v}|^2|u_i' - u_e'|^2$$

$$= m_i|\bar{v}|^2[\alpha T_i' + (1-\alpha)T_e' + \frac{\gamma}{m_i}|u_i' - u_e'|^2] =: |\bar{v}|^2 m_i T_{ie}', \tag{6.20}$$

$$M_{ie} = \frac{n_i'\bar{n}_i}{\sqrt{2\pi\bar{v}^2 T_{ie}'}} \exp\left(-\frac{|v' - u_{ie}'|^2}{2T_{ie}'}\right) =: \frac{\bar{n}_i}{\bar{v}} M_{ie}'.$$

With the same assumptions we obtain for u_{ei}, T_{ei} and M_{ei} in a similar way the expressions

$$u_{ei} = [(1 - \frac{m_i}{m_e}\varepsilon(1 - \delta))u_e' + \frac{m_i}{m_e}\varepsilon(1 - \delta)u_i']\bar{v} =: u_{ei}'\bar{v},$$

$$T_{ei} = [(1 - \varepsilon(1 - \alpha))T_e' + \varepsilon(1 - \alpha)T_i']\bar{T}$$
$$+ (\varepsilon m_i(1 - \delta)(\frac{m_i}{m_e}\varepsilon(\delta - 1) + \delta + 1) - \varepsilon\gamma)|u_i' - u_e'|^2|\bar{v}|^2$$
$$= [(1 - \varepsilon(1 - \alpha))T_e' + \varepsilon(1 - \alpha)T_i']|\bar{v}|^2 m_e\frac{m_i}{m_e}$$
$$+ (\varepsilon m_i(1 - \delta)(\frac{m_i}{m_e}\varepsilon(\delta - 1) + \delta + 1) - \varepsilon\gamma)|u_i' - u_e'|^2|\bar{v}|^2 =: |\bar{v}|^2 m_e\frac{m_i}{m_e}T_{ei}',$$

$$M_{ei} = \frac{\bar{n}_e}{\bar{v}}\frac{m_e}{m_i}\frac{n_e'}{\sqrt{2\pi T_{ei}'}}\exp(-\frac{|v' - u_{ei}'|^2}{2T_{ei}'}\frac{m_e}{m_i}) =: \frac{\bar{n}_e}{\bar{v}}M_{ei}'.$$

Now we replace all quantities in (6.6) by their non-dimensionalized expressions. For the left-hand side of the equation for the ions we obtain

$$\partial_t f_i + v\partial_x f_i + \frac{e}{m_i}E\partial_v f_i$$
$$= \frac{1}{\bar{t}}\frac{N}{\bar{x}\bar{v}}\partial_{t'}f_i' + \frac{1}{\bar{x}}\frac{N}{\bar{x}\bar{v}}\bar{v}v'\partial_{x'}f_i' + \frac{N}{\bar{x}\bar{v}}\frac{1}{\bar{v}}\bar{E}\frac{e}{m_i}E'\partial_{v'}f_i',$$
(6.21)

and for the right-hand side using that $\bar{n}_k = \frac{N}{\bar{x}}, k = i, e$, (6.8), (6.19) and (6.20), we get

$$\nu_{ii}n_i(M_i - f_i) + \nu_{ie}n_e(M_{ie} - f_i) = \nu_{ie}\beta_i n_i(M_i - f_i) + \nu_{ie}n_e(M_{ie} - f_i)$$
$$= \beta_i\bar{\nu}_{ie}\frac{N}{\bar{x}\bar{v}}\frac{N}{\bar{x}}\nu_{ie}'n_i'(M_i' - f_i') + \bar{\nu}_{ie}\frac{N}{\bar{x}\bar{v}}\frac{N}{\bar{x}}\nu_{ie}'n_e'(M_{ie}' - f_i').$$
(6.22)

Using that the left-hand side of (6.21) and the left-hand side of (6.22) coincide, we obtain that the right-hand side of (6.21) and the right-hand side of (6.22) are the same. Multiplying this obtained equality by $\frac{\bar{t}\bar{x}\bar{v}}{N}$ and dropping the primes in the variables leads to

$$\partial_t f_i + \frac{\bar{t}\bar{v}}{\bar{x}}v\partial_x f_i + \bar{t}\frac{\bar{E}}{\bar{v}}\frac{e}{m_i}E\partial_v f_i$$
$$= \beta_i\bar{\nu}_{ie}\bar{t}\frac{N}{\bar{x}}\nu_{ie}n_i(M_i - f_i) + \bar{\nu}_{ie}\bar{t}\frac{N}{\bar{x}}\nu_{ie}n_e(M_{ie} - f_i).$$

In a similar way we obtain for electrons

$$\partial_t f_e + \frac{\bar{t}\bar{v}}{\bar{x}}v\partial_x f_e - \bar{t}\frac{\bar{E}}{\bar{v}}\frac{e}{m_e}E\partial_v f_e$$
$$= \frac{\beta_e}{\varepsilon}\bar{\nu}_{ie}\bar{t}\frac{N}{\bar{x}}\nu_{ie}n_e(M_e - f_e) + \frac{1}{\varepsilon}\bar{\nu}_{ie}\bar{t}\frac{N}{\bar{x}}\nu_{ie}n_i(M_{ei} - f_e),$$

and the non-dimensionalized Maxwell distributions given by

$$M_i(x, v, t) = \frac{n_i}{\sqrt{2\pi T_i}} \exp\left(-\frac{|v - u_i|^2}{2T_i}\right),$$

$$M_e(x, v, t) = \frac{n_e}{\sqrt{2\pi T_e}} \left(\frac{m_e}{m_i}\right)^{\frac{1}{2}} \exp\left(-\frac{|v - u_e|^2}{2T_e}\frac{m_e}{m_i}\right),$$

$$M_{ie}(x, v, t) = \frac{n_i}{\sqrt{2\pi T_{ie}}} \exp\left(-\frac{|v - u_{ie}|^2}{2T_{ie}}\right),$$

$$M_{ei}(x, v, t) = \frac{n_e}{\sqrt{2\pi T_{ei}}} \left(\frac{m_e}{m_i}\right)^{\frac{1}{2}} \exp\left(-\frac{|v - u_{ei}|^2}{2T_{ei}}\frac{m_e}{m_i}\right),$$

(6.23)

with the non-dimensionalized macroscopic quantities

$$u_{ie} = \delta u_i + (1 - \delta)u_e, \tag{6.24}$$

$$T_{ie} = \alpha T_i + (1 - \alpha)T_e + \frac{\gamma}{m_i}|u_i - u_e|^2, \tag{6.25}$$

$$u_{ei} = (1 - \frac{m_i}{m_e}\varepsilon(1 - \delta))u_e + \frac{m_i}{m_e}\varepsilon(1 - \delta)u_i, \tag{6.26}$$

$$T_{ei} = [(1 - \varepsilon(1 - \alpha))T_e + \varepsilon(1 - \alpha)T_i]$$
$$+ (\varepsilon(1 - \delta)(\frac{m_i}{m_e}\varepsilon(\delta - 1) + \delta + 1) - \varepsilon\frac{\gamma}{m_i})|u_i - u_e|^2. \tag{6.27}$$

Defining dimensionless parameters

$$A = \frac{\bar{t}\bar{v}}{\bar{x}}, \quad B_i = \bar{t}\frac{\bar{E}}{\bar{v}}\frac{e}{m_i}, \quad B_e = \bar{t}\frac{\bar{E}}{\bar{v}}\frac{e}{m_e},$$
$$\frac{1}{\varepsilon_i} = \beta_i\bar{\nu}_{ie}\bar{t}\frac{N}{\bar{x}}, \quad \frac{1}{\tilde{\varepsilon}_i} = \bar{\nu}_{ie}\bar{t}\frac{N}{\bar{x}}, \quad \frac{1}{\varepsilon_e} = \frac{\beta_e}{\varepsilon}\bar{\nu}_{ie}\bar{t}\frac{N}{\bar{x}}, \quad \frac{1}{\tilde{\varepsilon}_e} = \frac{1}{\varepsilon}\bar{\nu}_{ie}\bar{t}\frac{N}{\bar{x}},$$

(6.28)

we get

$$\partial_t f_i + A\partial_x v f_i + B_i E\partial_v f_i = \frac{1}{\varepsilon_i}\nu_{ie}n_i(M_i - f_i) + \frac{1}{\tilde{\varepsilon}_i}\nu_{ie}n_e(M_{ie} - f_i),$$

$$\partial_t f_e + Av\partial_x f_e - B_e E\partial_v f_e = \frac{1}{\varepsilon_e}\nu_{ie}n_e(M_e - f_e) + \frac{1}{\tilde{\varepsilon}_e}\nu_{ie}n_i(M_{ei} - f_e).$$

(6.29)

In addition, we want to write the moments (6.5) in non-dimensionalized form. We can compute this in a similar way as for (6.6) and obtain after dropping the primes

$$\int f_k dv = n_k, \quad \int v f_k dv = n_k u_k, \quad k = i, e,$$

$$\frac{1}{n_i}\int |v - u_i|^2 f_i dv = T_i, \quad \frac{m_e}{m_i}\frac{1}{n_e}\int |v - u_e|^2 f_e dv = T_e.$$

(6.30)

For the non-dimensionalized form of the Maxwell equation (6.9) we obtain after dropping the primes

$$\frac{\bar{E}}{eN}\partial_x E = \rho.$$

We assume that $\frac{\bar{E}}{eN} = 1$. This means that we assume that the electric field is of the order of the number of particles times the elementary charge.

Remark 6.1.1. We described in section 5.2.1 that according to [11] there are the following relationships between the collision frequencies in the case of ions and electrons

$$\nu_{ee} = \nu_{ei} = \sqrt{\frac{m_i}{m_e}}\nu_{ii} = \frac{m_i}{m_e}\nu_{ie},$$

which means

$$\varepsilon = \frac{m_e}{m_i}, \quad \beta_e = 1, \quad \beta_i = \sqrt{\frac{m_i}{m_e}}.$$

6.1.4 Micro-macro decomposition for the two species model

In this section, we derive the micro-macro model which is equivalent to the kinetic equations (6.29). First, we take the dimensionless equations (6.29) and choose $A = B_e = \frac{m_i}{m_e}B_i = 1$. The choice $A = 1$ means $\bar{v} = \frac{\bar{x}}{\bar{t}}$. The choice $B_e = 1$ means that the reciprocal unit time scales are given by the cyclotron frequency of electrons in the $\bar{E}-$ field, that is $\frac{1}{\bar{t}} = \frac{\bar{E}}{\bar{v}}\frac{e}{m_e}$. Now, we propose to adapt the micro-macro decomposition presented in [12] and [28]. It is used for numerical methods to solve Boltzmann like equations for mixtures to capture the right compressible Navier-Stokes dynamics at small Knudsen numbers. The idea is to write each distribution function as the sum of its own equilibrium part verifying a fluid equation and a remainder of kinetic type. We decompose f_i and f_e as

$$f_i = M_i + g_{ii}, \quad f_e = M_e + g_{ee}. \tag{6.31}$$

Let us introduce $m(v) := \begin{pmatrix} 1 \\ v \\ |v|^2 \end{pmatrix}$ and the notation $\langle \cdot \rangle := \int \cdot \, dv$. Since f_i and M_i, and f_e and M_e, have the same moments: $\langle m(v)f_i \rangle = \langle m(v)M_i \rangle$ and $\langle m(v)f_e \rangle = \langle m(v)M_e \rangle$, the moments of g_{ii} and g_{ee} are zero and thus

$$\langle m(v)g_{ii} \rangle = \langle m(v)g_{ee} \rangle = 0. \tag{6.32}$$

With this decomposition we get from equation (6.29) for ions in dimensionless form

$$\partial_t M_i + \partial_t g_{ii} + v\partial_x M_i + v\partial_x g_{ii} + \frac{m_e}{m_i}E\partial_v M_i + \frac{m_e}{m_i}E\partial_v g_{ii}$$
$$= -\frac{1}{\varepsilon_i}\nu_{ie}n_i g_{ii} + \frac{1}{\tilde{\varepsilon}_i}\nu_{ie}n_e(M_{ie} - M_i - g_{ii}), \tag{6.33}$$

and a similar equation for electrons.

Now we consider the Hilbert spaces $L^2_{M_k} = \{\phi$ such that $\phi M_k^{-\frac{1}{2}} \in L^2(dv)\}$, $k = i, e$, with the weighted inner product $\langle \phi\psi M_k^{-1} \rangle$ of ϕ and ψ. We consider the subspace $\mathcal{N}_k$ =span $\{M_k, vM_k, |v|^2 M_k\}$, $k = i, e$. Let Π_{M_k} be the orthogonal projection in $L^2_{M_k}$ on this subspace $\mathcal{N}_k$. This subspace has the orthonormal basis

$$\tilde{B}_k = \{\frac{1}{\sqrt{n_k}}M_k, \frac{(v - u_k)}{\sqrt{T_k m_i/m_k}}\frac{1}{\sqrt{n_k}}M_k, (\frac{|v - u_k|^2}{2T_k m_i/m_k} - \frac{1}{2})\frac{1}{\sqrt{n_k}}M_k\} =: \{b_1^k, b_2^k, b_3^k\}.$$

Using this orthonormal basis of $\mathcal{N}_k$, one finds for any function $\phi \in L^2_{M_k}$ the following expression of $\Pi_{M_k}(\phi)$

$$\Pi_{M_k}(\phi) = \sum_{n=1}^{3}(\phi, b_n^k)b_n^k = \frac{1}{n_k}[\langle\phi\rangle + \frac{(v - u_k) \cdot \langle(v - u_k)\phi\rangle}{T_k m_i/m_k}$$

$$+ (\frac{|v - u_k|^2}{2T_k m_i/m_k} - \frac{1}{2})2\langle(\frac{|v - u_k|^2}{2T_k m_i/m_k} - \frac{1}{2})\phi\rangle]M_k. \qquad (6.34)$$

This orthogonal projection $\Pi_{M_k}(\phi)$ has some elementary properties.

Lemma 6.1.4 (Properties of Π_{M_k}). *We have, for $k = i, e$,*

$$(1 - \Pi_{M_k})(M_k) = (1 - \Pi_{M_k})(\partial_t M_k) = 0,$$
$$\Pi_{M_k}(g_{kk}) = \Pi_{M_k}(\partial_t g_{kk}) = (1 - \Pi_{M_k})(E\partial_v M_k) = 0,$$

and

$$\Pi_{M_i}(M_{ie}) = (1 + \frac{(v - u_i)(u_{ie} - u_i)}{T_i}$$

$$+ (\frac{|v - u_i|^2}{2T_i} - \frac{1}{2})(\frac{T_{ie}}{T_i} + \frac{|u_{ie} - u_i|^2}{T_i} - 1))M_i, \qquad (6.35)$$

$$\Pi_{M_e}(M_{ei}) = (1 + \frac{(v - u_e)(u_{ei} - u_e)}{T_e m_i/m_e}$$

$$+ (\frac{|v - u_e|^2}{2T_e m_i/m_e} - \frac{1}{2})(\frac{T_{ei}}{T_e} + \frac{|u_{ei} - u_e|^2}{T_e m_i/m_e} - 1))M_e. \qquad (6.36)$$

Proof. The proof of the first five equalities is analogue to the one species case and is given in [12]. For the convenience of the reader we will repeat it here.

- Since $\Pi_{M_k}(M_k) = M_k$, we have $(1 - \Pi_{M_k})(M_k) = 0$.

- Since

$$\partial_t M_k = \left(\frac{\partial_t n_k}{n_k} + \frac{v - u_k}{T_k \frac{m_k}{m_i}} \cdot \partial_t u_k + \left(\frac{|v - u_k|^2}{2(T_k \frac{m_k}{m_i})^2} - \frac{3}{2T_k \frac{m_k}{m_i}}\right)\partial_t T_k \frac{m_k}{m_i}\right)M_k, \qquad (6.37)$$

$\partial_t M_k$ is a linear combination of $M_k, vM_k, |v|^2 M_k$. So $\partial_t M_k \in \mathcal{N}_k$ and therefore $\Pi_{M_k}(\partial_t M_k) = \partial_t M_k$.

- With the definition of g_{kk} and the fact that f_k and M_k have the same moments we get

$$\langle m(v) g_{kk} \rangle = \langle m(v)(f_k - M_k) \rangle = 0.$$

Therefore g_{kk} is orthogonal to $\mathcal{N}_k$, so $\Pi_{M_k}(g_{kk}) = 0$.

- $\partial_t g_{kk}$ is also orthogonal to $\mathcal{N}_k$ since

$$\langle m(v) \partial_t g_{kk} \rangle = \partial_t \langle m(v) g_{kk} \rangle = 0.$$

- As in the case of $\partial_t M_k$, $\nabla_v \cdot (E M_k)$ is a linear combination of $M_k, v M_k, |v|^2 M_k$.

The last two equalities (6.35) and (6.36) are obtained using the explicit expression of Π_{M_k} given by (6.34) by direct computations.

$\square$

Now we apply the orthogonal projection $\mathbb{1} - \Pi_{M_i}$ to (6.33), use lemma 6.1.4 and obtain

$$\partial_t g_{ii} + (\mathbb{1} - \Pi_{M_i})(v \partial_x M_i) + (\mathbb{1} - \Pi_{M_i})(v \partial_x g_{ii}) + (\mathbb{1} - \Pi_{M_i})(\frac{m_e}{m_i} E \partial_v g_{ii})$$

$$= \frac{1}{\bar{\varepsilon}_i} \nu_{ie} n_e (M_{ie} - \Pi_{M_i}(M_{ie})) - (\frac{1}{\varepsilon_i} \nu_{ie} n_i + \frac{1}{\bar{\varepsilon}_i} \nu_{ie} n_e) g_{ii}.$$

Again with lemma 6.1.4 we replace $\Pi_{M_i}(M_{ie})$ by its explicit expression.

$$\partial_t g_{ii} + (\mathbb{1} - \Pi_{M_i})(v \partial_x M_i) + (\mathbb{1} - \Pi_{M_i})(v \partial_x g_{ii}) + (\mathbb{1} - \Pi_{M_i})(\frac{m_e}{m_i} E \partial_v g_{ii})$$

$$= \frac{1}{\bar{\varepsilon}_i} \nu_{ie} n_e (M_{ie} - (1 + \frac{(v - u_i)(u_{ie} - u_i)}{T_i}$$

$$+ (\frac{|v - u_i|^2}{2 T_i} - \frac{1}{2})(\frac{T_{ie}}{T_i} + \frac{1}{T_i}|u_{ie} - u_i|^2 - 1)) M_i) - (\frac{1}{\varepsilon_i} \nu_{ie} n_i + \frac{1}{\bar{\varepsilon}_i} \nu_{ie} n_e) g_{ii}.$$

We take the moments of equation (6.33) and get

$$\partial_t \langle m(v) M_i \rangle + \partial_t \langle m(v) g_{ii} \rangle + \partial_x \langle m(v)(v M_i) \rangle + \partial_x \langle m(v) v g_{ii} \rangle$$

$$+ \langle m(v) \frac{m_e}{m_i} E \partial_v M_i \rangle + \langle m(v) \frac{m_e}{m_i} E \partial_v g_{ii} \rangle$$

$$= -\frac{1}{\varepsilon_i} \nu_{ie} n_i \langle m(v) g_{ii} \rangle + \frac{1}{\bar{\varepsilon}_i} \nu_{ie} n_e (\langle m(v)(M_{ie} - M_i) \rangle - \langle m(v) g_{ii} \rangle).$$

With (6.32), we get

$$\partial_t \langle m(v) M_i \rangle + \partial_x \langle m(v) v M_i \rangle + \partial_x \langle m(v) v g_{ii} \rangle + \langle m(v) \frac{m_e}{m_i} E \partial_v M_i \rangle$$

$$+ \langle m(v) \frac{m_e}{m_i} E \partial_v g_{ii} \rangle = \frac{1}{\bar{\varepsilon}_i} \nu_{ie} n_e (\langle m(v)(M_{ie} - M_i) \rangle).$$

Using integration by parts and the fact that the moments of g_{ii} are zero we get that the term $\langle m(v)E\partial_v g_{ii}\rangle$ vanishes and so we have

$$\partial_t\langle m(v)M_i\rangle + \partial_x\langle m(v)vM_i\rangle + \partial_x\langle m(v)vg_{ii}\rangle + \langle m(v)\frac{m_e}{m_i}E\partial_v M_i\rangle$$
$$= \frac{1}{\tilde{\varepsilon}_i}\nu_{ie}n_e(\langle m(v)(M_{ie}-M_i)\rangle).$$

So altogether we get the following coupled system for the ions

$$\partial_t g_{ii} + (\mathbb{1}-\Pi_{M_i})(v\partial_x M_i) + (\mathbb{1}-\Pi_{M_i})(v\partial_x g_{ii}) + (\mathbb{1}-\Pi_{M_i})(\frac{m_e}{m_i}E\partial_v g_{ii})$$
$$= \frac{1}{\tilde{\varepsilon}_i}\nu_{ie}n_e(M_{ie}-(1+\frac{(v-u_i)(u_{ie}-u_i)}{T_i}+(\frac{|v-u_i|^2}{2T_i}-\frac{1}{2})(\frac{T_{ie}}{T_i}+\frac{1}{T_i}|u_{ie}-u_i|^2-1))M_i)$$
$$-(\frac{1}{\varepsilon_i}\nu_{ie}n_i+\frac{1}{\tilde{\varepsilon}_i}\nu_{ie}n_e)g_{ii}, \tag{6.38}$$

$$\partial_t\langle m(v)M_i\rangle + \partial_x\langle m(v)vM_i\rangle + \partial_x\langle m(v)vg_{ii}\rangle + \langle m(v)\frac{m_e}{m_i}E\partial_v M_i\rangle$$
$$= \frac{1}{\tilde{\varepsilon}_i}\nu_{ie}n_e(\langle m(v)(M_{ie}-M_i)\rangle). \tag{6.39}$$

In a similar way, we get an analogous coupled system for the electrons which is coupled with the system of the ions

$$\partial_t g_{ee} + (\mathbb{1}-\Pi_{M_e})(v\partial_x M_e) + (\mathbb{1}-\Pi_{M_e})(v\partial_x g_{ee}) - (\mathbb{1}-\Pi_{M_e})(E\partial_v g_{ee})$$
$$= \frac{1}{\tilde{\varepsilon}_e}\nu_{ie}n_i(M_{ei}-(1+\frac{(v-u_e)(u_{ei}-u_e)}{T_e}\frac{m_e}{m_i}$$
$$+(\frac{|v-u_e|^2}{2T_e}\frac{m_e}{m_i}-\frac{1}{2})(\frac{T_{ei}}{T_e}+\frac{m_e}{m_iT_e}|u_{ei}-u_e|^2-1))M_e)$$
$$-(\frac{1}{\varepsilon_e}\nu_{ie}n_e+\frac{1}{\tilde{\varepsilon}_e}\nu_{ie}n_i)g_{ee}, \tag{6.40}$$

$$\partial_t\langle m(v)M_e\rangle + \partial_x\langle m(v)vM_e\rangle + \partial_x\langle m(v)vg_{ee}\rangle - \langle m(v)E\partial_v M_e\rangle$$
$$= \frac{1}{\tilde{\varepsilon}_e}\nu_{ie}n_i(\langle m(v)(M_{ei}-M_e)\rangle). \tag{6.41}$$

Now we have obtained a system of two microscopic equations (6.38), (6.40) and two macroscopic equations (6.39), (6.41). One can show that this system is an equivalent formulation of the BGK equations for ions and electrons. This is analogous to what is done in [28].

6.2 Theoretical results to check the quality of numerical experiments

In this section we want to present two theoretical results which we can compare with numerical experiments. The first theoretical result are convergence rates to equilibrium in the space-homogeneous case. Since in the case of gas mixtures we expect that the distribution functions relax towards equilibrium distributions which are Maxwell distributions with a common equilibrium and temperature (see theorem 2.1.6), we proved convergence rates of the two velocities u_i and u_e and the two temperatures T_i and T_e to a common value and convergence rates of the two distribution functions to Maxwell distributions. This is presented in section 6.2.1. This is also presented in [29] by Crestetto, Klingenberg and Pirner. In addition in [29], we present numerical experiments using the micro-macro decomposition and verify these convergence rates numerically. The numerical discretization used in [28] uses a particle approximation for the kinetic part, the fluid part being always discretized by standard Finite Volume schemes. Furthermore, we study in [29] the influence on the speed of convergence under different choices of the collision frequencies.

The second theoretical result is an extension of the theory of Landau damping from one species presented in [84] to two species. This is done in section 6.2.2.

6.2.1 Space homogeneous case without electric field

We first propose to consider our model in the space-homogeneous case, without electric field, where we can show an estimate of the decay rate of $\|f_k(t) - M_k(t)\|_{L^1(dv)}$, $|u_i(t) - u_e(t)|^2$ and $|T_i(t) - T_e(t)|^2$. In the space-homogeneous case, without electric field, the BGK model for mixtures (6.6) simplifies to

$$\partial_t f_i = \frac{1}{\varepsilon_i}\nu_{ie}n_i(M_i - f_i) + \frac{1}{\tilde{\varepsilon}_i}\nu_{ie}n_e(M_{ie} - f_i),$$

$$\partial_t f_e = \frac{1}{\varepsilon_e}\nu_{ie}n_e(M_e - f_e) + \frac{1}{\tilde{\varepsilon}_e}\nu_{ie}n_i(M_{ei} - f_e),$$

(6.42)

and its micro-macro reformulation simplifies to

$$\partial_t g_{ii} = \frac{1}{\tilde{\varepsilon}_i}\nu_{ie}n_e(M_{ie} - (1 + \frac{(v - u_i)(u_{ie} - u_i)}{T_i}$$

$$+ (\frac{|v - u_i|^2}{2T_i} - \frac{1}{2})(\frac{T_{ie}}{T_i} + \frac{1}{T_i}|u_{ie} - u_i|^2 - 1))M_i) - (\frac{1}{\varepsilon_i}\nu_{ie}n_i + \frac{1}{\tilde{\varepsilon}_i}\nu_{ie}n_e)g_{ii},$$

$$\partial_t \langle m M_i \rangle = \frac{1}{\tilde{\varepsilon}_i}\nu_{ie}n_e(\langle m(M_{ie} - M_i) \rangle),$$

(6.43)

$$\partial_t g_{ee} = \frac{1}{\tilde{\varepsilon}_e}\nu_{ie}n_e(M_{ei} - (1 + \frac{(v - u_e)(u_{ei} - u_e)}{T_e}\frac{m_e}{m_i}$$

$$+ (\frac{|v - u_e|^2}{2T_e}\frac{m_e}{m_i} - \frac{1}{2})(\frac{T_{ei}}{T_e} + \frac{m_e}{m_i}\frac{1}{T_e}|u_{ei} - u_e|^2 - 1))M_e)$$

$$- (\frac{1}{\varepsilon_e}\nu_{ie}n_e + \frac{1}{\tilde{\varepsilon}_e}\nu_{ie}n_i)g_{ee},$$

$$\partial_t \langle m M_e \rangle = \frac{1}{\tilde{\varepsilon}_e}\nu_{ie}n_i(\langle m(M_{ei} - M_e) \rangle).$$

(6.44)

Convergence rate to Maxwell distributions

We denote by $H(f) = \int f \ln f dv$ the entropy of a function f and by $H(f|g) = \int f \ln \frac{f}{g} dv$ the relative entropy of f and g.

Theorem 6.2.1. *In the space homogeneous case without electric field (6.42), we have the following convergence rate of the distribution functions f_i and f_e:*

$$\|f_k - M_k\|_{L^1(dv)} \leq 4e^{-\frac{1}{2}Ct}[H(f_i^0|M_i^0) + H(f_e^0|M_e^0)]^{\frac{1}{2}}, \quad k = i, e,$$

where C is a constant.

Proof. We consider the entropy production of species i defined by

$$D_i(f_i, f_e) = -\int \frac{1}{\varepsilon_i} \nu_{ie} n_i \ln f_i \, (M_i - f_i) dv - \int \frac{1}{\tilde{\varepsilon}_i} \nu_{ie} n_e \ln f_i \, (M_{ie} - f_i) dv.$$

According to the proof of lemma 1.3.12 the function $h(x) := x \ln x - x$ satisfies $h'(x) = \ln x$, so we can deduce

$$D_i(f_i, f_e) = -\int \frac{1}{\varepsilon_i} \nu_{ie} n_i h'(f_i)(M_i - f_i) dv - \int \frac{1}{\tilde{\varepsilon}_i} \nu_{ie} n_e h'(f_i)(M_{ie} - f_i) dv.$$

Moreover, we have that h is convex according to lemma 1.3.12 and we obtain

$$D_i(f_i, f_e) \geq \int \frac{1}{\varepsilon_i} \nu_{ie} n_i (h(f_i) - h(M_i)) dv + \int \frac{1}{\tilde{\varepsilon}_i} \nu_{ie} n_e (h(f_i) - h(M_{ie})) dv$$

$$= \frac{1}{\varepsilon_i} \nu_{ie} n_i (H(f_i) - H(M_i)) + \frac{1}{\tilde{\varepsilon}_i} \nu_{ie} n_e (H(f_i) - H(M_{ie})). \tag{6.45}$$

In the same way we get a similar expression for $D_e(f_e, f_i)$ just exchanging the indices i and e.

If we use that $\ln M_i$ is a linear combination of $1, v$ and $|v|^2$, we see that $\int (M_i - f_i) \ln M_i dv = 0$ since f_i and M_i have the same moments. With this we can compute that

$$H(f_i|M_i) = H(f_i) - H(M_i). \tag{6.46}$$

Moreover, according to (2.17), we see that

$$\frac{1}{\tilde{\varepsilon}_i} \nu_{ie} n_e H(M_{ie}) + \frac{1}{\tilde{\varepsilon}_e} \nu_{ie} n_i H(M_{ei}) \leq \frac{1}{\tilde{\varepsilon}_i} \nu_{ie} n_e H(M_i) + \frac{1}{\tilde{\varepsilon}_e} \nu_{ie} n_i H(M_e). \tag{6.47}$$

With (6.46) and (6.47), we can deduce from (6.45) that

$$D_i(f_i, f_e) + D_e(f_e, f_i) \geq \left(\frac{1}{\varepsilon_i} \nu_{ie} n_i + \frac{1}{\tilde{\varepsilon}_i} \nu_{ie} n_e\right) H(f_i|M_i)$$

$$+ \left(\frac{1}{\varepsilon_e} \nu_{ie} n_e + \frac{1}{\tilde{\varepsilon}_e} \nu_{ie} n_i\right) H(f_e|M_e). \tag{6.48}$$

We want to relate the time derivative of the relative entropies

$$\frac{d}{dt}(H(f_i|M_i) + H(f_e|M_e)) = \frac{d}{dt}\left[\int f_i \ln \frac{f_i}{M_i} dv + \int f_e \ln \frac{f_e}{M_e} dv\right],$$

to the entropy production in the following. First, we use product rule and obtain

$$\frac{d}{dt}(H(f_i|M_i) + H(f_e|M_e)) = \int \partial_t f_i \left(\ln \frac{f_i}{M_i} + 1\right) dv - \int \frac{f_i}{M_i} \partial_t M_i dv$$

$$+ \int \partial_t f_e \left(\ln \frac{f_e}{M_e} + 1\right) dv - \int \frac{f_e}{M_e} \partial_t M_e dv. \tag{6.49}$$

By using the explicit expression of $\partial_t M_i$ given by (6.37), we can compute that

$$\int f_k \frac{\partial_t M_k}{M_k} dv = \partial_t n_k = 0, \quad k = i, e,$$

since n_k is constant in the space-homogeneous case. In the first term on the right-hand side of (6.49), we insert $\partial_t f_i$ and $\partial_t f_e$ from equation (6.42) and obtain

$$\frac{d}{dt}(H(f_i|M_i) + H(f_e|M_e)) = \int \left(\frac{1}{\varepsilon_i}\nu_{ie}n_i(M_i - f_i) + \frac{1}{\tilde{\varepsilon}_i}\nu_{ie}n_e(M_{ie} - f_i)\right) \ln f_i dv$$

$$+ \int \left(\frac{1}{\varepsilon_e}\nu_{ie}n_e(M_e - f_e) + \frac{1}{\tilde{\varepsilon}_e}\nu_{ie}n_i(M_{ei} - f_e)\right) \ln f_e dv.$$

Indeed, the terms with $\ln M_i$ and $\ln M_e$ vanish since $\ln M_i$ and $\ln M_e$ are a linear combination of $1, v$ and $|v|^2$ and our model satisfies the conservation of the number of particles, total momentum and total energy (see section 2.1.5). All in all, we obtain

$$\frac{d}{dt}(H(f_i|M_i) + H(f_e|M_e)) = -(D_i(f_i, f_e) + D_e(f_e, f_i)). \tag{6.50}$$

Using (6.48) we obtain

$$\frac{d}{dt}(H(f_i|M_i) + H(f_e|M_e))$$

$$\leq -\left[\left(\frac{1}{\varepsilon_i}\nu_{ie}n_i + \frac{1}{\tilde{\varepsilon}_i}\nu_{ie}n_e\right) H(f_i|M_i) + \left(\frac{1}{\varepsilon_e}\nu_{ie}n_e + \frac{1}{\tilde{\varepsilon}_e}\nu_{ie}n_i\right) H(f_e|M_e)\right]$$

$$\leq -\min\{\frac{1}{\varepsilon_i}\nu_{ie}n_i + \frac{1}{\tilde{\varepsilon}_i}\nu_{ie}n_e, \frac{1}{\varepsilon_e}\nu_{ie}n_e + \frac{1}{\tilde{\varepsilon}_e}\nu_{ie}n_i\}(H(f_i|M_i) + H(f_e|M_e)).$$

Define $C := \min\{\frac{1}{\varepsilon_i}\nu_{ie}n_i + \frac{1}{\tilde{\varepsilon}_i}\nu_{ie}n_e, \frac{1}{\varepsilon_e}\nu_{ie}n_e + \frac{1}{\tilde{\varepsilon}_e}\nu_{ie}n_i\}$, then we can deduce an exponential decay with Gronwall's inequality

$$H(f_k|M_k) \leq H(f_i|M_i) + H(f_e|M_e)$$

$$\leq e^{-Ct}[H(f_i^0|M_i^0) + H(f_e^0|M_e^0)], \quad k = i, e.$$

With the Ciszar-Kullback inequality (see appendix A.1) we get

$$\|f_k - M_k\|_{L^1(dv)} \leq \|f_i - M_i\|_{L^1(dv)} + \|f_e - M_e\|_{L^1(dv)}$$

$$\leq 4e^{-\frac{1}{2}Ct}[H(f_i^0|M_i^0) + H(f_e^0|M_e^0)]^{\frac{1}{2}}.$$

$$\square$$

Convergence rates for the velocities and temperatures

In this section we prove convergence rates for the velocities u_i, u_e and for the temperatures T_i, T_e, respectively, to a common value in the space-homogeneous case. We start with the decay of $|u_i - u_e|^2$.

Theorem 6.2.2. *Suppose that ν_{ie} is constant in time. In the space-homogeneous case without electric field (6.42), we have the following decay rate of the velocities*

$$|u_i(t) - u_e(t)|^2 = e^{-2\nu_{ie}(1-\delta)\left(\frac{1}{\bar{\varepsilon}_i}n_e + \frac{\varepsilon}{\bar{\varepsilon}_e}\frac{m_i}{m_e}n_i\right)t}|u_i(0) - u_e(0)|^2.$$

Proof. If we multiply the equations (6.42) by v and integrate with respect to v, we obtain by using (6.24), (6.26) and (6.28)

$$\partial_t(n_iu_i) = \frac{1}{\bar{\varepsilon}_i}\nu_{ie}n_en_i(u_{ie} - u_i) = \frac{1}{\bar{\varepsilon}_i}\nu_{ie}n_en_i(1-\delta)(u_e - u_i),$$

$$\partial_t(n_eu_e) = \frac{1}{\bar{\varepsilon}_e}\nu_{ie}n_en_i(u_{ei} - u_e) = \frac{1}{\bar{\varepsilon}_e}\nu_{ie}n_en_i\frac{m_i}{m_e}\varepsilon(1-\delta)(u_i - u_e).$$

Since in the space-homogeneous case the densities n_i and n_e are constant, we actually have

$$\partial_tu_i = \frac{1}{\bar{\varepsilon}_i}\nu_{ie}n_e(1-\delta)(u_e - u_i), \quad \partial_tu_e = \frac{1}{\bar{\varepsilon}_e}\nu_{ie}n_i\frac{m_i}{m_e}\varepsilon(1-\delta)(u_i - u_e).$$

With this we get

$$\frac{1}{2}\frac{d}{dt}|u_i - u_e|^2 = (u_i - u_e)\partial_t(u_i - u_e)$$

$$= (u_i - u_e)\nu_{ie}(1-\delta)\left(\frac{1}{\bar{\varepsilon}_i}n_e + \frac{\varepsilon}{\bar{\varepsilon}_e}\frac{m_i}{m_e}n_i\right)(u_e - u_i)$$

$$= -2\nu_{ie}(1-\delta)\left(\frac{1}{\bar{\varepsilon}_i}n_e + \frac{\varepsilon}{\bar{\varepsilon}_e}\frac{m_i}{m_e}n_i\right)|u_i - u_e|^2.$$

From this, we deduce

$$|u_i(t) - u_e(t)|^2 = e^{-2\nu_{ie}(1-\delta)\left(\frac{1}{\bar{\varepsilon}_i}n_e + \frac{\varepsilon}{\bar{\varepsilon}_e}\frac{m_i}{m_e}n_i\right)t}|u_i(0) - u_e(0)|^2.$$

$\square$

We continue with a decay rate of $|T_i(t) - T_e(t)|$.

Theorem 6.2.3. *Suppose ν_{ie} is constant in time. In the space-homogeneous case without electric field (6.42), we have the following decay rate of the temperatures*

$$|T_i(t) - T_e(t)| \le e^{-C_1t}\left[|T_i(0) - T_e(0)| + \frac{|C_2|}{C_1 - C_3}(e^{(C_1-C_3)t} - 1)|u_i(0) - u_e(0)|^2\right],$$

where the constants are defined by

$$C_1 = (1 - \alpha)\nu_{ie} \left(\frac{1}{\tilde{\varepsilon}_i} n_e + \frac{\varepsilon}{\tilde{\varepsilon}_e} n_i \right),$$

$$C_2 = \nu_{ie} \left(\frac{1}{\tilde{\varepsilon}_i} n_e \left((1 - \delta)^2 + \frac{\gamma}{m_i} \right) - \frac{\varepsilon}{\tilde{\varepsilon}_e} n_i \left(1 - \delta^2 - \frac{\gamma}{m_i} \right) \right),$$

$$C_3 = 2\nu_{ie}(1 - \delta) \left(\frac{1}{\tilde{\varepsilon}_i} n_e + \frac{\varepsilon}{\tilde{\varepsilon}_e} \frac{m_i}{m_e} n_i \right).$$

Proof. If we multiply the first equation of (6.42) by $\frac{1}{n_i}|v - u_i|^2$ and integrate with respect to v, we obtain

$$\int \frac{1}{n_i} |v - u_i|^2 \partial_t f_i dv = \frac{1}{\tilde{\varepsilon}_i} \nu_{ie} n_e \frac{1}{n_i} \int |v - u_i|^2 (M_{ie} - f_i) dv. \tag{6.51}$$

Indeed, the first relaxation term vanishes since M_i and f_i have the same temperature. We simplify the left-hand side of (6.51) to

$$\int \frac{1}{n_i} |v - u_i|^2 \partial_t f_i dv = \int \frac{1}{n_i} \partial_t (|v - u_i|^2 f_i) dv + 2 \int \frac{1}{n_i} f_i (v - u_i) \cdot \partial_t u_i dv$$

$$= \partial_t T_i,$$

and the right-hand side of (6.51) simplifies to

$$\frac{1}{\tilde{\varepsilon}_i} \nu_{ie} n_e \frac{1}{n_i} \int |v - u_i|^2 (M_{ie} - f_i) dv = \frac{1}{\tilde{\varepsilon}_i} \nu_{ie} n_e (T_{ie} + |u_{ie} - u_i|^2 - T_i)$$

$$= \frac{1}{\tilde{\varepsilon}_i} \nu_{ie} n_e \left((1 - \alpha)(T_e - T_i) + \left((1 - \delta)^2 + \frac{\gamma}{m_i} \right) |u_e - u_i|^2 \right).$$

For the second species we multiply the second equation of (6.42) by $\frac{m_e}{m_i} \frac{1}{n_e} |v - u_e|^2$. For the left-hand side, we obtain by using (6.30)

$$\int \frac{m_e}{m_i} \frac{1}{n_e} |v - u_e|^2 \partial_t f_e dv = \partial_t T_e$$

and for the right-hand side using (6.26), (6.27) and (6.28)

$$\frac{1}{\tilde{\varepsilon}_e} \nu_{ie} n_i \frac{m_e}{m_i} \frac{1}{n_e} \int |v - u_e|^2 (M_{ei} - f_e) dv = \frac{1}{\tilde{\varepsilon}_e} \nu_{ie} n_i (T_{ei} + \frac{m_e}{m_i} |u_{ei} - u_e|^2 - T_e)$$

$$= \frac{1}{\tilde{\varepsilon}_e} \nu_{ie} n_i \left[\varepsilon(1 - \alpha)(T_i - T_e) \right.$$

$$+ \left(\varepsilon(1 - \delta) \left(\frac{m_i}{m_e} \varepsilon(\delta - 1) + \delta + 1 \right) - \varepsilon \frac{\gamma}{m_i} + \varepsilon^2 (1 - \delta)^2 \frac{m_i}{m_e} \right) |u_i - u_e|^2 \right]$$

$$= \frac{1}{\tilde{\varepsilon}_e} \nu_{ie} n_i \left(\varepsilon(1 - \alpha)(T_i - T_e) + \varepsilon(1 - \delta^2 - \frac{\gamma}{m_i}) |u_i - u_e|^2 \right).$$

So, we obtain

$$\partial_t T_i = \frac{1}{\tilde{\varepsilon}_i} \nu_{ie} n_e \left((1-\alpha)(T_e - T_i) + \left((1-\delta)^2 + \frac{\gamma}{m_i} \right) |u_e - u_i|^2 \right),$$

$$\partial_t T_e = \frac{1}{\tilde{\varepsilon}_e} \nu_{ie} n_i \left(\varepsilon(1-\alpha)(T_i - T_e) + \varepsilon \left(1 - \delta^2 - \frac{\gamma}{m_i} \right) |u_i - u_e|^2 \right).$$

We deduce

$$\partial_t (T_i - T_e) = -(1-\alpha)\nu_{ie} \left(\frac{1}{\tilde{\varepsilon}_i} n_e + \frac{\varepsilon}{\tilde{\varepsilon}_e} n_i \right)(T_i - T_e)$$

$$+ \nu_{ie} \left(\frac{1}{\tilde{\varepsilon}_i} n_e \left((1-\delta)^2 + \frac{\gamma}{m_i} \right) - \frac{\varepsilon}{\tilde{\varepsilon}_e} n_i \left(1 - \delta^2 - \frac{\gamma}{m_i} \right) \right) |u_i - u_e|^2,$$

or with the constants defined in this theorem

$$\partial_t (T_i - T_e) = -C_1 (T_i - T_e) + C_2 |u_i - u_e|^2.$$

Duhamel's formula gives

$$T_i(t) - T_e(t) = e^{-C_1 t}(T_i(0) - T_e(0)) + C_2 e^{-C_1 t} \int_0^t e^{C_1 s} |u_i(s) - u_e(s)|^2 ds.$$

So we have the following inequality

$$|T_i(t) - T_e(t)| \le e^{-C_1 t}|T_i(0) - T_e(0)| + |C_2| e^{-C_1 t} \int_0^t e^{C_1 s} |u_i(s) - u_e(s)|^2 ds,$$

and by using theorem 6.2.2, we have

$$|T_i(t) - T_e(t)| \le e^{-C_1 t}|T_i(0) - T_e(0)| + |C_2| e^{-C_1 t} \int_0^t e^{C_1 s} e^{-C_3 s} ds |u_i(0) - u_e(0)|^2,$$

$$|T_i(t) - T_e(t)| \le e^{-C_1 t} \left(|T_i(0) - T_e(0)| + \frac{|C_2|}{C_1 - C_3} (e^{(C_1 - C_3)t} - 1)|u_i(0) - u_e(0)|^2 \right).$$

$\square$

Numerical simulation

In this section we sketch the idea of the numerical approximation of the two-species micro-macro system (6.38),(6.39),(6.40) and (6.41). Following the idea of [28], we propose to use a particle method to discretize both microscopic equations (6.38) and (6.40). The macroscopic equations (6.39) and (6.41) are solved by a classical Finite Volume method (see for example [85]).

In this section, we only present the main idea of the method and refer to [28] for the details.

For the microscopic parts, we use a Particle-In-Cell method (see for example [17]): we approach g_{ii} and g_{ee} by a set of N_{p_i} and N_{p_e} particles, with position $x_{i_k}(t)$

and $x_{e_k}(t)$, velocity $v_{i_k}(t)$ and $v_{e_k}(t)$ and weight $\omega_{i_k}(t)$ and $\omega_{e_k}(t)$, $k = 1, \ldots, N_{p_i}$ and $k = 1, \ldots, N_{p_e}$, respectively. Then we assume that the microscopic distribution functions have the following expression:

$$g_{ii}(x, v, t) = \sum_{k=1}^{N_{p_i}} \omega_{i_k}(t)\delta(x - x_{i_k}(t))\delta(v - v_{i_k}(t)),$$

$$g_{ee}(x, v, t) = \sum_{k=1}^{N_{p_e}} \omega_{e_k}(t)\delta(x - x_{e_k}(t))\delta(v - v_{e_k}(t)),$$

$$(6.52)$$

with δ the Dirac mass. Moreover, we have the following relations:

$$\omega_{i_k}(t) = g_{ii}(x_{i_k}(t), v_{i_k}(t), t)\frac{L_x L_v}{N_{p_i}}, \quad k = 1, \ldots, N_{p_i},$$

$$\omega_{e_k}(t) = g_{ee}(x_{e_k}(t), v_{e_k}(t), t)\frac{L_x L_v}{N_{p_e}}, \quad k = 1, \ldots, N_{p_e},$$

$$(6.53)$$

where $L_x \in \mathbb{R}$ and $L_v \in \mathbb{R}$ denote the length of the domain in the space and the length in the domain in velocity direction, respectively.

The method consists now in splitting the transport and the source parts of (6.38) and (6.40). Let us consider (6.38), the steps being the same for (6.40). The transport part

$$\partial_t g_{ii} + v\partial_x g_{ii} + E\partial_v g_{ii} = 0, \tag{6.54}$$

is solved in the following way. We evolve the positions x_{i_k} and the velocities v_{i_k} in (6.52) and (6.53) in the expression of g_{ii} by solving the characteristic equations

$$\frac{dx_{i_k}}{dt}(t) = v_{i_k}(t), \quad \frac{dv_{i_k}}{dt}(t) = E(x_{i_k}(t), t), \quad \forall k = 1, \ldots, N_{p_i}.$$

The source part

$$\partial_t g_{ii} = -(1 - \Pi_{M_i})(v\partial_x M_i) + \Pi_{M_i}(v\partial_x g_{ii}) + \Pi_{M_i}(E\partial_v g_{ii})$$

$$+ \frac{1}{\bar{\varepsilon}_i}\nu_{ie}n_e(M_{ie} - (1 + \frac{(v - u_i)(u_{ie} - u_i)}{T_i})$$

$$+ (\frac{|v - u_i|^2}{2T_i} - \frac{1}{2})(\frac{T_{ie}}{T_i} + \frac{1}{T_i}|u_{ie} - u_i|^2 - 1))M_i) - (\frac{1}{\varepsilon_i}\nu_{ie}n_i + \frac{1}{\bar{\varepsilon}_i}\nu_{ie}n_e)g_{ii},$$

$$(6.55)$$

is solved by evolving the weight $\omega_{i_k}(t)$. Let us denote by $S(x, v, t)$ the right-hand side such that $\partial_t g_{ii} = S(x, v, t)$. If we solve (6.53) for g_{ii}, compute $\partial_t g_{ii}$ of the obtained equation and use that $\partial_t g_{ii} = S$, we obtain

$$\partial_t \omega_{ik} = S(x_{i_k}, v_{i_k}, t)\frac{N_{p_i}}{L_x L_p}.$$

The strategy is the same as in paragraph 4.1.2 of [28], where only one species is considered (and so there are no coupling terms). The supplementary terms coming from the coupling of both species are treated in the source part as the other source terms. They do not add any particular difficulty.

Furthermore, we have to ensure that the moments of g_{ii} and g_{ee} remain zero at each time step at the discrete level. This is not guaranteed at the discrete level, since $\langle m(v)v\partial_x g_{kk}\rangle \neq 0$. This is called projection step where the moments of g_{kk} are matched to zero. Details are given in subsection 4.2 of [28]. The idea is the following. We construct at each time step a function $h_{kk}(x, v)$ which has the same macroscopic quantities as the discrete function $g_{kk}^{n+1}(x, v)$ at the time step $n+1$ by making the ansatz

$$h_{kk}(x, v) = \lambda_{kk}(x) \cdot m(v)M_k(x, v)$$

where M_k is the Maxwell distribution of f_k^{n+1}. The motivation for this ansatz is that we want to have a function which lies in $\mathcal{N}_k$. Then we solve the system

$$\langle m(v)h_{kk}(x, v)\rangle|_{X_i} = \langle m(v)g_{kk}^{n+1}\rangle|_{X_i}$$

on each position X_i on the uniform grid $(X_i)_i$, for the function $\lambda_{kk}(x) \in \mathbb{R}^3$. This is possible since we have three unknowns and three constraints. Then we correct g_{kk}^{n+1} by

$$g_{kk}^{n+1,new} = g_{kk}^{n+1} - h_{kk}$$

at each time step.

Finally, the macroscopic equations (6.39) and (6.41) are discretized on a grid in space and solved by a classical Finite Volume method. For the one species case, this is detailed in subsection 4.3 of [28]. The electric field is discretized on the same grid and computed at each time step by solving the Maxwell equation (6.9) with a Finite Difference method.

6.2.2 Landau damping for two species

The next theoretical result which can be used to compare the analytic behaviour of the solution with numerical results is the Landau damping. It is a qualitative behaviour of the Vlasov-Poisson equation. For one species this phenomena was developed by Landau in [63]. A more rigorous treatment of the derivation of Landau has been done by Villani in [84]. The derivation in this section is an extension of the result presented in [84] by Villani for one species to a mixture of ions and electrons.

The 1D Vlasov model for a mixture of two species

We consider a plasma consisting of electrons denoted by the index e and one species of ions denoted by the index i. The state of the two species will be described by two distribution functions $f_e(x, v, t), f_i(x, v, t) \geq 0$, $x \in [0, L], L > 0$, $v \in \mathbb{R}, t > 0$. We describe the time evolution by a system of Vlasov equations

$$\partial_t f_e + v \partial_x f_e - \frac{e}{m_e} E[f_e, f_i] \partial_v f_e = 0,$$
$$\partial_t f_i + v \partial_x f_i + \frac{e}{m_i} E[f_e, f_i] \partial_v f_i = 0,$$

(6.56)

with corresponding initial data. This set of equations describes the collision-less regime of equation (6.6) with the same notation as in section 6.1.3. Again we expect from physics that the electric field $E(x, t)$ can be written as a gradient of a potential $\phi(x, t)$. Together with the Maxwell equation $\partial_x E(x, t) = \rho(x, t)$, where

$$\rho(x, t) = e \int_{-\infty}^{\infty} (f_i(x, v, t) - f_e(x, v, t)) dv$$

(6.57)

describes the charge density, we get that ϕ solves a Poisson equation.

So now, we assume that E is of the form

$$E[f_e, f_i] = -\int \partial_x W(x - y) \rho(y, t) dy$$
$$= -\int \partial_x W(x - y) e \int_{-\infty}^{\infty} (f_i(y, v, t) - f_e(y, v, t)) dv dy$$
$$= -\int \int_{-\infty}^{\infty} \partial_x W(x - y) e (f_i(y, v, t) - f_e(y, v, t)) dv dy,$$

(6.58)

where W is a Green's function to the Poisson equation on $[0, L]$. The existence of the function W is argued in [84] and [75].

The advantage of the system (6.56) is the following property discovered by Landau. When one linearises the Vlasov equations around a homogeneous equilibrium, it is possible to analyse the stability and the asymptotic behaviour for the linearised equation. This analysis is carried out for the two species case in the following section in a more rigorous way inspired by the one species case done by Villani in [84]. We begin with the existence of equilibrium solutions in section 6.2.2. In section 6.2.2 we perform the linearisation around an equilibrium distribution. In section 6.2.2 we want to give a physical motivation of the properties we want to derive analytically from the linearised equations in section 6.2.2 to section 6.2.2.

Existence of equilibria

We want to find equilibrium solutions to (6.56).

Definition 6.2.1 (Equilibrium solution). We call a pair of functions (f_i^{equ}, f_e^{equ}) an equilibrium solution to (6.56) if and only if (f_i^{equ}, f_e^{equ}) satisfy (6.56) and $\partial_t f_i^{equ} = \partial_t f_e^{equ} = 0$.

We are able to prove that there exist an equilibrium solution to (6.56).

Theorem 6.2.4 (Existence of equilibrium solutions). *There exists at most one equilibrium solution to (6.56).*

Proof. We prove it by giving an example. Any pair of distributions $(f_i(x,v), f_e(x,v)) = (f_i^{equ}(v), f_e^{equ}(v))$ defines an equilibrium solution. Obviously the distributions are independent of t. So the only thing we have to prove is that they are a solution to (6.56). Since they are also independent of x, we have $v\partial_x f_i^{equ} = v\partial_x f_e^{equ} = 0$ for every $v \in \mathbb{R}$. Therefore the charge density ρ^{equ} given by (6.57) associated to f_i^{equ} and f_e^{equ} is constant in x. So the corresponding forces vanish since

$$\int \partial_x W(x-y)\rho^{equ}(y,t)dy = \int W(x-y)\partial_y \rho^{equ}(y,t)dy = 0.$$

$\square$

Linearisation of the Vlasov model

We assume that we are near an equilibrium, that means

$$f_e(x,v,t) = f_e^{equ}(v) + h_e(x,v,t),$$
$$f_i(x,v,t) = f_i^{equ}(v) + h_i(x,v,t),$$

(6.59)

where h_e, h_i are small deviations. The meaning of "small" will be specified in a moment. We assume that the equilibria are independent of x and they satisfy the condition of quasi-neutrality

$$\int_{-\infty}^{\infty} f_i^{equ}(v)dv = \int_{-\infty}^{\infty} f_e^{equ}(v)dv,$$

(6.60)

which means that in equilibrium the resulting charge ρ is equal to zero. Such an equilibrium exists according to theorem 6.2.4. Inserting ansatz (6.59) into (6.56) and (6.58) and neglecting quadratic terms in h_i and h_e, respectively, we obtain

$$\partial_t h_e + v\partial_x h_e - \frac{e}{m_e}E[h_e, h_i]\partial_v f_e = 0,$$
$$\partial_t h_i + v\partial_x h_i + \frac{e}{m_i}E[h_e, h_i]\partial_v f_i = 0,$$

(6.61)

where

$$E[h_e, h_i] = -\int\int_{-\infty}^{\infty} \partial_x W(x-y)[h_i(x,v,t) - h_e(x,v,t)]dvdy$$
$$= -\int\int_{-\infty}^{\infty} \partial_x W(x-y)h(x,v,t)dvdy.$$

(6.62)

The quantity h is defined as $h := h_i - h_e$. We now want to show the following. If a plasma is disturbed from equilibrium as in (6.59), we expect from physics that the plasma will relax back to equilibrium for $t \to \infty$. So the particles will be forced to go to the equilibrium configuration. Since we expect the particles to be inert by Newton's first law, we expect them to carry out oscillations. So we expect to see that the densities are oscillating in time and this oscillation is possibly damped depending on the physical configuration. We want to show that this property is contained in the model (6.61) and find criteria when these oscillations are damped. Since the electric field depends directly via (6.62) on the densities and the electric field is related to the electric energy W_{el} via $W_{el} = \frac{1}{2}E^2$, we also expect to observe a damping by considering the electric energy or E^2.

Physical motivation of Landau damping in plasmas

In this section we want to give a physical motivation of Landau damping, before we start to derive its qualitative behaviour in the following sections. We consider an electrically quasi-neutral plasma according to (6.60) in equilibrium consisting of positively charged ions and negatively charged electrons. Now we assume that we disturb the position of the electrons or the ions away from the equilibrium configuration. Then we expect to have a force which forces the electrons back to the equilibrium configuration. Similar as in the case of a spring, this can lead to an oscillation around the equilibrium configuration. This oscillation can be damped or not. Since an oscillation around the equilibrium configuration has to do with a change of the position of the particles, we expect that the oscillations are reflected somehow in the corresponding density.

Methods of characteristics

In this section and the following ones we want to find criteria for damping and show the property of damping in an analytical way from equations (6.56). Similar as in [84] in step 1, we apply the methods of characteristics on (6.61) and obtain

$$h_e(x, v, t) = h_{e,in}(x - vt, v) + \frac{e}{m_e} \int_0^t E[h_i, h_e](\tau, x - v(t - \tau)) \partial_v f_e^{equ}(v) d\tau,$$

$$h_i(x, v, t) = h_{i,in}(x - vt, v) - \frac{e}{m_i} \int_0^t E[h_i, h_e](\tau, x - v(t - \tau)) \partial_v f_i^{equ}(v) d\tau. \tag{6.63}$$

This is similar to the definition of a mild solution treating the force term as the inhomogeneity of the transport equation, see section 4.1.1.

Fourier transform of the distribution functions

Similar as in [84] in step 2, we take the Fourier transform (see appendix A.2) in both x and v of both equations (6.63) and do the substitutions $y = x - vt$ in the part with the initial data and $y = x - v(t - \tau)$ in the term with the time integration (source term). We use that the Fourier transform of the source terms separates into the product of the Fourier transform in x of the force term $E[h_i, h_e]$ and the Fourier transform in v of the derivative of the equilibrium function $\partial_v f_k^{equ}$, $k = 1, 2$. Then we use that the Fourier transform of a convolution leads to a product of two Fourier transforms. We get rid of the derivatives doing integration by parts in v. All in all, we obtain

$$\tilde{h}_e(k,\eta,t) = \tilde{h}_{e,in}(k,\eta + kt)$$

$$- \frac{e}{m_e} 4\pi^2 \hat{W}(k) \int_0^t \hat{\rho}^h(k,\tau) \tilde{f}_e^{equ}(\eta + k(t-\tau))k \cdot [\eta + k(t-\tau)]d\tau,$$

$$\tilde{h}_i(k,\eta,t) = \tilde{h}_{i,in}(k,\eta + kt)$$

$$+ \frac{e}{m_i} 4\pi^2 \hat{W}(k) \int_0^t \hat{\rho}^h(k,\tau) \tilde{f}_i^{equ}(\eta + k(t-\tau))k \cdot [\eta + k(t-\tau)]d\tau,$$

(6.64)

where k and η denote the new variables after the transformations and $\rho^h = \rho_i^{h_i} - \rho_e^{h_e}$, $\rho_i^{h_i} = \int h_i dv$, $\rho_e^{h_e} = \int h_e dv$. The notation $\sim$ denotes the Fourier transform in x and v whereas $\wedge$ denotes the Fourier transform only in the x variable.

Fourier transform of the densities

If we set $\eta = 0$, we obtain the Fourier transform of the densities. Since the exponential in the Fourier transform in v reduces to 1 and it remains the integration with respect to v over the function h_k, $k = e, i$ which is the density.

$$\hat{\rho}_e^{h_e}(k,t) = h_{e,in}(k,kt) + \int_0^t K_e^0(t-\tau,k)[\hat{\rho}_i^{h_i} - \hat{\rho}_e^{h_e}](k,\tau)d\tau,$$

$$\hat{\rho}_i^{h_i}(k,t) = h_{i,in}(k,kt) + \int_0^t K_i^0(t-\tau,k)[\hat{\rho}_i^{h_i} - \hat{\rho}_e^{h_e}](k,\tau)d\tau,$$

(6.65)

where

$$K_e^0(t-\tau,k) = -\frac{e}{m_e} 4\pi^2 \hat{W}(k) f_e^{equ}(kt)|k|^2 t,$$

$$K_i^0(t-\tau,k) = \frac{e}{m_i} 4\pi^2 \hat{W}(k) f_i^{equ}(kt)|k|^2 t.$$

Convergence rates of the densities

Assume that the equilibrium is a global one which means $f_i^{equ} = f_e^{equ}$. It exists according to the proof of theorem 6.2.4. Then

$$K_e^0(t-\tau,k) = -\frac{m_i}{m_e} K_i^0(t-\tau,k) =: -K(t-\tau,k).$$

Then we have two equations of the form

$$\phi_e(t) = a_e(t) - \int_0^t K(t-\tau)[\phi_i(\tau) - \phi_e(\tau)]d\tau,$$

$$\phi_i(t) = a_i(t) + \frac{m_e}{m_i} \int_0^t K(t-\tau)[\phi_i(\tau) - \phi_e(\tau)]d\tau.$$

(6.66)

Theorem 6.2.5. *Let* $K = K(t)$ *be a kernel defined for* $t \geq 0$, *such that*

(i) $|K(t)| \leq C_0 e^{-2\pi\lambda_0 t}$ *for a constant* $\lambda_0 > 0$.

(ii) $|(1 + \frac{m_e}{m_i})K^L(\xi) - 1| \geq \kappa > 0$ *for* $0 \leq Re(\xi) \leq \Lambda$ *where the index* L *denotes the complex Laplace transform. For the definition of the Laplace transform see appendix A.2. Note, that property* (i) *in this theorem corresponds to the property* (ii) *in the definition of the Laplace transform as a required property such that the Laplace transform is well-defined.*

Let further $a_i = a_i(t), a_e = a_e(t)$ *satisfy*

(iii) $|\frac{m_e}{m_i} a_e(t) + a_i(t)| \leq \alpha_+ e^{-2\pi\lambda_+ t}$,

(iv) $|a_i(t) - a_e(t)| \leq \alpha_- e^{-2\pi\lambda_- t}$

and let ϕ_i, ϕ_e *solve (6.66). Then for any* $\lambda' < \min(\lambda_+, \lambda_-, \Lambda, \lambda_0)$

$$|\phi_e(t)| \leq C_e e^{-2\pi\lambda' t},$$

$$|\phi_i(t)| \leq C_i e^{-2\pi\lambda' t},$$

where C_i, C_e *are constants depending on* $\lambda_+, \lambda_-, \Lambda, \lambda_0, \kappa, C_0, \lambda'$.

Proof. Let us write $\Phi_e(t) = e^{2\pi\lambda' t}\phi_e(t), \Phi_i(t) = e^{2\pi\lambda' t}\phi_i(t), A_e(t) = e^{2\pi\lambda' t}a_e(t),$ $A_i(t) = e^{2\pi\lambda' t}a_i(t)$. Now similar as in the proof of lemma 3.5 in [84], we multiply (6.66) by $e^{2\pi\lambda' t}$ and obtain

$$\Phi_e(t) = A_e(t) - \int_0^t K(t - \tau)e^{2\pi\lambda'(t-\tau)}[\Phi_i(\tau) - \Phi_e(\tau)]d\tau,$$

$$\Phi_i(t) = A_i(t) + \frac{m_e}{m_i} \int_0^t K(t - \tau)e^{2\pi\lambda'(t-\tau)}[\Phi_i(\tau) - \Phi_e(\tau)]d\tau. \tag{6.67}$$

The functions Φ_e, Φ_i, A_e, A_i and K are defined only for $t \geq 0$. We extend the domain of these functions for negative times by setting them equal to zero for $t < 0$. Then take the Fourier transform in the time variable. This leads to

$$\hat{\Phi}_e(\omega) = \hat{A}_e(\omega) - \int_{-\infty}^{\infty} e^{-i\omega t} \int_0^t K(t - \tau)e^{2\pi\lambda'(t-\tau)}[\Phi_i(\tau) - \Phi_e(\tau)]d\tau dt$$

$$= \hat{A}_e(\omega) - \int_{-\infty}^{\infty}\int_{-\infty}^{\infty} K(s)e^{2\pi\lambda' s}e^{-i\omega(s+\tau)}[\Phi_i(\tau) - \Phi_e(\tau)d\tau ds$$

$$= \hat{A}_e(\omega) - K^L(\lambda' + i\omega)[\hat{\Phi}_i(\omega) - \hat{\Phi}_e(\omega)],$$

$$\hat{\Phi}_i(\omega) = \hat{A}_i(\omega) + \frac{m_e}{m_i}K^L(\lambda' + i\omega)[\hat{\Phi}_i(\omega) - \hat{\Phi}_e(\omega)],$$

where K^L denotes the complex Laplace transform, see appendix A.2. Subtract the first equation from the second one. Then

$$\hat{\Phi}_i(\omega) - \hat{\Phi}_e(\omega) = \hat{A}_i(\omega) - \hat{A}_e(\omega) + \left(\frac{m_e}{m_i} + 1\right)K^L(\lambda' + i\omega)[\hat{\Phi}_i(\omega) - \hat{\Phi}_e(\omega)],$$

which is equivalent to

$$\hat{\Phi}_i(\omega) - \hat{\Phi}_e(\omega) = \frac{\hat{A}_i(\omega) - \hat{A}_e(\omega)}{1 - (\frac{m_e}{m_i} + 1)K^L(\lambda' + i\omega)},$$

since we assumed $(\frac{m_e}{m_i} + 1)K^L(\lambda' + i\omega) \neq 1$. Actually, we assumed that $|(1 + \frac{m_e}{m_i})K^L(\xi) - 1| \geq \kappa > 0$ and so we obtain

$$||\hat{\Phi}_i - \hat{\Phi}_e||_{L^2(\omega)} \leq \frac{||\hat{A}_i - \hat{A}_e||_{L^2(\omega)}}{\kappa}.$$

Therefore by Plancherel's identity (see appendix A.2) and the decay assumption (iv), we get

$$||\Phi_i - \Phi_e||_{L^2(t)} \leq \frac{||A_i - A_e||_{L^2(t)}}{\kappa} \leq \frac{\alpha_-}{\kappa\sqrt{4\pi(\lambda_- - \lambda')}},$$

since the integral $\int_0^\infty e^{-2\pi\lambda_- t}e^{2\pi\lambda' t}dt$ is equal to $\frac{1}{\sqrt{4\pi(\lambda_- - \lambda')}}$. We plug this into the following system equivalent to (6.67)

$$\frac{m_e}{m_i}\Phi_e(t) + \Phi_i(t) = \frac{m_e}{m_i}A_e(t) + A_i(t),$$

$$\Phi_i(t) - \Phi_e(t) = A_i(t) - A_e(t) + \left(\frac{m_e}{m_i} + 1\right)\int_0^t K(t-\tau)e^{2\pi\lambda'(t-\tau)}[\Phi_i(\tau) - \Phi_e(\tau)]d\tau,$$

$$(6.68)$$

and obtain by using the estimates (iii) and (iv), Hölder inequality and estimate (ii)

$$||\frac{m_e}{m_i}\Phi_e + \Phi_i||_{L^\infty(t)} \leq ||\frac{m_e}{m_i}A_e + A_i||_{L^\infty(t)} \leq \alpha_+,$$

$$||\Phi_i - \Phi_e||_{L^\infty(t)} \leq ||A_i - A_e||_{L^\infty(t)} + (\frac{m_e}{m_i} + 1)||(Ke^{2\pi\lambda' t}) * [\Phi_i - \Phi_e]||_{L^\infty(dt)}$$

$$\leq \alpha_- + \left(\frac{m_e}{m_i} + 1\right)||Ke^{2\pi\lambda' t}||_{L^2(dt)}||\Phi_i - \Phi_e||_{L^2(dt)}$$

$$\leq \alpha_- + \left(\frac{m_e}{m_i} + 1\right)\frac{C_0}{\sqrt{4\pi(\lambda_0 - \lambda')}} \frac{\alpha_-}{\kappa\sqrt{4\pi(\lambda_- - \lambda')}}.$$

$$(6.69)$$

From this we obtain the estimates

$$||\Phi_i||_{L^\infty(t)} = ||\frac{1}{1 + \frac{m_e}{m_i}}(\frac{m_e}{m_i}\Phi_e + \Phi_i) + \frac{1}{1 + \frac{m_e}{m_i}}\frac{m_e}{m_i}(\Phi_i - \Phi_e)||_{L^\infty(t)}$$

$$\leq \frac{1}{1 + \frac{m_e}{m_i}}||\frac{m_e}{m_i}\Phi_e + \Phi_i||_{L^\infty(t)} + \frac{1}{1 + \frac{m_e}{m_i}}\frac{m_e}{m_i}||\Phi_i - \Phi_e||_{L^\infty(t)}$$

$$\leq \frac{1}{1 + \frac{m_e}{m_i}}\alpha_+ + \frac{1}{1 + \frac{m_e}{m_i}}\frac{m_e}{m_i}\left[\alpha_- + \left(\frac{m_e}{m_i} + 1\right)\frac{C_0}{\sqrt{4\pi(\lambda_0 - \lambda')}}\frac{\alpha_-}{\kappa\sqrt{4\pi(\lambda_- - \lambda')}}\right],$$

$$||\Phi_e||_{L^\infty(t)} = ||\frac{1}{1 + \frac{m_e}{m_i}}\left(\frac{m_e}{m_i}\Phi_e + \Phi_i\right) - \frac{1}{1 + \frac{m_e}{m_i}}(\Phi_i - \Phi_e)||_{L^\infty(t)}$$

$$\leq \frac{1}{1 + \frac{m_e}{m_i}}\alpha_+ + \frac{1}{1 + \frac{m_e}{m_i}}\left[\alpha_- + (\frac{m_e}{m_i} + 1)\frac{C_0}{\sqrt{4\pi(\lambda_0 - \lambda')}}\frac{\alpha_-}{\kappa\sqrt{4\pi(\lambda_- - \lambda')}}\right].$$

We also have to check if the Fourier transforms of Φ_e, Φ_i exist and if they are in $L^2(\mathbb{R})$. We replace λ' by a parameter α varying from $-\varepsilon$ to λ'. By the integrability of K and Gronwall's lemma, $\frac{m_e}{m_i}\Phi_e + \Phi_i$ and $\Phi_i - \Phi_e$ are bounded as a function of t. Therefore Φ_e and Φ_i are bounded as a function of t. So $\phi_i(k,t)e^{-\varepsilon|k|t}$, $\phi_e(k,t)e^{-\varepsilon|k|t}$ are integrable for any $\varepsilon > 0$ and continuous as $\varepsilon \to 0$. Then assumption (ii) guarantees that the bounds are uniform in the strip $0 \leq \mathrm{Re}(\xi) \leq \lambda'$. □

From (6.69) we see that the difference of the densities is damped which results in a damping of the electric field and the electric energy what we wanted to show. In [29] Crestetto, Klingenberg and Pirner also numerically observed a damped electric field in the non-collision-less regime, see section 6 in [29].

The Landau-Penrose stability criterion

In order to have damping, we have to ensure that

$$|(1 + \frac{m_e}{m_i})K^L(\xi) - 1| \geq \kappa > 0 \quad \text{for} \quad 0 \leq \mathrm{Re}(\xi) \leq \Lambda.$$

[84] computes that

$$(K)^L((\lambda + i\omega)|k|) = \hat{W}(k) \int \frac{(f^{equ})'(v)}{v - w + i\lambda} dv,$$

using integration by parts and the fact that we have

$$\int_0^\infty e^{-2i\pi|k|tv} e^{2\pi(\lambda+iw)|k|t} |k| dv dt = \frac{1}{2i\pi} \frac{1}{v - w + i\lambda},$$

assuming $(f^{equ})'$ decays fast enough at infinity such that the integral exists. Then [84] considers the limit $\lambda \to 0^+$ (and extends the statement later to a strip $0 \leq \lambda \leq \Lambda$ by continuity arguments). In the limit [84] obtains

$$Z(k,\omega) := \hat{W}(k)[\int \frac{(f^{equ})'(v) - (f^{equ})'(\omega)}{v - \omega} dv - i\pi(f^{equ})'(\omega)].$$

So the aim is to find a condition such that Z does not approach 1. If the imaginary part of Z stays away from zero, then Z does not approach the real value 1. So Z only approaches zero, if $(f^{equ})'(v)$ approaches zero in the imaginary part. But when the imaginary part will approach zero or equivalently if ω approaches zero of $(f^{equ})'$, the real part has to stay away from one. This leads to the Penrose stability criterion

$$\forall_{\omega \in \mathbb{R}}, \ (f^{equ})'(\omega) = 0 \ \Rightarrow \ \hat{W}(k) \int \frac{(f^{equ})'(v)}{v - \omega} dv < 1.$$

In our case there is a factor $(1 + \frac{m_e}{m_i})$ in front of K^L, so we get

$$\forall_{\omega \in \mathbb{R}}, \ (f^{equ})'(\omega) = 0 \ \Rightarrow \ (1 + \frac{m_e}{m_i})\hat{W}(k) \int \frac{(f^{equ})'(v)}{v - \omega} dv < 1.$$

Example 6.2.1. In the case of Coulomb interactions, we have $\hat{W}(k) = \frac{1}{|k|^2}$. If f^{equ} has only one maximum at the origin meaning $\omega = 0$, and is non-decreasing for $v > 0$ meaning $(f^{equ})'(v) < 0$ for $v > 0$, and non-increasing for $v > 0$ meaning $(f^{equ})'(v) > 0$ for $v > 0$, then obviously

$$\int \frac{(f^{equ})'(v)}{v} dv < 0,$$

and so the Penrose stability criterion trivially holds true.

Example 6.2.2. If f^{equ} is a small perturbation of the function described in the previous example, so that it has a slight secondary bump, then the Penrose criterion will still be satisfied. If the bump becomes larger, there will be instability.

Chapter 7

Extension to an ES-BGK model for a multi-component gas mixture

In the previous chapters we concerned ourselves with a kinetic description of gases using the BGK approximation. It has the advantage of being less complicated than the full Boltzmann equation and at the same time fulfils the main properties of the Boltzmann equation namely conservation of mass, momentum and energy and the H-theorem with its entropy inequality and Maxwell distributions in equilibrium. However, the drawback of the BGK approximation is its incapability of reproducing the correct Boltzmann hydrodynamic regime in the asymptotic continuum limit to the Navier-Stokes equations. Therefore, a modified version called ES-BGK approximation was suggested by Holway in the case of one species [54]. The H-theorem of this model then was proven in [2] and existence and uniqueness of solutions in [87].

Here we shall focus on gas mixtures modelled via an ES-BGK approach. We saw that in the literature there is a BGK model for gas mixtures suggested by Andries, Aoki and Perthame in [1] which contains only one collision term on the right-hand side. One extension of this model to an ES-BGK model for gas mixtures is given by Brull in [21]. His extension is based on an entropy minimization problem and leads to a correct Prandtl number in the Navier-Stokes equations. Another extension is given by Groppi, Monica and Spiga in [48]. They noticed that in the case of a gas mixture, besides the Prandtl number, the diffusion coefficient and the thermal diffusion parameter need to be fixed, too. With the proposed model in [48], they are merely able to fix the diffusion parameter and the Prandtl number but not the thermal diffusion parameter.

In this chapter we are interested in an extension of the BGK model to an ES-BGK model for gas mixtures presented in chapter 2 which just like the Boltzmann equation for gas mixtures contains a sum of collision terms on the right-hand side. The advantage of the model presented in chapter 2 is that we have free parameters. With these free parameters, we are possibly able to determine all macroscopic physical constants like diffusion parameter, Prandtl number and the thermal diffusion parameter when taking the limit to the Navier-Stokes equations.

The outline of this chapter is as follows: in section 7.1 we want to motivate the ES-BGK model for one species. In section 7.2 we want to briefly repeat the main issues of the BGK model for gas mixtures in chapter 2. In section 7.3, we want to introduce the macroscopic equations and quantities which Groppi, Monica and Spiga [48] expect in the case of gas mixtures. In section 7.4 the Chapman-Enskog expansion for the BGK model for mixtures presented in chapter 2 is performed and the differences in the case of the ES-BGK extension for gas mixtures is illustrated in order to see that we are able to capture the right hydrodynamic regime. In section 7.5 we introduce the Brunn-Minkowski inequality which will be needed to prove the H-theorem of ES-BGK models. In section 7.6, we suggest extensions to ES-BGK models for mixtures and prove the corresponding H-theorems.

7.1   Motivation of the model for one species

For one species the BGK equation is given by

$$\partial_t f + v \cdot \nabla_x f = \nu n(M(f) - f),$$

with the distribution function $f(x, v, t) > 0$ and the collision frequency $\nu(x, t) > 0$ where $x \in \mathbb{R}^3$, $v \in \mathbb{R}^3$ are the phase space variables and $t \geq 0$ the time. The Maxwell distribution M is given by

$$M = \frac{n}{\sqrt{2\pi \frac{T}{m}}^3} \exp\left(-\frac{|v - u|^2}{2T/m}\right),$$

with mass m and the moments

$$\int f(v) \begin{pmatrix} 1 \\ v \\ m|v - u|^2 \end{pmatrix} dv =: \begin{pmatrix} n \\ nu \\ 3nT \end{pmatrix}.$$

We now replace the Maxwell distribution $M(f)$ by another relaxation operator and consider the new equation

$$\partial_t f + v \cdot \nabla_x f = \nu n(G(f) - f),$$

where $G(f)$ is no longer a Maxwell distribution. Instead of the scalar temperature in the Maxwell distribution we take a linear combination of the temperature and the pressure tensor $\mathbb{P} = m \int (v - u) \otimes (v - u) dv$ in the following way

$$G(f) = \frac{n}{\sqrt{\det(2\pi \frac{\mathcal{T}}{m})}} e^{-\frac{1}{2}(v-u)\cdot(\frac{\mathcal{T}}{m})^{-1}\cdot(v-u)}, \tag{7.1}$$

where

$$\mathcal{T} = (1 - \tilde{\mu})T\mathbb{1} + \tilde{\mu}\frac{\mathbb{P}}{n}, \quad \text{with} \quad -\frac{1}{2} \leq \tilde{\mu} \leq 1,$$

being a free parameter. We see that for $\tilde{\mu} = 0$ we regain the BGK model because then $\mathcal{T}^{-1} = \frac{1}{T}\mathbb{1}$ and $\det(2\pi\frac{\mathcal{T}}{m}\mathbb{1}) = (2\pi\frac{T}{m})^3$ in three space dimensions.

   For writing $\mathcal{T}^{-1}$ we have to ensure that $\mathcal{T}$ is invertible. However, the next lemma and theorem provide that the matrix $\mathcal{T}$ is invertible for all parameters $\tilde{\nu}$ in the range $[-\frac{1}{2}, 1]$.

Lemma 7.1.1. *Assume that $f > 0$. Then $\frac{\mathbb{P}}{n}$ has strictly positive eigenvalues.*

We skip the proof since we will show it later in the two species case. The one species case is a special case of the two species case by taking one species which does not interact with the other one. Then the corresponding collision frequency is zero or the density of the second species is zero.

Theorem 7.1.2. *Assume that $f > 0$ and $-\frac{1}{2} \leq \tilde{\mu} \leq 1$. Then $\mathcal{T}$ has strictly positive eigenvalues. Especially $\mathcal{T}$ is invertible.*

The proof is given in [2]. We skip the prove since we also show it in the two species case later.

7.1.1 Minimization of the entropy

Another motivation of the ES-BGK model comes from a minimization problem for the entropy. In section 1.4.1 we motivated the BGK model coming from a minimization problem of the entropy. We considered a model of the form

$$\partial_t f + v \cdot \nabla_x f = \nu n (G(f) - f), \tag{7.2}$$

for a function $G(f)$ which we determined as the minimizer to the problem

$$S(n, u, T) = \min_{g \in \chi} \int H(g) dv, \tag{7.3}$$

where $H(g)$ is given by $H(g) = g \ln g$ and χ is the following set

$$\chi = \{g \geq 0, (1 + |v|^2) g \in L^1(dv), \int g dv = n, \int v g dv = nu, \int v \otimes v g dv = n(u \otimes u + \frac{T}{m} \mathbf{1})\}.$$

The set χ was motivated in the following way. We expect that during the relaxation process the density n, the momentum nu and the energy $\frac{1}{2} n (m|u|^2 + T)$ should be conserved, so the solution to (7.3) should have the same density, momentum and energy as f. We observe that we also determined the off-diagonal terms of $\int v \otimes v g dv$. The physical meaning of the choice

$$\int v \otimes v g dv = nu \otimes u + n \frac{T}{m} \mathbf{1},$$

was the following. If we compute the integral $m \int v \otimes v f dv$ we get $mnu \otimes u + \mathbb{P}$. Therefore the restriction on $g \in \chi$ means that we expect that in equilibrium the tensor $\mathbb{P}$ becomes diagonal, which means that we have no friction in the equilibrium. Now we also allow for off-diagonal terms and choose a different set χ_{new} of the following form

$$\chi_{new} = \{g \geq 0, (1 + |v|^2) g \in L^1(dv), \int g dv = n, \int v g dv = nu, \int v \otimes v g dv = nu \otimes u + \frac{n}{m} \mathcal{T}\},$$

with $\mathcal{T}$ given by

$$\mathcal{T} = (1 - \tilde{\mu}) T \mathbf{1} + \tilde{\mu} \frac{\mathbb{P}}{n}, \quad -\frac{1}{2} \leq \tilde{\mu} \leq 1.$$

Theorem 7.1.3. *The unique minimizer to*

$$\min_{g \in \chi_{new}} \int H(g) dv,$$

is the function G given by (7.1).

Proof. The proof is analogous to the proof of theorem 1.4.1 for the set χ. □

7.1.2 Chapman-Enskog expansion

The essential motivation to extend the BGK model to an ES-BGK model is the following. The BGK model has the disadvantage that if one performs the Chapman-Enskog expansion, the obtained transport coefficients, which are the viscosity and the heat conductivity, obtained in the Navier-Stokes level are not satisfactory, as their ratio, the Prandtl number is equal to 1. The meaning of the Prandtl number is the following. The viscosity of a fluid describes its resistance to shearing flows. Imagine parallel plates with a fluid in between. When we move the top plate the fluid region next to the top plate will move with the same speed due to friction. The fluid next to the bottom plate has speed zero since the plate on the bottom does not move. The speed of the fluid decreases from the bottom to the top (Couette flow [38]). In physical experiments, one can measure that the force which is needed to move the top plate is proportional to the area of the plate and to the reciprocal of the distance of the two plates. The proportionality constant is given by the viscosity. In most of the fluids the viscosity has a temperature dependence via $e^{C/T}$ with a constant $C > 0$. The heat conductivity describes the ability of conducting heat. So the reciprocal of the heat conductivity has the meaning of a heat resistance. It increases with the temperature. Since the Prandtl number Pr is the ratio of the viscosity and the heat conductivity, it compares the momentum transfer with the heat transfer in the gas or equivalently the resistance to shearing with the resistance of heat transfer. For most gases Pr is approximately constant. For most gases one of the two effects dominates, so for most gases, we observe $Pr \neq 1$. The physical background described here is given in a lot of introductory physics or chemistry books for example in [52].

Chapman-Enskog expansion of the BGK equation

We write the BKG equation in dimensionless form

$$\partial_t f_{\tilde{\varepsilon}} + v \cdot \nabla_x f_{\tilde{\varepsilon}} = \frac{1}{\tilde{\varepsilon}} (M(f_{\tilde{\varepsilon}}) - f_{\tilde{\varepsilon}}). \tag{7.4}$$

Now we expand $f_{\tilde{\varepsilon}}$ into a series in the parameter $\tilde{\varepsilon}$

$$f_{\tilde{\varepsilon}} = f_0 + \tilde{\varepsilon} f_1 + \tilde{\varepsilon}^2 f_2 + \cdots. \tag{7.5}$$

Now consider (7.4) in the limit $\tilde{\varepsilon} \to 0$. In this limit $\frac{1}{\tilde{\varepsilon}}(M(f_{\tilde{\varepsilon}}) - f_{\tilde{\varepsilon}})$ must remain finite, because the left-hand side of (7.4) remains finite, too. Therefore $Q(f_{\tilde{\varepsilon}})$ has to converge to zero as $\tilde{\varepsilon} \to 0$. This means that in the limit the remaining distribution function f_0, has to coincide with its Maxwell distribution, so $f_0 = M(f_0)$. Due to conservation of the number of particles, conservation of momentum and conservation of energy, in this limit $\tilde{\varepsilon} \to 0$, the distribution function needs to have the macroscopic quantities n, u and T as $f_{\tilde{\varepsilon}}$. This means that f_0 is a Maxwell distribution with the same density, mean velocity and temperature as $f_{\tilde{\varepsilon}}$, so $f_0 = M(f_{\tilde{\varepsilon}})$. Therefore the density, mean velocity and temperature of the other components are zero

$$0 = m \int f_k dv = m \int v f_k dv = \frac{m}{2} \int |v|^2 f_k dv \quad \text{for} \quad k \geq 1.$$

Consider the macroscopic equations by taking moments from theorem 2.1.9 for one species

$$\partial_t n + \nabla_x \cdot (nu) = 0,$$
$$m \partial_t (nu) + m \nabla_x \cdot (nu \otimes u) + \nabla_x \cdot \mathbb{P} = 0, \tag{7.6}$$
$$\partial_t \left(\frac{3}{2} nT + \frac{1}{2} mn|u|^2 \right) + \nabla_x \cdot Q = 0,$$

where $\mathbb{P} = m \int f(v-u) \otimes (v-u) dv$ denotes the pressure tensor and $Q = \frac{m}{2} \int f|v|^2 v dv$ the heat flux. Since f_0 already has the same number density, mean velocity and temperature as $f_{\tilde{\varepsilon}}$, higher order terms in the expansion of $f_{\tilde{\varepsilon}}$ only influence $\mathbb{P}$ and Q. From (7.4) we get

$$f_{\tilde{\varepsilon}} = M(f_{\tilde{\varepsilon}}) - \tilde{\varepsilon}(\partial_t f_{\tilde{\varepsilon}} + v \cdot \nabla_x f_{\tilde{\varepsilon}}). \tag{7.7}$$

We insert our expansion (7.5) for $f_{\tilde{\varepsilon}}$ with $f_0 = M(f_{\tilde{\varepsilon}})$ into the right-hand side and obtain

$$f_{\tilde{\varepsilon}} = M(f_{\tilde{\varepsilon}}) - \tilde{\varepsilon}(\partial_t M(f_{\tilde{\varepsilon}}) + v \cdot \nabla_x M(f_{\tilde{\varepsilon}})) + O(\tilde{\varepsilon}^2).$$

Then

$$\mathbb{P} = m \int f_{\tilde{\varepsilon}}(v - u) \otimes (v - u) dv$$

$$= nT\mathbb{1} - \tilde{\varepsilon}m \left(\int \partial_t M(f_{\tilde{\varepsilon}})(v - u) \otimes (v - u) dv + \int v \cdot \nabla_x M(f_{\tilde{\varepsilon}})(v - u) \otimes (v - u) dv \right)$$
$$+ O(\tilde{\varepsilon}^2)$$

$$= nT\mathbb{1} - \tilde{\varepsilon}T \left(\nabla_x u + (\nabla_x u)^T - \frac{2}{3}(\nabla_x \cdot u)\mathbb{1} \right) + O(\tilde{\varepsilon}^2).$$

The last equality follows by replacing the time derivative by the zeroth order of the equations (7.6) and then computing the Gaussian integrals. In a similar way, we can compute

$$Q = -\tilde{\varepsilon}T\nabla_x \left(\frac{5}{2} \frac{T}{m} \right) + O(\tilde{\varepsilon}^2).$$

The term of order 1 in $\tilde{\varepsilon}$ in front of the brackets in the pressure tensor is expected from physics to be the viscosity $\mu(T) := \tilde{\varepsilon}T$, and the corresponding term in the heat flux $\kappa(T) := \tilde{\varepsilon}T$ is expected to be the heat conductivity. In this case the ratio $\frac{\mu}{\kappa}$, the Prandtl number, is one which is a contradiction to experiments.

Chapman-Enskog expansion of the ES-BGK equation

Consider $\tilde{\mu} < 1$. We start with the non-dimensional ES-BGK equation

$$\partial_t f_{\tilde{\varepsilon}} + v \cdot \nabla_x f_{\tilde{\varepsilon}} = \frac{1}{\tilde{\varepsilon}}(G(f_{\tilde{\varepsilon}}) - f_{\tilde{\varepsilon}}). \tag{7.8}$$

We expand $f_{\tilde{\varepsilon}}$ into a series in $\tilde{\varepsilon}$:

$$f_{\tilde{\varepsilon}} = f_0 + \tilde{\varepsilon} f_1 + \tilde{\varepsilon}^2 f_2 + \cdots . \tag{7.9}$$

Then the equilibrium distribution is still a Maxwell distribution. This is argued in the following. Suppose we are in equilibrium, then

$$f_{\tilde{\varepsilon}} = G(f_{\tilde{\varepsilon}}). \tag{7.10}$$

If we take the moment of the pressure tensor of equation (7.10), we obtain

$$\frac{\mathbb{P}}{n} = \mathcal{T} = (1 - \tilde{\mu})T\mathbf{1} + \tilde{\mu}\frac{\mathbb{P}}{n},$$

which is equivalent to

$$\frac{\mathbb{P}}{n} = T\mathbf{1},$$

for $\tilde{\mu} < 1$. This means that in equilibrium the tensor is diagonal and the function $G(f_{\tilde{\varepsilon}})$ reduces to a Maxwell distribution. Since the equilibrium function of $f_{\tilde{\varepsilon}}$ is $M(f_{\tilde{\varepsilon}})$, we get in the limit $\tilde{\varepsilon} \to 0$ that $f_0 = M(f_0)$. Since in this case f_0 has to have the density n, the mean velocity u and the temperature T, we have $f_0 = M(f_{\tilde{\varepsilon}})$. So f_0 has the same density, mean velocity and temperature as $f_{\tilde{\varepsilon}}$ and higher orders of $f_{\tilde{\varepsilon}}$ have zero density, mean velocity and temperatures. Combining this with (7.8) and (7.9), we get

$$f_{\tilde{\varepsilon}} = G(f_{\tilde{\varepsilon}}) - \tilde{\varepsilon}(\partial_t M(f_{\tilde{\varepsilon}}) + v \cdot \nabla_x M(f_{\tilde{\varepsilon}})) + O(\tilde{\varepsilon}^2).$$

So now, if we compute the pressure tensor of this expression, we obtain

$$\mathbb{P} = n(1-\tilde{\mu})T\mathbf{1} + \tilde{\mu}\mathbb{P} + \tilde{\varepsilon}nT\nabla_x \cdot \left(m \int (\partial_t M(f_{\tilde{\varepsilon}}) + v \cdot \nabla_x M(f_{\tilde{\varepsilon}}))(v-u) \otimes (v-u)dv \right) + O(\tilde{\varepsilon}^2),$$

which is equivalent to

$$\frac{\mathbb{P}}{n} = T\mathbf{1} + \tilde{\varepsilon}\frac{1}{1-\tilde{\mu}}T\nabla_x \cdot \left(m \int (\partial_t M(f_{\tilde{\varepsilon}}) + v \cdot \nabla_x M(f_{\tilde{\varepsilon}}))(v-u) \otimes (v-u)dv \right) + O(\tilde{\varepsilon}^2)$$

and similarly

$$Q = -\tilde{\varepsilon}T\nabla_x \left(\frac{5}{2}\frac{T}{m} \right) + O(\tilde{\varepsilon}^2).$$

So in this case the viscosity, which is the coefficient in the term of order 1 in $\tilde{\varepsilon}$ in front of ∇_x, is $\tilde{\varepsilon}\frac{T}{1-\tilde{\mu}}$, whereas the heat conductivity, which is the parameter in front of ∇_x in Q, is still $\tilde{\varepsilon}T$, so the Prandtl number, which is the ratio, is $\frac{1}{1-\tilde{\mu}}$. Now, we can choose $\tilde{\mu}$ such that it fits to experiments. Note that it seems that the viscosity coefficient has not the right dependence on the temperature. One ansatz to correct this, is to treat the collision frequency as a free parameter which can be chosen such that the viscosity has the correct physical value, see [48].

7.1.3 The theory of persistence of the velocity

Later, in section 7.6 we want to propose possible extensions to an ES-BGK model for gas mixtures. One of these extensions is based on the following physical theory. It is called theory of the persistence of velocities. It is described in [55, 56, 53, 26]. The theory of persistence of velocities after a collision is a physical phenomenon developed by Jeans in [55]. It states the following. After a collision with another particle the velocity of a given molecule will, on the average, still retain a component in the direction of its original motion. This is explained in the following. Jeans computes a mean speed $\bar{c}'_1(c_1, c_2)$ after collision of a particle with mass m_1 which has the speed c_1 before collision with another particle 2 via

$$\bar{c}'_1(c_1, c_2) = \frac{\int_{S^2} c'_1 \tilde{\nu}_{12}(c_2, \omega) d\omega}{\int_{S^2} \tilde{\nu}_{12}(c_2, \omega) d\omega}. \tag{7.11}$$

The index 2 denotes the other particle with mass m_2 and a fixed but arbitrary value of the second absolute value of the velocity c_2. The average is taken over all possible deflection angles $\omega \in S^2$ and $\tilde{\nu}_{12}(c_2, \omega)$ denotes the probability of a collision.

The case of equal masses

In the case of hard balls and equal masses one can compute the integral in (7.11). For details of this computation see [55, 56, 26]. We obtain

$$\bar{c}'_1(c_1, c_2) = \begin{cases} \frac{15c_1^4 + c_2^4}{10c_1(3c_1^2 + c_2^2)} & c_1 > c_2, \\ \frac{c_1(5c_2^2 + 3c_1^2)}{5(3c_2^2 + c_1^2)} & c_1 < c_2. \end{cases}$$

Then the expression $\frac{\bar{c}'_1(c_1, c_2)}{c_1}$ is called the measure of persistence of the velocity of the first particle. We observe that $\frac{\bar{c}'_1(c_1, c_2)}{c_1}$ depends only on the ratio $\kappa := \frac{c_1}{c_2}$, namely

$$\frac{\bar{c}'_1}{c_1}(\kappa) = \begin{cases} \frac{15\kappa^4 + 1}{10\kappa^2(3\kappa^2 + 1)} & \kappa > 1, \\ \frac{3\kappa^2 + 5}{5(\kappa^2 + 3)} & \kappa < 1. \end{cases}$$

We observe the following estimate:

Lemma 7.1.4. $\frac{\bar{c}'_1}{c_1}$ *satisfies the following estimate*

$$\frac{1}{4} \leq \frac{\bar{c}'_1}{c_1}(\kappa) \leq 1,$$

for all $\kappa \in \mathbb{R}^+$.

Proof. First, let $\kappa > 1$. Then $\frac{\bar{c}'_1}{c_1}(\kappa)$ is bounded from above by a number smaller or equal to 1 since

$$\frac{15\kappa^4 + 1}{10\kappa^2(3\kappa^2 + 1)} \leq \frac{15\kappa^4 + 1}{30\kappa^4} = \frac{1}{2} + \frac{1}{30}\frac{1}{\kappa^4} \leq \frac{1}{2} + \frac{1}{30},$$

and from below by $\frac{1}{3}$ since

$$\frac{15\kappa^4 + 1}{10\kappa^2(3\kappa^2 + 1)} \geq \frac{15\kappa^4 + 1}{10\kappa^2(3\kappa^2 + \kappa^2)} = \frac{15\kappa^4}{40\kappa^4} = \frac{3}{8} \geq \frac{1}{3}.$$

Next, let $\kappa < 1$, then we have

$$\frac{3\kappa^2 + 5}{5(\kappa^2 + 3)} \leq \frac{3\kappa^2 + 5}{15} \leq 1,$$

and

$$\frac{3\kappa^2 + 5}{5(\kappa^2 + 3)} \geq \frac{5}{20} = \frac{1}{4}.$$

$\square$

We observe that for any κ we expect a positive persistence of the velocity before collision which is larger than $\frac{1}{4}$.

Remark 7.1.1. In the case of different masses we obtain from the integral in (7.11) the expression

$$\frac{\bar{c}_1'}{c_1} = \frac{m_1 - m_2}{m_1 + m_2} + \frac{2m_2}{m_1 + m_2}\left(\frac{\bar{c}_1'}{c_1}\right)_e,$$

where $\left(\frac{\bar{c}_1'}{c_1}\right)_e$ denotes the persistence when the masses are equal. So we obtain the following inequality

$$\frac{\bar{c}_1'}{c_1} \geq \frac{m_1 - \frac{1}{2}m_2}{m_1 + m_2},$$

if we use lemma 7.1.4 for the persistence when the masses are equal. So in the case of different masses we observe that it is dependent on the masses, whether we have a persistence or not.

Consequences for the choice of the ES-BGK operator

The choice of the tensor $\mathcal{T} = (1 - \tilde{\mu})T\mathbf{1} + \tilde{\mu}\mathbb{P}$ can be motivated with the theory of persistence of the velocities as follows. This was done by Holway in [54] who is the physicist who invented the ES-BGK model. The theory of persistence of the velocity argues that in the post-collisional speed there is a memory of the pre-collisional speed of the particle. In the single species BGK equation this yields the choice of

$$\mathcal{T} = (1 - \tilde{\mu})T\mathbf{1} + \tilde{\mu}\mathbb{P}, \quad -\frac{1}{2} \leq \tilde{\mu} \leq 1,$$

the tensor chosen in the well-known ES-BGK model, where $\tilde{\mu}\mathbb{P}$ preserves the memory of the off-equilibrium content of the pre-collisional velocity. This can be rewritten as

$$\mathcal{T} = T\mathbf{1} + \tilde{\mu} \, \text{traceless}[\mathbb{P}],$$

where traceless$[\mathbb{P}]$ denotes the traceless part of $\mathbb{P}$. So the off-equilibrium part is contained in $\tilde{\mu} \, \text{traceless}[\mathbb{P}]$.

7.2 The BGK approximation for gas mixtures

In this section we want to briefly repeat the BGK model for a mixture of two species such that it is possible to read this chapter independently. For simplicity we consider in the following a mixture composed of two different species, but the discussion can be generalized to multi-species mixtures. Thus, our kinetic model has two distribution functions $f_1(x, v, t) > 0$ and $f_2(x, v, t) > 0$ where $x \in \Lambda \subset \mathbb{R}^3$ and $v \in \mathbb{R}^3$ are the phase space variables and $t \geq 0$ the time.

Furthermore, for any $f_1, f_2 : \Lambda \times \mathbb{R}^3 \times \mathbb{R}_0^+ \to \mathbb{R}$, $\Lambda \subset \mathbb{R}^3$ with $(1 + |v|^2)f_1, (1 + |v|^2)f_2 \in L^1(dv), f_1, f_2 \geq 0$ we relate the distribution functions to macroscopic quantities by mean-values of f_k, $k = 1, 2$

$$
\int f_k(v) \begin{pmatrix} 1 \\ v \\ m_k|v - u_k|^2 \\ m_k(v - u_k) \otimes (v - u_k) \end{pmatrix} dv =: \begin{pmatrix} n_k \\ n_k u_k \\ 3 n_k T_k \\ \mathbb{P}_k \end{pmatrix}, \quad k = 1, 2, \tag{7.12}
$$

where n_k is the number density, u_k the mean velocity, T_k the temperature and $\mathbb{P}_k$ the pressure tensor of species k, $k = 1, 2$. Note that in this section we shall write T_k instead of $k_B T_k$, where k_B is the Boltzmann constant.

Then we consider the following BGK model for gas mixtures presented in chapter 2:

$$
\begin{aligned}
\partial_t f_1 + v \cdot \nabla_x f_1 &= \nu_{11} n_1 (M_1 - f_1) + \nu_{12} n_2 (M_{12} - f_1), \\
\partial_t f_2 + v \cdot \nabla_x f_2 &= \nu_{22} n_2 (M_2 - f_2) + \nu_{21} n_1 (M_{21} - f_2),
\end{aligned} \tag{7.13}
$$

with the Maxwell distributions

$$
M_k(x, v, t) = \frac{n_k}{\sqrt{2\pi \frac{T_k}{m_k}}^3} \exp\left(-\frac{|v - u_k|^2}{2\frac{T_k}{m_k}}\right), \quad k = 1, 2,
$$

$$
\tag{7.14}
$$

$$
M_{kj}(x, v, t) = \frac{n_{kj}}{\sqrt{2\pi \frac{T_{kj}}{m_k}}^3} \exp\left(-\frac{|v - u_{kj}|^2}{2\frac{T_{kj}}{m_k}}\right), \quad k, j = 1, 2, \; k \neq j,
$$

where $\nu_{11} n_1$ and $\nu_{22} n_2$ are the collision frequencies of the particles of each species with itself, while ν_{12} and ν_{21} are related to interspecies collisions. To be flexible in choosing the relationship between the collision frequencies, we now assume the relationship

$$
\nu_{12} = \varepsilon \nu_{21}, \qquad\qquad 0 < \varepsilon \leq 1, \tag{7.15}
$$

$$
\nu_{11} = \beta_1 \nu_{12}, \quad \nu_{22} = \beta_2 \nu_{21}, \qquad \beta_1, \beta_2 > 0. \tag{7.16}
$$

If we assume that

$$
n_{12} = n_1 \quad \text{and} \quad n_{21} = n_2, \tag{7.17}
$$

$$u_{12} = \delta u_1 + (1 - \delta)u_2, \quad \delta \in \mathbb{R}, \tag{7.18}$$

and

$$T_{12} = \alpha T_1 + (1 - \alpha)T_2 + \gamma |u_1 - u_2|^2, \quad 0 \leq \alpha \leq 1, \gamma \geq 0, \tag{7.19}$$

we have conservation of the number of particles, of total momentum and total energy provided that

$$u_{21} = u_2 - \frac{m_1}{m_2}\varepsilon(1 - \delta)(u_2 - u_1), \tag{7.20}$$

and

$$T_{21} = \left[\frac{1}{3}\varepsilon m_1(1 - \delta)\left(\frac{m_1}{m_2}\varepsilon(\delta - 1) + \delta + 1\right) - \varepsilon\gamma\right]|u_1 - u_2|^2 \tag{7.21}$$
$$+ \varepsilon(1 - \alpha)T_1 + (1 - \varepsilon(1 - \alpha))T_2,$$

see theorem 2.1.1, theorem 2.1.2 and theorem 2.1.3.

We see that without using an ES-BGK extension, we already have three free parameters in (7.18) and (7.19) in order to possibly match coefficients like the Fick constant or the heat conductivity in the Navier-Stokes equations. In order to ensure the positivity of all temperatures, we need to impose restrictions on δ and γ,

$$0 \leq \gamma \leq \frac{m_1}{3}(1 - \delta)\left[(1 + \frac{m_1}{m_2}\varepsilon)\delta + 1 - \frac{m_1}{m_2}\varepsilon\right], \tag{7.22}$$

and

$$\frac{\frac{m_1}{m_2}\varepsilon - 1}{1 + \frac{m_1}{m_2}\varepsilon} \leq \delta \leq 1, \tag{7.23}$$

see theorem 2.1.4 in chapter 2. This summarizes our kinetic model (7.13) of two species that contains three free parameters. More details can be found in chapter 2.

7.3 Coefficients on the Navier-Stokes level

In section 7.1.2 our aim was to fix the right values of the viscosity coefficient and the heat conductivity in the case of one species. In this section we want to present what parameters we want to fix in the two species case. This is described by Groppi, Monica and Spiga in [48]. They expect the following expansion of the velocities according to Fick's law

$$u_s^\varepsilon = \bar{u} - \sum_{r=1}^{2} D_{sr}\nabla_x n_r, \quad s = 1, 2,$$

with the four diffusion coefficients coupled with the three independent constraints

$$D_{12} = D_{21}, \quad \sum_{s=1}^{2} D_{sr} m_s n_s = 0, \quad r = 1, 2,$$

so the full Fick matrix is determined by only one of its entries. The physical meaning of Fick's law is the following. It relates the flux of the fluid to the density. It postulates that the flux goes from regions of high densities to regions of low densities. In the case of the pressure tensor, [48] expects the following expansion

$$\mathbb{P} := \mathbb{P}_1 + \mathbb{P}_2 = (n_1 + n_2)\bar{T}\mathbf{1} - \mu(\nabla_x \bar{u} + \nabla_x \bar{u}^T - \frac{2}{3}\nabla_x \cdot \bar{u}\mathbf{1}),$$

where μ denotes the viscosity coefficient of the gas mixture. In the case of the derivation of the heat flux, they expect the following

$$\tilde{Q} := \tilde{Q}_1 + \tilde{Q}_2 = -\lambda(m_1 + m_2) \sum_{s=1}^{2} \frac{n_s}{m_s} \nabla_x \bar{T} + C \sum_{s=1}^{2} n_s(u_s - \bar{u}),$$

where $\tilde{Q}_k$ is $\tilde{Q}_k = m_k \int (v - u_k)|v - u_k|^2 f_k dv$, $k = 1, 2$ and λ denotes the heat conductivity and C the thermal diffusion parameter. The thermal diffusion parameter indicates the rate of transfer of the heat of a fluid from hot regions to cold regions.

7.4 Chapman-Enskog expansion for the mixture

In this section we will partly perform the Chapman-Enskog expansion for the BGK equation for mixtures (7.13) in order to see where the free parameters α, δ and γ will show up at the Navier-Stokes level. This expansion is also presented by Klingenberg and Pirner in [59].

7.4.1 Dimensionless form of the two species BGK model

In section 6.1.3 we derived the dimensionless form of (7.13) in the one dimensional case in position and velocity. One can also carry out the non-dimensionalization in the three dimensional case in the same way. One obtains

$$\partial_t f_1 + v \cdot \nabla_x f_1 = \frac{1}{\varepsilon_1}(M_1 - f_1) + \frac{1}{\tilde{\varepsilon}_1}(M_{12} - f_1),$$

$$\partial_t f_2 + v \cdot \nabla_x f_2 = \frac{1}{\varepsilon_2}(M_2 - f_2) + \frac{1}{\tilde{\varepsilon}_2}(M_{21} - f_2),$$

(7.24)

where

$$\frac{1}{\tilde{\varepsilon}_1} = \frac{1}{\varepsilon_1} \frac{1}{\beta_1} \frac{n_2}{n_1},$$

$$\frac{1}{\tilde{\varepsilon}_2} = \frac{1}{\varepsilon_1} \frac{1}{\beta_1} \frac{1}{\varepsilon},$$

$$\frac{1}{\varepsilon_2} = \frac{1}{\varepsilon_1} \frac{\beta_2}{\beta_1 \varepsilon} \frac{n_2}{n_1},$$

if we assume $\nu_{11} = \beta_1 \nu_{12}$ and $\nu_{22} = \beta_2 \nu_{21}$ and $\nu_{12} = \varepsilon \nu_{21}$. The parameter ε_1 coming from the non-dimensionalization is given by $\varepsilon_1 = \frac{1}{\beta_1 \bar{\nu}_{12} \bar{t} N / \bar{x}}$. Here, $\bar{t}$ is a typical order of magnitude of the time, $\bar{x}$ a typical order of magnitude of the length, $\bar{\nu}_{12}$ the typical order of magnitude of the collision frequency and N the typical number of particles in the volume $\bar{x}^3$. In the following, we assume that ε_1 is small and consider the limit $\varepsilon_1 \to 0$. The non-dimensionalized Maxwell distributions are given by

$$M_1(x, v, t) = \frac{n_1}{\sqrt{2\pi T_1}^3} \exp\left(-\frac{|v - u_1|^2}{2T_1}\right),$$

$$M_2(x, v, t) = \frac{n_2}{\sqrt{2\pi T_2}^3} \left(\frac{m_2}{m_1}\right)^{\frac{1}{2}} \exp\left(-\frac{|v - u_2|^2}{2T_2} \frac{m_2}{m_1}\right),$$

$$M_{12}(x, v, t) = \frac{n_1}{\sqrt{2\pi T_{12}}^3} \exp\left(-\frac{|v - u_{12}|^2}{2T_{12}}\right), \tag{7.25}$$

$$M_{21}(x, v, t) = \frac{n_2}{\sqrt{2\pi T_{21}}^3} \left(\frac{m_2}{m_1}\right)^{\frac{1}{2}} \exp\left(-\frac{|v - u_{21}|^2}{2T_{21}} \frac{m_2}{m_1}\right),$$

with the non-dimensionalized macroscopic quantities

$$u_{12} = \delta u_1 + (1 - \delta) u_2, \tag{7.26}$$

$$T_{12} = \alpha T_1 + (1 - \alpha) T_2 + \frac{\gamma}{m_1} |u_1 - u_2|^2, \tag{7.27}$$

$$u_{21} = (1 - \frac{m_1}{m_2} \varepsilon (1 - \delta)) u_2 + \frac{m_1}{m_2} \varepsilon (1 - \delta) u_1, \tag{7.28}$$

$$T_{21} = [(1 - \varepsilon(1 - \alpha)) T_2 + \varepsilon(1 - \alpha) T_1]$$
$$+ \left(\frac{1}{3} \varepsilon (1 - \delta) \left(\frac{m_1}{m_2} \varepsilon (\delta - 1) + \delta + 1\right) - \varepsilon \frac{\gamma}{m_1}\right) |u_1 - u_2|^2. \tag{7.29}$$

The macroscopic quantities n_k, u_k and T_k in non-dimensionalized form are given by

$$n_k = \int f_k dv,$$

$$n_k u_k = \int v f_k dv, \quad k = 1, 2,$$

$$T_1 = \frac{1}{3} \frac{1}{n_1} \int |v - u_1|^2 f_1 dv, \tag{7.30}$$

$$T_2 = \frac{1}{3} \frac{m_2}{m_1} \frac{1}{n_2} \int |v - u_2|^2 f_2 dv.$$

7.4.2 Expansion of the distribution functions

Now, we want to do the Chapman-Enskog expansion presented in section 7.1.2 for one species in the two species case. We now expand both f_1 and f_2 in terms of ε_1

$$f_1 = f_1^0 + \varepsilon_1 f_1^1 + \varepsilon_1^2 f_1^2 + \cdots,$$
$$f_2 = f_2^0 + \varepsilon_1 f_2^1 + \varepsilon_1^2 f_2^2 + \cdots.$$

From (7.24) we get in the limit $\varepsilon_1 \to 0$

$$f_1 = \frac{1}{1 + \frac{1}{\tilde{\beta}_1}} \left(M_1 + \frac{1}{\tilde{\beta}_1} M_{12} \right),$$

$$f_2 = \frac{1}{1 + \frac{1}{\tilde{\beta}_2}} \left(M_2 + \frac{1}{\tilde{\beta}_2} M_{21} \right),$$

from which we can deduce that both distribution functions are Maxwell distributions with equal mean velocity and temperature (see theorem 2.1.8).

From (7.24) we get

$$f_1 = \frac{1}{\frac{1}{\varepsilon_1} + \frac{1}{\tilde{\varepsilon}_1}} \left(\frac{1}{\varepsilon_1} M_1 + \frac{1}{\tilde{\varepsilon}_1} M_{12} \right) - \frac{1}{\frac{1}{\varepsilon_1} + \frac{1}{\tilde{\varepsilon}_1}} (\partial_t f_1 + v \cdot \nabla_x f_1),$$

$$f_2 = \frac{1}{\frac{1}{\varepsilon_2} + \frac{1}{\tilde{\varepsilon}_2}} \left(\frac{1}{\varepsilon_2} M_2 + \frac{1}{\tilde{\varepsilon}_2} M_{21} \right) - \frac{1}{\frac{1}{\varepsilon_2} + \frac{1}{\tilde{\varepsilon}_2}} (\partial_t f_2 + v \cdot \nabla_x f_2).$$

$$(7.31)$$

The zeroth order terms of the expansion have the same number density as the distribution functions itself so the number densities are independent of ε_1. But the velocities and temperatures do not coincide, since we expect a common value in equilibrium, so they depend on ε_1. This means the first term in the expansion $\frac{1}{\frac{1}{\varepsilon_1} + \frac{1}{\tilde{\varepsilon}_1}} (\frac{1}{\varepsilon_1} M_1 + \frac{1}{\tilde{\varepsilon}_1} M_{12})$ is not the zeroth order in ε_1 but also contains higher orders. In the one species case we directly obtained that the zeroth order is given by $M(f)$. This helped a lot in the expansion since is was possible to insert $M(f)$ as zeroth order in (7.7). Since we are not able to explicitly specify the zeroth order here, we are not able to do this. But we can find a linear combination of f_1 and f_2 such that the mean velocity and the temperature coincide with the zeroth order of $\frac{1}{\frac{1}{\varepsilon_1} + \frac{1}{\tilde{\varepsilon}_1}} (\frac{1}{\varepsilon_1} M_1 + \frac{1}{\tilde{\varepsilon}_1} M_{12})$ which is similar to equation (7.5) in the one species case. This is done in the next section.

7.4.3 Combination of distribution functions whose macroscopic quantities correspond to the macroscopic quantities of the zeroth order

We now want to find a linear combination of the distribution functions whose mean velocity and temperature are independent of ε_1.

The zeroth order of f_1 is contained in

$$\frac{1}{\frac{1}{\varepsilon_1} + \frac{1}{\tilde{\varepsilon}_1}} \left(\frac{1}{\varepsilon_1} M_1 + \frac{1}{\tilde{\varepsilon}_1} M_{12} \right),$$

and the zeroth order of f_2 is contained in

$$\frac{1}{\frac{1}{\varepsilon_2} + \frac{1}{\tilde{\varepsilon}_2}} \left(\frac{1}{\varepsilon_2} M_2 + \frac{1}{\tilde{\varepsilon}_2} M_{21} \right).$$

So a combination of the distribution functions whose mean velocity and temperature are of zeroth order is obtained if the mean velocity and the temperature of $A f_1 + B f_2$ are equal to the mean velocity and temperature of $\frac{A}{\frac{1}{\varepsilon_1} + \frac{1}{\tilde{\varepsilon}_1}} (\frac{1}{\varepsilon_1} M_1 + \frac{1}{\tilde{\varepsilon}_1} M_{12}) + \frac{B}{\frac{1}{\varepsilon_2} + \frac{1}{\tilde{\varepsilon}_2}} (\frac{1}{\varepsilon_2} M_2 + \frac{1}{\tilde{\varepsilon}_2} M_{21})$. By taking moments we get conditions on A and B. From the velocities we get

$$A n_1 u_1 + B n_2 u_2 = \frac{A}{\frac{1}{\varepsilon_1} + \frac{1}{\tilde{\varepsilon}_1}} \left(\frac{1}{\varepsilon_1} n_1 u_1 + \frac{1}{\tilde{\varepsilon}_1} n_1 u_{12} \right) + \frac{B}{\frac{1}{\varepsilon_2} + \frac{1}{\tilde{\varepsilon}_2}} \left(\frac{1}{\varepsilon_2} n_2 u_2 + \frac{1}{\tilde{\varepsilon}_2} n_2 u_{21} \right)$$

$$= \frac{A}{\frac{1}{\varepsilon_1} + \frac{1}{\tilde{\varepsilon}_1}} \left(\frac{1}{\varepsilon_1} n_1 u_1 + \frac{1}{\tilde{\varepsilon}_1} n_1 \delta u_1 + \frac{1}{\tilde{\varepsilon}_1} n_1 (1 - \delta) u_2 \right)$$

$$+ \frac{B}{\frac{1}{\varepsilon_2} + \frac{1}{\tilde{\varepsilon}_2}} \left(\frac{1}{\varepsilon_2} n_2 u_2 + \frac{1}{\tilde{\varepsilon}_2} n_2 u_2 - \frac{1}{\tilde{\varepsilon}_2} n_2 \frac{m_1}{m_2} \varepsilon (1 - \delta)(u_2 - u_1) \right)$$

$$= \left[\frac{A}{\frac{1}{\varepsilon_1} + \frac{1}{\tilde{\varepsilon}_1}} \left(\frac{1}{\varepsilon_1} + \frac{1}{\tilde{\varepsilon}_1} \delta \right) n_1 + \frac{B}{\frac{1}{\varepsilon_2} + \frac{1}{\tilde{\varepsilon}_2}} \frac{1}{\tilde{\varepsilon}_2} n_2 \frac{m_1}{m_2} \varepsilon (1 - \delta) \right] u_1$$

$$+ \left[\frac{A}{\frac{1}{\varepsilon_1} + \frac{1}{\tilde{\varepsilon}_1}} \left(\frac{1}{\tilde{\varepsilon}_1} (1 - \delta) n_1 + \frac{B}{\frac{1}{\varepsilon_2} + \frac{1}{\tilde{\varepsilon}_2}} \left(\left(\frac{1}{\varepsilon_2} + \frac{1}{\tilde{\varepsilon}_2} \right) n_2 - \frac{1}{\tilde{\varepsilon}_2} n_2 \right) \frac{m_1}{m_2} \varepsilon (1 - \delta) \right) \right] u_2.$$

A comparison of the coefficients in front of u_1 and u_2 leads to

$$A = B \frac{\tilde{\varepsilon}_1}{\tilde{\varepsilon}_2} \frac{m_1}{m_2} \varepsilon \frac{n_2}{n_1} \frac{\frac{1}{\varepsilon_1} + \frac{1}{\tilde{\varepsilon}_1}}{\frac{1}{\varepsilon_2} + \frac{1}{\tilde{\varepsilon}_2}}.$$

If we do the same with the temperatures, we will obtain the same result. We can simplify this expression by using the relationships between ε_1, ε_2, $\tilde{\varepsilon}_1$ and $\tilde{\varepsilon}_2$. First we get

$$\frac{\tilde{\varepsilon}_1}{\tilde{\varepsilon}_2} = \frac{1}{\varepsilon^2} \frac{n_1}{n_2},$$

and secondly

$$\frac{\frac{1}{\varepsilon_1} + \frac{1}{\tilde{\varepsilon}_1}}{\frac{1}{\varepsilon_2} + \frac{1}{\tilde{\varepsilon}_2}} = \frac{\beta_1 \frac{n_1}{n_2} + 1}{\beta_2 \frac{n_2}{n_1} + 1} \varepsilon^2.$$

Multiplying the two expressions we get

$$\frac{\beta_1 \frac{n_1}{n_2} + 1}{\beta_2 \frac{n_2}{n_1} + 1} \frac{n_1}{n_2}.$$

So all in all we get the following simplification

$$A = B \frac{m_1}{m_2} \varepsilon \frac{\beta_1 \frac{n_1}{n_2} + 1}{\beta_2 \frac{n_2}{n_1} + 1}.$$

We choose $B = 1$ and $A = \frac{m_1}{m_2} \varepsilon \frac{\beta_1 \frac{n_1}{n_2} + 1}{\beta_2 \frac{n_2}{n_1} + 1}$ and obtain

$$\frac{m_1}{m_2} \varepsilon \frac{\beta_1 \frac{n_1}{n_2} + 1}{\beta_2 \frac{n_2}{n_1} + 1} f_1 + f_2. \tag{7.32}$$

So

$$\frac{m_1}{m_2} \varepsilon \frac{\beta_1 \frac{n_1}{n_2} + 1}{\beta_2 \frac{n_2}{n_1} + 1} f_1 + f_2 = \bar{f}^0 + \varepsilon_1 \bar{f}^1 + \cdots, \tag{7.33}$$

where $\bar{f}^0$ has the same density, velocity and temperature according to (7.30) as the left hand-side of (7.33) and the moments of $\bar{f}^k$, $k \geq 1$ are zero as in the one species case. We can explicitly compute the macroscopic quantities of $\bar{f}^0$. This is done in the next section.

7.4.4 Macroscopic quantities of the distribution $\bar{f}^0$

The density of $\bar{f}^0$ is given by

$$\bar{n}^0 = \int \bar{f}^0 dv = \int \left(\frac{m_1}{m_2} \varepsilon \frac{\beta_1 \frac{n_1}{n_2} + 1}{\beta_2 \frac{n_2}{n_1} + 1} f_1 + f_2 \right) dv = \frac{m_1}{m_2} \varepsilon \frac{\beta_1 \frac{n_1}{n_2} + 1}{\beta_2 \frac{n_2}{n_1} + 1} n_1 + n_2.$$

Therefore, the mean velocity is given by

$$\bar{u}^0 = \frac{1}{\bar{n}^0} \int \bar{f}^0 v dv = \frac{\frac{m_1}{m_2} \varepsilon \frac{\beta_1 + 1}{\beta_2 \frac{n_2}{n_1} + 1} n_1 u_1 + n_2 u_2}{\frac{m_1}{m_2} \varepsilon \frac{\beta_1 \frac{n_1}{n_2} + 1}{\beta_2 + 1} n_1 + n_2},$$

and the energy is given by

$$\frac{1}{2} \bar{n}^0 |\bar{u}^0|^2 + \frac{3}{2} \bar{n}^0 \frac{\bar{T}^0}{\bar{m}^0} = \int \bar{f}^0 \frac{1}{2} |v|^2 dv$$

$$= \frac{1}{2} A n_1 |u_1|^2 + \frac{1}{2} n_2 |u_2|^2 + \frac{3}{2} n_1 A T_1 + \frac{3}{2} n_2 T_2 \frac{m_1}{m_2},$$

and solving this for $\frac{\bar{T}^0}{\bar{m}^0}$ leads to the temperature

$$\frac{\bar{T}^0}{\bar{m}^0} = \frac{1}{3} \frac{A n_1 n_2}{(A n_1 + n_2)^2} |u_1 - u_2|^2 + \frac{A n_1}{A n_1 + n_2} T_1 + \frac{n_2}{A n_1 + n_2} T_2 \frac{m_1}{m_2}.$$

7.4.5 Combination of the distribution functions whose moments are zero

We observe that in section 7.4.3 we regained a property of the one species case. But since we are in the two species case it is not enough to have one equation for the sum of the two distribution functions. We need a second equation. Since we expect that in equilibrium the mean velocities and the temperatures of the two distribution functions are the same we know that the zeroth order of $\frac{n_2}{n_1} f_1 - f_2$ has zero mean velocity and the zeroth order of the combination $\frac{n_2}{n_1} \frac{m_1}{m_2} f_1 - f_2$ has zero temperature. Therefore, we have

$$\frac{n_2}{n_1} f_1 - f_2 = \tilde{f}^0 + \varepsilon_1 \tilde{f}^1 + \cdots , \tag{7.34}$$

$$\frac{n_2}{n_1} \frac{m_1}{m_2} f_1 - f_2 = \tilde{\tilde{f}}^0 + \varepsilon_1 \tilde{\tilde{f}}^1 + \cdots , \tag{7.35}$$

where $\tilde{f}^0$ has zero mean velocity and $\tilde{\tilde{f}}^0$ has zero temperature.
Solving (7.33) and (7.34) for f_1 and f_2 leads to

$$f_1 = \frac{1}{A + \frac{n_2}{n_1}} \left(\bar{f}^0 + \tilde{f}^0 \right) + \varepsilon_1 \frac{1}{A + \frac{n_2}{n_1}} \left(\bar{f}^1 + \tilde{f}^1 \right) + \varepsilon_1^2 \frac{1}{A + \frac{n_2}{n_1}} \left(\bar{f}^2 + \tilde{f}^2 \right) + O(\varepsilon_1^3),$$
$$\tag{7.36}$$

$$f_2 = \frac{1}{\frac{n_2}{n_1} + A} \left(\frac{n_2}{n_1} \bar{f}^0 - A \tilde{f}^0 \right) + \varepsilon_1 \frac{1}{\frac{n_2}{n_1} + A} \left(\frac{n_2}{n_1} \bar{f}^1 - A \tilde{f}^1 \right)$$
$$+ \varepsilon_1^2 \frac{1}{\frac{n_2}{n_1} + A} \left(\frac{n_2}{n_1} \bar{f}^2 - A \tilde{f}^2 \right) + O(\varepsilon_1^3), \tag{7.37}$$

and solving (7.33) and (7.35) for f_1 and f_2 leads to

$$f_1 = \frac{1}{A + \frac{n_2}{n_1} \frac{m_1}{m_2}} \left(\bar{f}^0 + \tilde{\tilde{f}}^0 \right) + \varepsilon_1 \frac{1}{A + \frac{n_2}{n_1} \frac{m_1}{m_2}} \left(\bar{f}^1 + \tilde{\tilde{f}}^1 \right) + O(\varepsilon_1^2), \tag{7.38}$$

$$f_2 = \frac{1}{\frac{n_2}{n_1} \frac{m_1}{m_2} + A} \left(\frac{n_2}{n_1} \frac{m_1}{m_2} \bar{f}^0 - A \tilde{\tilde{f}}^0 \right) + \varepsilon_1 \frac{1}{\frac{n_2}{n_1} \frac{m_1}{m_2} + A} \left(\frac{n_2}{n_1} \frac{m_1}{m_2} \bar{f}^1 - A \tilde{\tilde{f}}^1 \right)$$
$$+ \varepsilon_1^2 \frac{1}{\frac{n_2}{n_1} \frac{m_1}{m_2} + A} \left(\frac{n_2}{n_1} \frac{m_1}{m_2} \bar{f}^2 - A \tilde{\tilde{f}}^2 \right) + O(\varepsilon_1^2). \tag{7.39}$$

So for $\varepsilon_1 \to 0$ we see from (7.36) and (7.37) that the order zero in ε_1 are two Maxwell distributions f_1^M and f_2^M with the common velocity

$$\bar{u} := \frac{1}{An_1 + n_2} (An_1 u_1 + n_2 u_2),$$

and from (7.38) and (7.39) that the zeroth order are two Maxwell distributions f_1^M and f_2^M with the equal temperature

$$\bar{T} := \frac{1}{An_1 + n_2\frac{m_1}{m_2}}(An_1 + n_2)\frac{\bar{T}^0}{\bar{m}^0}.$$

Remember from the remark below (7.31) that f_1^M has density n_1 and f_2^M has density n_2. Using this we can deduce from (7.31) by inserting the expansions of f_1 and f_2 that

$$f_1 = \frac{1}{\frac{1}{\varepsilon_1} + \frac{1}{\tilde{\varepsilon}_1}}\left(\frac{1}{\varepsilon_1}M_1 + \frac{1}{\tilde{\varepsilon}_1}M_{12}\right) - \frac{1}{\frac{1}{\varepsilon_1} + \frac{1}{\tilde{\varepsilon}_1}}\int(\partial_t f_1^M + v \cdot \nabla_x f_1^M)dv + O(\varepsilon_1^2),$$

$$f_2 = \frac{1}{\frac{1}{\varepsilon_2} + \frac{1}{\tilde{\varepsilon}_2}}\left(\frac{1}{\varepsilon_2}M_2 + \frac{1}{\tilde{\varepsilon}_2}M_{21}\right) - \frac{1}{\frac{1}{\varepsilon_2} + \frac{1}{\tilde{\varepsilon}_2}}\int(\partial_t f_2^M + v \cdot \nabla_x f_2^M)dv + O(\varepsilon_1^2).$$

$$(7.40)$$

7.4.6 Expansion of the velocities, the pressure tensors and the heat fluxes

The exact macroscopic conservation equations that need to be closed in the two species case compared to the one species case in (7.6) read as

$$\partial_t n_1 + \nabla_x \cdot (n_1 u_1) = 0,$$

$$\partial_t n_2 + \nabla_x \cdot (n_2 u_2) = 0,$$

$$\partial_t(n_1 u_1) + \nabla_x \cdot \mathbb{P}_1 + \nabla_x \cdot (n_1 u_1 \otimes u_1) = \frac{1}{\tilde{\varepsilon}_1}(1-\delta)(u_2 - u_1),$$

$$\partial_t(n_2 u_2) + \nabla_x \cdot \mathbb{P}_2\frac{m_1}{m_2} + \nabla_x \cdot (n_2 u_2 \otimes u_2) = \frac{1}{\tilde{\varepsilon}_1}(1-\delta)(u_1 - u_2),$$

$$\partial_t\left(\frac{1}{2}n_1|u_1|^2 + \frac{3}{2}n_1 T_1\right) + \nabla_x \cdot Q_1$$

$$= \frac{1}{2}\frac{1}{\tilde{\varepsilon}_1}((\delta^2-1)|u_1|^2+(1-\delta)^2|u_2|^2+2\delta(1-\delta)u_1 \cdot u_2)+\frac{3}{2}\frac{1}{\tilde{\varepsilon}_1}((1-\alpha)(T_2-T_1)+\frac{\gamma}{m_1}|u_1-u_2|^2),$$

$$\partial_t\left(\frac{1}{2}n_2|u_2|^2 + \frac{3}{2}n_2 T_2\frac{m_1}{m_2}\right) + \nabla_x \cdot Q_2$$

$$= \frac{1}{2}\frac{1}{\tilde{\varepsilon}_1}((\delta^2-1)|u_1|^2+(1-\delta)^2|u_2|^2+2\delta(1-\delta)u_1 \cdot u_2)+\frac{3}{2}\frac{1}{\tilde{\varepsilon}_1}((1-\alpha)(T_2-T_1)+\frac{\gamma}{m_1}|u_1-u_2|^2),$$

according to theorem 2.1.9, where we need expressions for the velocities, pressure tensors and heat fluxes

$$u_k(x,t) = \frac{1}{n_k(x,t)}\int v f_k(x,v,t)dv, \tag{7.41}$$

$$\mathbb{P}_k(x,t) = \frac{m_k}{m_1}\int(v-u_k(x,t)) \otimes (v-u_k(x,t))f_k(x,v,t)dv, \tag{7.42}$$

$$Q_k(x,t) = \frac{m_k}{m_1}\frac{1}{2}\int|v|^2 v f_k(x,v,t)dv, \tag{7.43}$$

for $k = 1, 2$. In the following we want to insert the expansions for $f_1^{\varepsilon_1}$ and $f_2^{\varepsilon_2}$ in these three integrals in order to see if we are able to fix the diffusion coefficient, the viscosity, the heat conductivity and the thermal diffusion parameter described in section 7.3.

Expansion of the velocities

First, we want to consider the expansion of the velocities $u_k^{\varepsilon_1}(x, t)$. If we just insert (7.40) into (7.41), this will lead to expansions of the form

$$
\begin{aligned}
u_1^{\varepsilon_1} &= u_2^{\varepsilon_1} + O(\varepsilon_1), \\
u_2^{\varepsilon_1} &= u_1^{\varepsilon_1} + O(\varepsilon_1).
\end{aligned}
\tag{7.44}
$$

This is in accordance to our expectation that for $\varepsilon_1 \to 0$ the two velocities converge to a common value, but the expansion (7.44) cannot be used to solve it for the two velocities $u_1^{\varepsilon_1}$ and $u_2^{\varepsilon_1}$. In order to do this we need an additional trick. This is done in the following. The velocity of the ion expansion of the first term in (7.40) using the expression of u_{12} is given by

$$
\frac{1 + \frac{1}{\beta_1}\delta}{1 + \frac{1}{\beta_1}} u_1^{\varepsilon_1} + \frac{\frac{1}{\beta_1}(1 - \delta)}{1 + \frac{1}{\beta_1}} u_2^{\varepsilon_1}.
$$

We can split this expression into

$$
\bar{u} + \frac{-An_1(1 - \delta) + n_2(\beta_1 + \delta)}{(\beta_1 + 1)(An_1 + n_2)} u_1^{\varepsilon_1} + \frac{An_1(1 - \delta) - n_2(\beta_1 + \delta)}{(\beta_1 + 1)(An_1 + n_2)} u_2^{\varepsilon_1}.
$$

We denote

$$
c_1 := \frac{-An_1(1 - \delta) + n_2(\beta_1 + \delta)}{(\beta_1 + 1)(An_1 + n_2)},
$$

so

$$
\frac{1 + \frac{1}{\beta_1}\delta}{1 + \frac{1}{\beta_1}} u_1^{\varepsilon_1} + \frac{\frac{1}{\beta_1}(1 - \delta)}{1 + \frac{1}{\beta_1}} u_2^{\varepsilon_1} = \bar{u} + c_1 u_1^{\varepsilon_1} - c_1 u_2^{\varepsilon_1}.
$$

So we see from (7.40) that we get

$$
u_1^{\varepsilon_1} = \bar{u} + c_1 u_1^{\varepsilon_1} - c_1 u_2^{\varepsilon_1} - \frac{1}{\frac{1}{\varepsilon_1} + \frac{1}{\bar{\varepsilon}_1}} \frac{1}{n_1} \int v \left(\partial_t f_1^M + v \cdot \nabla_x f_1^M \right) dv + O(\varepsilon_1^2).
$$

Solving this for $u_1^{\varepsilon_1}$ leads to

$$
\begin{aligned}
u_1^{\varepsilon_1} &= -\frac{c_1}{1 - c_1} u_2^{\varepsilon_1} + \frac{1}{1 - c_1} \bar{u} \\
&\quad - \frac{1}{1 - c_1} \frac{1}{\frac{1}{\varepsilon_1} + \frac{1}{\bar{\varepsilon}_1}} \frac{1}{n_1} \int v \left(\partial_t f_1^M + v \cdot \nabla_x f_1^M \right) dv + O(\varepsilon_1^2).
\end{aligned}
\tag{7.45}
$$

Similarly, we get for the electrons

$$
u_2^{\varepsilon_1} = -\frac{A\frac{n_1}{n_2}c_1}{1 - A\frac{n_1}{n_2}c_1}u_1^{\varepsilon_1} + \frac{1}{1 - A\frac{n_1}{n_2}c_1}\bar{u}
$$
$$
- \frac{1}{1 - A\frac{n_1}{n_2}c_1}\frac{1}{\frac{1}{\varepsilon_2} + \frac{1}{\bar{\varepsilon}_2}}\frac{1}{n_2}\int v\left(\partial_t f_2^M + v\cdot\nabla_x f_2^M\right)dv + O(\varepsilon_1^2).
$$

(7.46)

Solving (7.45) and (7.46) for $u_1^{\varepsilon_1}$ and u_2^{ε} we get

$$
u_1^{\varepsilon_1} = \bar{u} - \frac{1 - A\frac{n_1}{n_2}c_1}{1 - A\frac{n_1}{n_2}c_1 - c_1}\frac{1}{\frac{1}{\varepsilon_1} + \frac{1}{\bar{\varepsilon}_1}}\frac{1}{n_1}\int v\left(\partial_t f_1^M + v\cdot\nabla_x f_1^M\right)dv
$$
$$
+ \frac{c_1}{1 - A\frac{n_1}{n_2}c_1 - c_1}\frac{1}{\frac{1}{\varepsilon_2} + \frac{1}{\bar{\varepsilon}_2}}\frac{1}{n_2}\int v\left(\partial_t f_2^M + v\cdot\nabla_x f_2^M\right)dv + O(\varepsilon_1^2),
$$
$$
u_2^{\varepsilon_1} = \bar{u} - \frac{A\frac{n_1}{n_2}c_1}{-1 + c_1 + A\frac{n_1}{n_2}c_1}\frac{1}{\frac{1}{\varepsilon_1} + \frac{1}{\bar{\varepsilon}_1}}\frac{1}{n_1}\int v\left(\partial_t f_1^M + v\cdot\nabla_x f_1^M\right)dv
$$
$$
+ \frac{1 - c_1}{-1 + c_1 + A\frac{n_1}{n_2}c_1}\frac{1}{\frac{1}{\varepsilon_2} + \frac{1}{\bar{\varepsilon}_2}}\frac{1}{n_2}\int v\left(\partial_t f_2^M + v\cdot\nabla_x f_2^M\right)dv + O(\varepsilon_1^2).
$$

(7.47)

Remark 7.4.1. At this point we observe that we cannot use the parameter γ from (7.27) because $|u_1^{\varepsilon_1} - u_2^{\varepsilon_1}|^2$ is of order ε_1^2 and has no influence on the order ε_1.

Remark 7.4.2. According to section 7.3 we expect to have a diffusion coefficient in front of the integrals in (7.47). We want to be able to fix it to the value which one obtains from experiments. We observe that we have a free parameter in the expression in front of the integrals in (7.47), since the constant c_1 depends on the undetermined parameter δ.

Expansion of the temperatures

In the case of the temperature we do the same trick as in the case of the velocities. The temperature of the first term on the right-hand side in the ion expansion in (7.40) can be written as

$$
\frac{1 + \frac{1}{\beta_1}\alpha}{1 + \frac{1}{\beta_1}}T_1^{\varepsilon_1} + \frac{\frac{1}{\beta_1}(1 - \alpha)}{1 + \frac{1}{\beta_1}}T_2^{\varepsilon_1},
$$

by using the expression of T_{12}. We can split this expression into

$$
\bar{T} + \frac{-An_1(1-\alpha) + \frac{m_1}{m_2}n_2(\beta_1+\alpha)}{(\beta_1+1)(An_1 + \frac{m_1}{m_2}n_2)}T_1^{\varepsilon_1} + \frac{An_1(1-\alpha) - \frac{m_1}{m_2}n_2(\beta_1+\alpha)}{(\beta_1+1)(An_1 + \frac{m_1}{m_2}n_2)}T_2^{\varepsilon_1}.
$$

We denote

$$
c_2 := \frac{-An_1(1-\alpha) + \frac{m_1}{m_2}n_2(\beta_1+\alpha)}{(\beta_1+1)(An_1 + \frac{m_1}{m_2}n_2)},
$$

so

$$\frac{1+\frac{1}{\beta_1}\alpha}{1+\frac{1}{\beta_1}}T_1^{\varepsilon_1} + \frac{\frac{1}{\beta_1}(1-\alpha)}{1+\frac{1}{\beta_1}}T_2^{\varepsilon_1} = \bar{T} + c_2 T_1^{\varepsilon_1} - c_2 T_2^{\varepsilon_1}.$$

We see from (7.40) that

$$T_1^{\varepsilon_1} = \bar{T} + c_2 T_1^{\varepsilon_1} - c_2 T_2^{\varepsilon_1} - \frac{1}{\frac{1}{\varepsilon_1}+\frac{1}{\bar{\varepsilon}_1}}\frac{1}{3n_1}\int |v-\bar{u}|^2 (\partial_t f_1^M + \nabla_x \cdot (v f_1^M)) dv + O(\varepsilon_1^2).$$

Solving this for $T_1^{\varepsilon_1}$ leads to

$$T_1^{\varepsilon_1} = \frac{1}{1-c_2}\bar{T} - \frac{c_2}{1-c_2}T_2^{\varepsilon_1}$$
$$- \frac{1}{\frac{1}{\varepsilon_1}+\frac{1}{\bar{\varepsilon}_1}}\frac{1}{3n_1}\frac{1}{1-c_2}\int |v-\bar{u}|^2 (\partial_t f_1^M + \nabla_x \cdot (v f_1^M)) dv + O(\varepsilon_1^2). \tag{7.48}$$

Similarly, we get for the electrons

$$T_2^{\varepsilon_1} = \frac{1}{1-A\frac{n_1}{n_2}\frac{m_2}{m_1}c_2}\bar{T} - \frac{A\frac{n_1}{n_2}\frac{m_2}{m_1}c_2}{1-A\frac{n_1}{n_2}\frac{m_2}{m_1}c_2}T_1^{\varepsilon_1}$$
$$- \frac{1}{\frac{1}{\varepsilon_2}+\frac{1}{\bar{\varepsilon}_2}}\frac{m_2}{m_1}\frac{1}{3n_2}\frac{1}{1-A\frac{n_1}{n_2}\frac{m_2}{m_1}c_2}\int |v-\bar{u}|^2 (\partial_t f_2^M + \nabla_x \cdot (v f_2^M)) dv + O(\varepsilon_1^2). \tag{7.49}$$

Solving (7.48) and (7.49) for $T_1^{\varepsilon_1}$ and T_2^{ε} we get

$$T_1^{\varepsilon_1} = \bar{T} - \frac{1}{\frac{1}{\varepsilon_1}+\frac{1}{\bar{\varepsilon}_1}}\frac{1-A\frac{n_1}{n_2}\frac{m_2}{m_1}c_2}{1-A\frac{n_1}{n_2}\frac{m_2}{m_1}c_2 - c_2}\frac{1}{3n_1}\int |v-\bar{u}|^2 (\partial_t f_1^M + \nabla_x \cdot (v f_1^M)) dv$$
$$- \frac{1}{\frac{1}{\varepsilon_2}+\frac{1}{\bar{\varepsilon}_2}}\frac{m_2}{m_1}\frac{1}{3n_2}\frac{c_2}{1-A\frac{n_1}{n_2}\frac{m_2}{m_1}c_2 - c_2}\int |v-\bar{u}|^2 (\partial_t f_2^M + \nabla_x \cdot (v f_2^M)) dv + O(\varepsilon_1^2), \tag{7.50}$$

$$T_2^{\varepsilon_1} = \bar{T} - \frac{1}{\frac{1}{\varepsilon_2}+\frac{1}{\bar{\varepsilon}_2}}\frac{1-c_2}{-1+c_2+A\frac{n_1}{n_2}\frac{m_2}{m_1}c_2}\frac{m_2}{m_1}\frac{1}{3n_2}\int |v-\bar{u}|^2 (\partial_t f_2^M + \nabla_x \cdot (v f_2^M)) dv$$
$$- \frac{1}{\frac{1}{\varepsilon_1}+\frac{1}{\bar{\varepsilon}_1}}\frac{1}{3n_1}\frac{A\frac{n_1}{n_2}\frac{m_2}{m_1}c_2}{-1+c_2+A\frac{n_1}{n_2}\frac{m_2}{m_1}c_2}\int |v-\bar{u}|^2 (\partial_t f_1^M + \nabla_x \cdot (v f_1^M)) dv + O(\varepsilon_1^2). \tag{7.51}$$

If we compute the expansion of the non-diagonal elements of the pressure tensor, the zeroth order vanishes and we obtain

$$\int (v_l - u_{1,l})(v_m - u_{1,m}) f_1 dv =$$
$$- \frac{1}{\frac{1}{\varepsilon_1}+\frac{1}{\bar{\varepsilon}_1}}\int (v_l - u_{1,l})(v_m - u_{1,m})(\partial_t f_1^M + \nabla_x \cdot (v f_1^M)) dv + O(\varepsilon_1^2),$$

$$\int (v_l - u_{2,l})(v_m - u_{2,m}) f_2 dv =$$

$$-\frac{1}{\frac{1}{\varepsilon_2} + \frac{1}{\tilde{\varepsilon}_2}} \int (v_l - u_{2,l})(v_m - u_{2,m})(\partial_t f_2^M + \nabla_x \cdot (v f_2^M)) dv + O(\varepsilon_1^2),$$

for $l, m = 1, 2, 3, l \neq m$.

Remark 7.4.3. According to section 7.3 we expect to have a viscosity coefficient in front of the integrals in the expansion of the pressure tensor which we want to be able to fix it to the value which one obtains from experiments. We observe that we have a free parameter in the expression in front of the integrals in (7.50) and (7.51), since the constant c_2 depends on the undetermined parameter α from (7.27).

Expansion of the heat fluxes

If we insert the expansion (7.40) into (7.43) and use the definition of the mixture Maxwell distributions (7.14), we get

$$\frac{1}{2} \int |v|^2 v f_1 dv = \frac{5}{2} \frac{1}{1 + \beta_1 \frac{n_1}{n_2}} n_1$$

$$[(\beta_1 \frac{n_1}{n_2} + \alpha \delta) T_1^{\varepsilon_1} u_1^{\varepsilon_1} + \alpha(1 - \delta) T_1^{\varepsilon_1} u_2^{\varepsilon_1} + (1 - \alpha)\delta T_2^{\varepsilon_1} u_1^{\varepsilon_1} + (1 - \alpha)(1 - \delta) T_2^{\varepsilon_1} u_2^{\varepsilon_1}]$$

$$+ \frac{1}{2} \frac{1}{1 + \beta_1 \frac{n_1}{n_2}} n_1 [(\beta_1 \frac{n_1}{n_2} |u_1^{\varepsilon_1}|^2 u_1^{\varepsilon_1} + |\delta u_1^{\varepsilon_1} + (1 - \delta) u_2^{\varepsilon_1}|^2 (\delta u_1^{\varepsilon_1} + (1 - \delta) u_2^{\varepsilon_1})]$$

$$- \frac{1}{2} \frac{1}{\frac{1}{\varepsilon_1} + \frac{1}{\tilde{\varepsilon}_1}} \int |v|^2 v (\partial_t f_1^M + \nabla_x \cdot (v f_1^M)) dv + O(\varepsilon_1^2),$$

in which we can insert the expansions for the velocities and the temperatures. The zeroth order is given by

$$\frac{5}{2} n_1 (\bar{T}\bar{u} + |\bar{u}|^2 \bar{u}).$$

The heat flux for species 2 is given by

$$\frac{m_2}{m_1} \frac{1}{2} \int |v|^2 v f_2 dv = \frac{5}{2} \frac{1}{1 + \beta_1 \frac{n_1}{n_2}} n_2$$

$$[\beta_2 \frac{n_2}{n_1} T_2^{\varepsilon_1} u_2^{\varepsilon_1} + (\alpha T_1^{\varepsilon_1} + (1 - \alpha) T_2^{\varepsilon_1})(u_2^{\varepsilon_1} - \frac{m_1}{m_2} \varepsilon(1 - \delta)(u_2^{\varepsilon_1} - u_1^{\varepsilon_1}))]$$

$$+ \frac{1}{2} \frac{1}{1 + \beta_2 \frac{n_2}{n_1}} n_1 [\beta_2 \frac{n_2}{n_1} |u_2^{\varepsilon_1}|^2 u_2^{\varepsilon_1}$$

$$+ |u_2^{\varepsilon_1} - \frac{m_1}{m_2} \varepsilon(1 - \delta)(u_2^{\varepsilon_1} - u_1^{\varepsilon_1})|^2 (u_2^{\varepsilon_1} - \frac{m_1}{m_2} \varepsilon(1 - \delta)(u_2^{\varepsilon_1} - u_1^{\varepsilon_1}))]$$

$$- \frac{1}{2} \frac{1}{\frac{1}{\varepsilon_2} + \frac{1}{\tilde{\varepsilon}_2}} \int |v|^2 v (\partial_t f_2^M + \nabla_x \cdot (v f_2^M)) dv + O(\varepsilon_1^2).$$

Here, the zeroth order is given by

$$\frac{5}{2} n_2 (\bar{T}\bar{u} + |\bar{u}|^2 \bar{u}).$$

Remark 7.4.4. According to section 7.3 we expect to have the thermal conductivity and the thermal diffusion parameter in front of the integrals in the expansion of the heat flux which we want to be able to fix to the value which one obtains from experiments. We observe that we do not have more free parameters since α and δ are already fixed in order to obtain the right viscosity and diffusion coefficient. But if we perform the extension of the BGK to an ES-BGK model we will obtain an additional free parameter in the pressure tensor similar as it is done in the one species case. Instead of fixing α as described in remark 7.4.3, this allows to fix the additional parameter such that we get the right expression of the viscosity coefficient. Then the parameter α remains undetermined and we can use it to fix the thermal conductivity in the heat flux. A fourth free parameter is gained if we treat the one collision frequency as a free parameter as it is done in [48]. With this we can determine the diffusion coefficient such that the parameter δ remains undetermined and we can fix it in the heat flux expansion such that the thermal diffusion parameter has the right physical value.

7.5 The Brunn-Minkowski inequality

In the following we want to introduce ES-BGK models for mixtures. First of all we want to prove consistency of these models meaning that they satisfy the conservation properties and the H-theorem. For the H-theorem we will need the next lemma which is proven in [2]. For the convenience of the reader we will repeat it here.

Lemma 7.5.1 (Brunn-Minkowski inequality). *Let $0 \leq b \leq 1$ and A, B be two positive symmetric matrices with the same basis of eigenvectors. Then*

$$\det(bA + (1 - b)B) \geq (\det A)^b (\det B)^{1-b}. \tag{7.52}$$

Proof. The proof is given in [2]. For the convenience of the reader we want to repeat it here. If A and B both have an eigenvalue zero, the inequality is satisfied. So without loss of generality all eigenvalues of A are different from zero. Then A is invertible, and we have

$$\det(bA + (1 - b)B) = \det A \det(b\mathbb{1} + (1 - b)A^{-1}B). \tag{7.53}$$

Denote $C = A^{-1}B$. Then C is diagonalizable and in particular it has the same set of eigenvectors than A and B, since A and B are simultaneously diagonalizable.. We denote the eigenvalues of C by c_i. We observe that all c_i are strictly positive since they are the quotients of the strictly positive eigenvalues of A and B. Then (7.52) is equivalent to

$$\det(b\mathbb{1} + (1 - b)C) \geq (\det C)^{1-b} \tag{7.54}$$

using (7.53). So we want to prove (7.54). This is equivalent to

$$\prod_i (b + (1 - b)c_i) \geq \prod_i c_i^{1-b}.$$

We take the logarithm, which is possible, since b and c_i are strictly positive:

$$\sum_i \ln(b + (1-b)c_i) \geq (1-b) \sum_i \ln c_i.$$

This is satisfied, since ln is a concave function. So (7.54) is true. Replacing C by $A^{-1}B$, we get

$$\det(b\mathbb{1} + (1-b)A^{-1}B) \geq (\det A^{-1}B)^{1-b} = (\det A)^{-(1-b)}(\det B)^{1-b}$$
$$= (\det A)^{b-1}(\det B)^{1-b},$$

which is the claimed inequality. □

Actually, we will need the following extension of the Brunn-Minkowski inequality.

Lemma 7.5.2 (Extension of the Brunn-Minkowski inequality). *Let $-\frac{1}{2} \leq a \leq 1$ and A, E, D positive symmetric matrices. Then*

$$\det\left(\frac{1+2a}{3}A + \frac{1-a}{3}E + \frac{1-a}{3}D\right) \geq (\det A)^{\frac{1+2a}{3}}(\det E)^{\frac{1-a}{3}}(\det D)^{\frac{1-a}{3}}.$$

Proof. We have

$$\det\left(\frac{1+2a}{3}A + \frac{1-a}{3}E + \frac{1-a}{3}D\right) = \det\left(\frac{1+2a}{3}A + 2\frac{1-a}{3}\frac{1}{2}(E+D)\right).$$

Choose $b = \frac{1+2a}{3}$ and $B = \frac{1}{2}(E+D)$. Since $-\frac{1}{2} \leq a \leq 1$, b is restricted to $0 \leq b \leq 1$. Then by the Brunn-Minkowski inequality (7.52), we get

$$\det\left(\frac{1+2a}{3}A + \frac{1-a}{3}E + \frac{1-a}{3}D\right) \geq (\det A)^{\frac{1+2a}{3}}\left(\det\left(\frac{1}{2}(E+D)\right)\right)^{2\frac{1-a}{3}}.$$

Again by the Brunn-Minkowski inequality for $b = \frac{1}{2}$ on the second term, we obtain

$$\det\left(\frac{1+2a}{3}A + \frac{1-a}{3}E + \frac{1-a}{3}D\right) \geq (\det A)^{\frac{1+2a}{3}}(\det E)^{\frac{1-a}{3}}(\det D)^{\frac{1-a}{3}}.$$

□

7.6 Extensions to an ES-BGK approximation

In this section we want to present three possible extensions to an ES-BGK model for gas mixtures. The first one has the attempt to keep it as simple as possible and extend only the Maxwell distribution in the single relaxation term describing the relaxation of the distribution function to an equilibrium distribution due to interactions of the species with itself. The two other extensions try to do it more symmetrically and extend every Maxwell distribution. The first ansatz in this case is to extend it exactly analogously as in the one species case and the second ansatz proposes a different extension which is motivated by the physical intuition of the physicist Holway who invented the ES-BGK model in [54]. These extensions are also presented by Klingenberg, Pirner and Puppo in [62].

7.6.1 Extension of the single relaxation terms

Motivated by the need to find a two species kinetic model that allows us to model physical parameters better we extend the above model by generalizing the Maxwell distributions. The simplest choice is to replace only the collision operators which represent the collisions of a species with itself by the ES-BGK collision operator for one species as suggested in [3]. Then the model can be written as:

$$\partial_t f_k + v \cdot \nabla_x f_k = \nu_{kk} n_k (G_k - f_k) + \nu_{kj} n_j (M_{kj} - f_k), \quad k, j = 1, 2, \ j \neq k, \quad (7.55)$$

with the modified Maxwell distributions

$$G_k(x, v, t) = \frac{n_k}{\sqrt{\det(2\pi \frac{T_k}{m_k})}} \exp\left(-\frac{1}{2}(v - u_k) \cdot \left(\frac{T_k}{m_k}\right)^{-1} \cdot (v - u_k)\right), \quad k = 1, 2, \quad (7.56)$$

and M_{12}, M_{21} the Maxwell distributions described in the previous sections. G_1 and G_2 have the same densities, velocities and pressure tensors as f_1 and f_2, respectively, so we still guarantee the conservation of mass, momentum and energy in interactions of one species with itself. Since the first term describes the interactions of a species with itself, it should correspond to the single ES-BGK collision operator suggested in section 7.1. So we choose T_1 and T_2 as

$$T_k = (1 - \tilde{\mu}_k) T_k \mathbf{1} + \tilde{\mu}_k \frac{\mathbb{P}_k}{n_k}, \quad (7.57)$$

with $\tilde{\mu}_k \in \mathbb{R}$, $k = 1, 2$ being free parameters which we can choose in a way to fix physical parameters in the Navier-Stokes equations. So, all in all, together with the parameters in the mixture Maxwell distributions (7.18) and (7.19) we now have five free parameters, see (7.18) and (7.19) for the other free parameters.

Since we wrote T_k^{-1} we have to check whether T_k is invertible. Otherwise the model is not well-posed. For the one species tensor this is done by the following theorem proven in [3].

Theorem 7.6.1. *Assume that $f_k > 0$. Then $\frac{\mathbb{P}_k}{n_k}$ has strictly positive eigenvalues. If we further assume that $-\frac{1}{2} \leq \mu_k \leq 1$, then T_k has strictly positive eigenvalues and therefore T_k is invertible.*

Equilibrium of the single extension

In global equilibrium when f_1 and f_2 are independent of x and t, the right-hand side of (7.55) has to be zero. In this case we get

$$f_1 = \frac{1}{\nu_{11} n_1 + \nu_{12} n_2} (\nu_{11} n_1 G_1 + \nu_{12} n_2 M_{12}).$$

If we compute the velocities of this expression, we can deduce $u_1 = u_2$ for $\delta \neq 1$. If we compute the temperatures of this expression using $u_1 = u_2$, we get

$$T_1 = \frac{1}{\nu_{11}n_1 + \nu_{12}n_2}(\nu_{11}n_1 T_1 + \nu_{12}n_2(\alpha T_1 + (1-\alpha)T_2)),$$

which is equivalent to $T_1 = T_2$ for $\alpha \neq 1$. So let $T := T_1 = T_2$ and use $u_1 = u_2$. If we compute the pressure tensors, we get

$$(\nu_{11}n_1 + \nu_{12}n_2)\mathbb{P}_1 = \nu_{11}n_1 \mathcal{T}_1 + \nu_{12}n_2 T_{12}$$
$$= \nu_{11}n_1(1 - \tilde{\mu}_1)T\mathbf{1} + \nu_{11}n_1\tilde{\mu}_1\mathbb{P}_1 + \nu_{12}n_2 T\mathbf{1},$$

which is equivalent to

$$(\nu_{11}n_1 + \nu_{12}n_2 - \nu_{11}n_1\tilde{\mu}_1)\mathbb{P}_1 = (\nu_{11}n_1 + \nu_{12}n_2 - \nu_{11}n_1\tilde{\mu}_1)T\mathbf{1},$$

which is $\mathbb{P}_1 = T\mathbf{1}$ for $\delta, \alpha \neq 1$, $\tilde{\mu}_1 \leq 1$. This means that the pressure tensors of f_1 and f_2 are diagonal and f_1, f_2 are Maxwell distributions with equal mean velocity and temperature. $\delta = 1$ or $\alpha = 1$ are cases in which the mixture Maxwell distributions do not contain the velocity or the temperature of the other species, see (7.18) and (7.19). In this case the two gases do not exchange information and a global equilibrium cannot be reached.

Entropy inequality of the single extension

Theorem 7.6.2 (H-theorem for the mixture). *Assume $\delta, \alpha \neq 1$. Assume that $f_1, f_2 > 0$ are solutions to (7.13). Assume the relationship between the collision frequencies (7.16), the conditions for the interspecies Maxwell distributions (7.18), (7.20), (7.19) and (7.21) and the positivity of the temperatures (7.22). Denote the collision terms on the right-hand side of (7.13) by $Q_{11}(f_1, f_1), Q_{12}(f_1, f_2), Q_{21}(f_2, f_1)$ and $Q_{22}(f_2, f_2)$. Then*

$$\int \ln f_1 \, Q_{11}(f_1, f_1) + \ln f_1 \, Q_{12}(f_1, f_2)dv + \int \ln f_2 \, Q_{22}(f_2, f_2) + \ln f_2 \, Q_{21}(f_2, f_1)dv \leq 0,$$

with equality if and only if f_1 and f_2 are Maxwell distributions with equal mean velocity and temperature.

Proof. The fact that $\int \ln f_k \, Q_{kk}(f_k, f_k)dv \leq 0$, $k = 1, 2$ with a criteria for equality follows from the H-theorem of the ES-BGK model for one species, see [3]. The fact that $\int \ln f_1 \, Q_{12}(f_1, f_2)dv + \int \ln f_2 \, Q_{21}(f_1, f_2)dv \leq 0$ with a corresponding criteria for equality follows from the H-theorem of the BGK model for mixtures, see the proof of theorem 2.1.6. $\qquad\square$

7.6.2 Alternative extensions to an ES-BGK model

In this section we also want to replace the scalar temperatures in the mixture Maxwell distributions by a tensor. In the first model the terms $(v_j - u_{kj})f_k(v_i - u_{ki})$ for $i \neq j$

do not appear in the relaxation operator. To obtain a more detailed description of the viscous effects in the mixture we take into account these cross terms during the relaxation process. Then the model can be written as:

$$\partial_t f_k + v \cdot \nabla_x f_k = \nu_{kk} n_k (G_k - f_k) + \nu_{kj} n_j (G_{kj} - f_k), \quad k = 1, 2, k \neq j, \quad (7.58)$$

with the modified Maxwell distributions

$$G_k(x,v,t) = \frac{n_k}{\sqrt{\det(2\pi \frac{\mathcal{T}_k}{m_k})}} \exp\left(-\frac{1}{2}(v-u_k)\cdot\left(\frac{\mathcal{T}_k}{m_k}\right)^{-1}\cdot(v-u_k)\right) \quad k=1,2,$$

$$G_{kj}(x,v,t) = \frac{n_k}{\sqrt{\det(2\pi \frac{\mathcal{T}_{kj}}{m_k})}} \exp\left(-\frac{1}{2}(v-u_{kj})\cdot\left(\frac{\mathcal{T}_{kj}}{m_k}\right)^{-1}\cdot(v-u_{kj})\right) \quad k,j=1,2,k\neq j,$$

$$(7.59)$$

with $\mathcal{T}_k$ defined by (7.57). Again, the conservation of mass, momentum and energy in interactions of one species with itself is ensured by this choice of the modified Maxwell distributions G_1 and G_2 which have the same densities, velocities and pressure tensors as f_1 and f_2, respectively. In addition, the choice of the densities in G_{12} and G_{21} also guarantees conservation of mass in interactions of one species with the other one.

If we extend T_{12} and T_{21} in the same fashion to a tensor as in the case of one species, we obtain

$$\mathcal{T}_{12} = (1-\tilde{\mu}_{12})(\alpha T_1 + (1-\alpha)T_2)\mathbf{1} + \tilde{\mu}_{12}\frac{\alpha \mathbb{P}_1 + (1-\alpha)\mathbb{P}_2}{n_1} + \gamma|u_1-u_2|^2\mathbf{1}, \quad (7.60)$$

$$\mathcal{T}_{21} = (1-\tilde{\mu}_{21})((1-\varepsilon(1-\alpha))T_2 + \varepsilon(1-\alpha)T_1)\mathbf{1} + \tilde{\mu}_{21}\frac{(1-\varepsilon(1-\alpha))\mathbb{P}_2 + \varepsilon(1-\alpha)\mathbb{P}_1}{n_2}$$

$$+ \left(\frac{1}{3}\varepsilon m_1(1-\delta)\left(\frac{m_1}{m_2}\varepsilon(\delta-1)+\delta+1\right)-\varepsilon\gamma\right)|u_1-u_2|^2\mathbf{1}. \quad (7.61)$$

If we check the equilibrium distributions as in section 7.6.1, we obtain the following restrictions on $\tilde{\mu}_{12}$ and $\tilde{\mu}_{21}$ given by

$$\tilde{\mu}_{12} = 1 + (1-\tilde{\mu}_1)\frac{n_1}{n_2}\frac{\nu_{11}}{\nu_{12}}, \quad (7.62)$$

and

$$\frac{1}{n_1^2}[-(\alpha-1)^2\tilde{\mu}_{12}^2 n_2^2\nu_{12}^2 + \frac{n_1}{n_2^2}((\frac{\tilde{\mu}_{21}}{\varepsilon} - \tilde{\mu}_{21} + \alpha\tilde{\mu}_{21})n_1\nu_{12} + (\tilde{\mu}_2-1)n_2\nu_{22})$$

$$\cdot(n_1((\alpha-1)\tilde{\mu}_{21}n_1 + \frac{1}{\varepsilon}(\mu_{21}-1)n_2)\nu_{12} + (\tilde{\mu}_2-1)n_2^2\nu_{22})] = 0. \quad (7.63)$$

An alternative choice to (7.60),(7.61) is given by

$$\mathcal{T}_{12} = \alpha\frac{\mathbb{P}_1}{n_1} + (1-\alpha)T_2\mathbf{1} + \gamma|u_1-u_2|^2\mathbf{1}, \quad (7.64)$$

$$\mathcal{T}_{21} = (1-\varepsilon(1-\alpha))\frac{\mathbb{P}_2}{n_2} + \varepsilon(1-\alpha)T_1\mathbf{1}$$

$$+ \left(\frac{1}{3}\varepsilon m_1(1-\delta)\left(\frac{m_1}{m_2}\varepsilon(\delta-1)+\delta+1\right)-\varepsilon\gamma\right)|u_1-u_2|^2\mathbf{1}. \quad (7.65)$$

This choice of $\mathcal{T}_{12}$ still contains the temperature of gas 1, since the trace of the pressure tensor $\frac{\mathbb{P}_1}{n_1}$ is the temperature T_1.

In (7.64) compared to (7.60) we replace only the temperature T_1 of species 1 by the pressure tensor $\mathbb{P}_1$ while we keep the temperature T_2. This asymmetric choice can be motivated by the theory of persistence of velocity described by Jeans in [55], [56], [53] and chapter 7.1.3. Jeans argues that in the post-collisional speed of particle 1 there is a memory of the pre-collisional speed of particle 1. In the single species BGK equation this yields to the choice of

$$\mathcal{T} = (1 - \tilde{\mu})T\mathbf{1} + \tilde{\mu}\mathbb{P}, \quad -\frac{1}{2} \le \tilde{\mu} \le 1,$$

the tensor chosen in the well-known ES-BGK model, where $\tilde{\mu}\,\mathbb{P}$ preserves the memory of the off-equilibrium content of the pre-collisional velocity. This can be rewritten as

$$\mathcal{T} = T\mathbf{1} + \tilde{\mu}\,\text{traceless}[\mathbb{P}],$$

where traceless$[\mathbb{P}]$ denotes the traceless part of $\mathbb{P}$. So the off-equilibrium part is contained in $\tilde{\mu}\,\text{traceless}[\mathbb{P}]$. Doing this analogously for two species we arrive at

$$\mathcal{T}_{12} = T_{12}\mathbf{1} + \frac{\alpha}{n_1}\text{traceless}[\mathbb{P}_1].$$

The quantity T_{12} is defined in (7.19). If we plug this definition in the equation above, we end up with (7.64).

With the second choice the model is well-defined, because $\mathcal{T}_{12}$ and $\mathcal{T}_{21}$ are invertible as a combination of strictly positive matrices as soon as all coefficients in front of these matrices are positive, which is the case due to (7.22) and (7.23). The first choice needs additional conditions coming from the restrictions on $\tilde{\mu}_{12}$ and $\tilde{\mu}_{21}$ given by (7.62) and (7.63). The first one leads to

$$\tilde{\mu}_1 \le \frac{n_2}{n_1}\frac{\tilde{\nu}_{12}}{\nu_{11}} + 1,$$

such that $\tilde{\mu}_{12}$ given by (7.62) is positive. The requirement of positivity of $\tilde{\mu}_{21}$ leads to a corresponding restriction on $\tilde{\mu}_2$ using (7.63).

Equilibrium of the alternative extensions

The aim of this section is to discuss the property of the equilibrium and the entropy inequality for the alternative extensions described in section 7.6.2 with the tensors (7.60), (7.61) and (7.64), (7.65), respectively. For the tensors (7.60), (7.61) we proved the property of equilibrium and the H-theorem in section 7.6.1 in the particular case of $\tilde{\mu}_{12} = \tilde{\mu}_{21} = 0$ for simplicity, but we can also prove it in the general case. In this section we will prove an entropy inequality for the alternative model (7.64),(7.65). First we will check that the equilibrium distributions are Maxwell distributions. In global equilibrium when f_1 and f_2 are independent of x and t, the right-hand side of (7.58) has to be zero. In this case we get

$$f_1 = \frac{1}{1 + \frac{1}{\beta_1^2}\frac{n_2}{n_1}}\left(G_1 + \frac{1}{\beta_1^2}\frac{n_2}{n_1}G_{12}\right).$$

If we compute the temperatures of this expression, we get

$$T_1 = \frac{1}{1 + \frac{1}{\beta_1^2}\frac{n_2}{n_1}}(T_1 + \frac{1}{\beta_1^2}\frac{n_2}{n_1}(\alpha T_1 + (1-\alpha)T_2)),$$

which is equivalent to $T_1 = T_2$ for $\alpha \neq 1$. So denote $T := T_1 = T_2$. If we compute pressure tensors, we get

$$(1 + \frac{1}{\beta_1^2}\frac{n_2}{n_1})\mathbb{P}_1 = \mathcal{T}_1 + \frac{1}{\beta_1^2}\frac{n_2}{n_1}\mathcal{T}_{12}$$

$$= (1 - \nu_1)T + \nu_1\mathbb{P}_1 + \frac{1}{\beta_1^2}\frac{n_2}{n_1}\alpha\mathbb{P}_1 + \frac{1}{\beta_1^2}\frac{n_2}{n_1}(1-\alpha)T\mathbf{1},$$

which is equivalent to

$$((1 - \nu_1) + \frac{1}{\beta_1^2}\frac{n_2}{n_1}(1-\alpha))\mathbb{P}_1 = ((1 - \nu_1) + \frac{1}{\beta_1^2}\frac{n_2}{n_1}(1-\alpha))T\mathbf{1},$$

which leads to $\mathbb{P}_1 = T\mathbf{1}$ for $\nu_1, \alpha \neq 1$. That means that the pressure tensors of f_1 and f_2 are diagonal and they are Maxwell distributions with equal mean velocity and temperature.

Entropy inequality of the alternative extensions

Next, we want to prove the H-theorem of the model (7.64) and (7.65).

Lemma 7.6.3. *Assuming (7.64) and (7.65) and the positivity of all temperatures and pressure tensors (7.22), we have the following inequality*

$$S := (\det \mathcal{T}_{12})^\varepsilon(\det \mathcal{T}_{21}) \geq \left(\det \frac{\mathbb{P}_1}{n_1}\right)^\varepsilon \det \frac{\mathbb{P}_2}{n_2}.$$

Proof. Using the definition of $\mathcal{T}_{12}$ we get

$$\det \mathcal{T}_{12} = \det \left(\alpha\frac{\mathbb{P}_1}{n_1} + (1-\alpha)T_2\mathbf{1} + \gamma|u_1 - u_2|^2\mathbf{1}\right).$$

Since γ is non-negative, we can estimate the expression by dropping the positive term on the diagonal $\gamma|u_1 - u_2|^2\mathbf{1}$

$$\det \mathcal{T}_{12} \geq \det \left(\alpha\frac{\mathbb{P}_1}{n_1} + (1-\alpha)T_2\mathbf{1}\right).$$

With the Brunn-Minkowski-inequality (7.52) presented in chapter 7.5 we obtain

$$\det \mathcal{T}_{12} \geq \left(\det \frac{\mathbb{P}_1}{n_1}\right)^\alpha (\det T_2\mathbf{1})^{1-\alpha}.$$

In a similar way, we can show a similar inequality for $\mathcal{T}_{21}$, so all in all we get

$$S \geq \left(\det \frac{\mathbb{P}_1}{n_1} \right)^{\alpha \varepsilon} (\det T_2 \mathbf{1})^{\varepsilon(1-\alpha)} \left(\det \frac{\mathbb{P}_2}{n_2} \right)^{1-\varepsilon(1-\alpha)} (\det T_1 \mathbf{1})^{\varepsilon(1-\alpha)}.$$

Consider the logarithm of this equation

$$\ln S \geq \varepsilon \alpha \ln \left(\det \left(\frac{\mathbb{P}_1}{n_1} \right) \right) + \varepsilon(1-\alpha) \ln \left(\det \left(T_2 \mathbf{1} \right) \right)$$

$$+ (1 - \varepsilon(1-\alpha)) \ln \left(\det \left(\frac{\mathbb{P}_2}{n_2} \right) \right) + \varepsilon(1-\alpha) \ln \left(\det \left(T_1 \mathbf{1} \right) \right).$$

We use that $\ln (\det (T_i \mathbf{1})) = \text{Tr}(\ln (T_i \mathbf{1}))$ and denote the eigenvalues of $\frac{\mathbb{P}_i}{n_i}$ by $\lambda_{i,1}, \lambda_{i,2}$ and $\lambda_{i,3}$. Since the pressure tensors are symmetric, we can diagonalize them and use that $T_i = \text{Tr} \frac{\mathbb{P}_i}{3n_i} = \frac{1}{3}(\lambda_{i,1} + \lambda_{i,2} + \lambda_{i,3})$.

$$\ln S \geq \varepsilon \alpha (\ln \lambda_{1,1} + \ln \lambda_{1,2} + \ln \lambda_{1,3}) + \varepsilon(1-\alpha) 3 \ln \frac{1}{3}(\lambda_{1,1} + \lambda_{1,2} + \lambda_{1,3})$$

$$+ (1 - \varepsilon(1-\alpha))(\ln \lambda_{2,1} + \ln \lambda_{2,2} + \ln \lambda_{2,3}) + \varepsilon(1-\alpha) 3 \ln \frac{1}{3}(\lambda_{2,1} + \lambda_{2,2} + \lambda_{2,3}).$$

Since $\ln$ is concave, we can estimate $\ln \frac{1}{3}(\lambda_{1,1} + \lambda_{1,2} + \lambda_{1,3})$ from below by $\frac{1}{3}(\ln \lambda_{1,1} + \ln \lambda_{1,2} + \ln \lambda_{1,3})$ and obtain

$$\ln S \geq \varepsilon \ln \left(\det \left(\frac{\mathbb{P}_1}{n_1} \right) \right) + \varepsilon(1-\alpha) \ln \left(\det \left(\frac{\mathbb{P}_2}{n_2} \right) \right).$$

This is equivalent to the required inequality. $\qquad \square$

Remark 7.6.1. From the case of one species ES-BGK model it follows that

$$\int G_k \ln G_k dv \leq \int G_{k,\tilde{\mu}_k=1} \ln G_{k,\tilde{\mu}_k=1} dv \leq \int f_k \ln f_k dv,$$

for $k = 1, 2$, see [3], where $G_{k,\tilde{\mu}_k=1}$ denotes the modified Maxwell distribution where in the case of $\tilde{\mu}_k = 1$ the tensor $\mathcal{T}_k$ is given by (7.57).

Theorem 7.6.4 (H-theorem for the mixture). *Assume $\alpha, \delta \neq 1$. Assume $f_1, f_2 > 0$. Assume the relationship between the collision frequencies (7.16), the conditions for the interspecies Maxwell distribution (7.18), (7.20), (7.64) and (7.65) and the positivity of the temperatures (7.22), then*

$$\int \ln f_1 \, Q_{11}(f_1, f_1) + \ln f_1 \, Q_{12}(f_1, f_2) dv + \int \ln f_2 \, Q_{22}(f_2, f_2) + \ln f_2 \, Q_{21}(f_2, f_1) dv \leq 0,$$

with equality if and only if f_1 and f_2 are Maxwell distributions with equal mean velocity and temperature.

Proof. The fact that $\int \ln f_k Q_{kk}(f_k, f_k) dv \leq 0$, $k = 1, 2$ is shown in the proof of the H-theorem of the single ES-BGK-model, for example in [3]. In both cases we have equality if and only if $f_1 = M_1$ and $f_2 = M_2$.
Let us define

$$S(f_1, f_2) := \nu_{12} n_2 \int \ln f_1 (G_{12} - f_1) dv + \nu_{21} n_1 \int \ln f_2 (G_{21} - f_2) dv.$$

The task is to prove that $S(f_1, f_2) \leq 0$. Consider now $S(f_1, f_2)$ and apply the inequality in lemma 1.3.12 to each of the two terms in S.

$$S(f_1, f_2) \leq \nu_{12} n_2 \left[\int G_{12} \ln G_{12} dv - \int f_1 \ln f_1 dv - \int G_{12} dv + \int f_1 dv \right]$$

$$+ \nu_{21} n_1 \left[\int G_{21} \ln G_{21} dv - \int f_2 \ln f_2 dv - \int G_{21} dv + \int f_2 dv \right],$$

with equality if and only if $f_1 = G_{12}$ and $f_2 = G_{21}$. If we compute the velocities of $f_1 = G_{12}$ and $f_2 = G_{21}$, we can deduce $u_1 = u_{12}$ and $u_2 = u_{21}$ which leads to $u_1 = u_2$ using the definitions of u_{12}, u_{21} given by (7.18) and (7.20). Analogously, computing the temperatures, we get $T_{12} = T_{21} = T_1 = T_2 =: T$. Finally, computing the pressure tensors, we obtain $\frac{\mathbb{P}_1}{n_1} = \frac{\mathbb{P}_2}{n_2} = T\mathbf{1}$, which means that we have equality if and only if f_1 and f_2 are Maxwell distributions with equal temperatures and velocities.
Since G_{12} and f_1 have the same density and G_{21} and f_2 have the same density too, the right-hand side reduces to

$$\nu_{12} n_2 \left(\int G_{12} \ln G_{12} dv - \int f_1 \ln f_1 dv \right) + \nu_{21} n_1 \left(\int G_{21} \ln G_{21} dv - \int f_2 \ln f_2 dv \right).$$

Since

$$\int G \ln G dv = n \ln \left(\frac{n}{\sqrt{\det(\frac{2\pi T}{m})}} \right) - \frac{3}{2} n \quad \text{for} \quad G = \frac{n}{\sqrt{\det(\frac{2\pi T}{m})}} e^{-(v-u)\cdot(\frac{T}{m})^{-1}\cdot(v-u)},$$

we will have that

$$\nu_{12} n_2 \int G_{12} \ln G_{12} dv + \nu_{21} n_1 \int G_{21} \ln G_{21} dv$$

$$\leq \nu_{21} n_1 \int G_{2,\tilde{\mu}_2=1} \ln M_{2,\tilde{\mu}_2=1} dv + \nu_{12} n_2 \int G_{1,\tilde{\mu}_1=1} \ln G_{1,\tilde{\mu}_1=1} dv,$$

provided that

$$\nu_{12} n_2 n_1 \ln \frac{n_1}{\sqrt{\det(2\pi \frac{T_{12}}{m_1})}} + \nu_{21} n_2 n_1 \ln \frac{n_2}{\sqrt{\det(2\pi \frac{T_{21}}{m_2})}}$$

$$\leq \nu_{12} n_2 n_1 \ln \frac{n_1}{\sqrt{\det(2\pi \frac{\mathbb{P}_1}{m_1})}} + \nu_{21} n_2 n_1 \ln \frac{n_2}{\sqrt{\det(2\pi \frac{\mathbb{P}_2}{m_2})}},$$

which is equivalent to the condition

$$(\det \mathcal{T}_{12})^{\varepsilon}(\det \mathcal{T}_{21}) \geq \left(\det \frac{\mathbb{P}_1}{n_1}\right)^{\varepsilon} \det \frac{\mathbb{P}_2}{n_2},$$

proven in lemma 7.6.3.
With this inequality we get

$$S(f_1, f_2) \leq \nu_{12} n_2 \left[\int G_{1,\tilde{\mu}_1=1} \ln G_{1,\tilde{\mu}_1=1} dv - \int f_1 \ln f_1 dv \right]$$

$$+ \nu_{21} n_1 \left[\int G_{2,\tilde{\mu}_2=1} \ln G_{2,\tilde{\mu}_2=1} dv - \int f_2 \ln f_2 dv \right] \leq 0.$$

The last inequality follows from remark 7.6.1. Here we also have equality if and only if $f_1 = M_1$ and $f_2 = M_2$, but since we already noticed that equality also implies $f_1 = G_{12}$ and $f_2 = G_{21}$. $\qquad\square$

Define the total entropy $H(f_1, f_2) = \int (f_1 \ln f_1 + f_2 \ln f_2) dv$. We can compute

$$\partial_t H(f_1, f_2) + \nabla_x \cdot \int (f_1 \ln f_1 + f_2 \ln f_2) v dv = S(f_1, f_2),$$

by multiplying the ES-BGK equation for the species 1 by $\ln f_1$, the ES-BGK equation for the species 2 by $\ln f_2$ and integrating the sum with respect to v.

Corollary 7.6.5 (Entropy inequality for mixtures). *Assume $f_1, f_2 > 0$. Assume a fast enough decay of f to zero for $v \to \infty$. Assume relationship (7.16), the conditions (7.18), (7.20), (7.64) and (7.65) and the positivity of the temperatures (7.22), then we have the following entropy inequality*

$$\partial_t \left(\int f_1 \ln f_1 dv + \int f_2 \ln f_2 dv \right) + \nabla_x \cdot \left(\int v f_1 \ln f_1 dv + \int v f_2 \ln f_2 dv \right) \leq 0,$$

with equality if and only if f_1 and f_2 are Maxwell distributions with equal mean velocity and temperature.

In summary the ES-BGK models (7.55) and (7.58) have the five free parameters $\alpha, \delta, \tilde{\mu}_1, \tilde{\mu}_2$ and ν_{12}. Let us summarize our result concerning what we expect from section 7.3. As we have seen in section 7.4, the parameter ν_{12} will show up in the expansion of the velocities (7.47) where we expect the diffusion coefficient. The parameters α and δ will show up in the expansion of the heat fluxes (7.43) where we expect the heat conductivity and the heat flux. From the motivation in the case of one species in section 7.1, we observe that the ES-BGK extension has the effect that the parameters $\tilde{\mu}_1$ and $\tilde{\mu}_2$ will show up in the expansion of the pressure tensors (7.50) and (7.51) where we expect the viscosity coefficient.

Chapter 8

Application to polyatomic mixtures

In this chapter we shall concern ourselves with a kinetic description of gas mixtures for polyatomic molecules. Evolution of a polyatomic gas is very important in applications, for instance air consists of a gas mixture of polyatomic molecules. But, most kinetic models modelling air deal with the case of a mono-atomic gas consisting of only one species.

In the literature one can find two types of models for one species of polyatomic molecules. There are models which contain a sum of collision terms on the right-hand side corresponding to the elastic and inelastic collisions. Examples are the models of Rykov [76], Holway [54] and Morse [68]. The other type of models contain only one collision term on the right-hand side taking into account both elastic and inelastic interactions. Examples for this are Bernard, Iollo, Puppo [14] and the model by Bisi and Caceres [18]. In this chapter we want to extend the model of Bernard, Iollo and Puppo [14] from one species of molecules to a gas mixture of polyatomic molecules. In contrast to mono-atomic molecules, in a polyatomic gas, energy is not entirely stored in the kinetic energy of its molecules but also in their rotational and vibrational modes. For simplification we present the model in the case of two species. We allow the two species to have different degrees of freedom in internal energy. For example, we may consider a mixture consisting of a mono-atomic and a diatomic gas. In addition, we want to model it via an ES-BGK approach in order to reproduce the correct Boltzmann hydrodynamic regime close to the asymptotic continuum limit. The ES-BGK approximation was suggested by Holway in the case of one species [54]. The H-theorem of this model then was proven in [2]. Brull and Schneider relate this model to a minimization problem in [23]. This model presented here for polyatomic molecules is also described in [60] by Klingenberg, Pirner and Puppo.

The outline of this chapter is as follows: in section 8.1 we will present the extension of the BGK model for one species of polyatomic molecules from [14] to two species of polyatomic molecules. In section 8.2, we extend it to an ES-BGK model and check if it is well-defined. In sections 8.2.1 to 8.2.4 we prove the conservation properties and the H-theorem. We show the positivity of all temperatures and quantify the structure of the equilibrium. In section 8.3, we compare our model reduced to one species with an other model presented in the literature from [2] which considers an ES-BGK model for one species of polyatomic molecules. In section 8.4.1 we apply the method of Chu reduction to our model in order to reduce the complexity of the variables for the rotational and vibrational energy degrees of freedom for numerical purposes. In section 8.4.2 we give an application in the case of a mono-atomic and a diatomic molecule. In section 8.5 we show that with a polyatomic model we are able to capture an equation of state which is different to the ideal gas law. In section 8.6 we conclude with existence and uniqueness results.

8.1   The BGK approximation for polyatomic mixtures

For simplicity in the following we consider a mixture composed of two different species. Let $x \in \mathbb{R}^d$ and $v \in \mathbb{R}^d, d \in \mathbb{N}$ be the phase space variables and $t \geq 0$ the time. Let M be the total number of different rotational and vibrational degrees of freedom and l_k the number of rotational and vibrational degrees of freedom of species k, $k = 1, 2$. Note that the sum $l_1 + l_2$ is not necessarily equal to M, because M counts only the different degrees of freedom in the internal energy, $l_1 + l_2$ counts all degrees of freedom in the internal energy. For example, consider two species consisting of diatomic molecules which have two rotational degrees of freedom. In addition, the second species has one vibrational degree of freedom. Then we have $M = 3, l_1 = 2, l_2 = 3$. Further, $\eta \in \mathbb{R}^M$ is the variable for the internal energy degrees of freedom, $\eta_{l_k} \in \mathbb{R}^M$ coincides with η in the components corresponding to the internal degrees of freedom of species k and is zero in the other components.

Since we want to describe two different species, our kinetic model has two distribution functions $f_1(x, v, \eta_{l_1}, t) > 0$ and $f_2(x, v, \eta_{l_2}, t) > 0$. Furthermore, we relate the distribution functions to macroscopic quantities by mean-values of f_k, $k = 1, 2$ as follows

$$\int f_k(v, \eta_{l_k}) \begin{pmatrix} 1 \\ v \\ \eta_{l_k} \\ m_k |v - u_k|^2 \\ m_k |\eta_{l_k} - \bar{\eta}_k|^2 \\ m_k (v - u_k) \otimes (v - u_k) \end{pmatrix} dv d\eta_{l_k} =: \begin{pmatrix} n_k \\ n_k u_k \\ n_k \bar{\eta}_k \\ d n_k T_k^{trans} \\ l_k n_k T_k^{rot} \\ \mathbb{P}_k \end{pmatrix}, \quad k = 1, 2, \quad (8.1)$$

where n_k is the number density, u_k the mean velocity and T_k^{trans} the temperature of the translation, T_k^{rot} the temperature of the rotation and vibration and $\mathbb{P}_k$ the pressure tensor of species k, $k = 1, 2$. Note that in this chapter we shall write T_k^{trans} and T_k^{rot} instead of $k_B T_k^{trans}$ and $k_B T_k^{rot}$, where k_B is Boltzmann's constant. In the following, we will require $\bar{\eta}_k = 0$, which means that the energy in rotations clockwise is the same as in rotations counter clockwise. Similar for vibrations.

First, we are interested in a BGK approximation of the interaction terms and write the model as:

$$\begin{aligned} \partial_t f_1 + v \cdot \nabla_x f_1 &= \nu_{11} n_1 (M_1 - f_1) + \nu_{12} n_2 (M_{12} - f_1), \\ \partial_t f_2 + v \cdot \nabla_x f_2 &= \nu_{22} n_2 (M_2 - f_2) + \nu_{21} n_1 (M_{21} - f_2), \end{aligned} \quad (8.2)$$

with the Maxwell distributions

$$M_k(x, v, \eta_{l_k}, t) = \frac{n_k}{\sqrt{2\pi \frac{\Lambda_k}{m_k}}^d} \frac{1}{\sqrt{2\pi \frac{\Theta_k}{m_k}}^{l_k}} \exp\left(-\frac{|v - u_k|^2}{2\frac{\Lambda_k}{m_k}} - \frac{|\eta_{l_k}|^2}{2\frac{\Theta_k}{m_k}}\right), \quad k = 1, 2,$$

$$M_{12}(x, v, \eta_{l_1}, t) = \frac{n_1}{\sqrt{2\pi \frac{\Lambda_{12}}{m_1}}^d} \frac{1}{\sqrt{2\pi \frac{\Theta_{12}}{m_1}}^{l_1}} \exp\left(-\frac{|v - u_{12}|^2}{2\frac{\Lambda_{12}}{m_1}} - \frac{|\eta_{l_1}|^2}{2\frac{\Theta_{12}}{m_1}}\right),$$

$$M_{21}(x, v, \eta_{l_2}, t) = \frac{n_2}{\sqrt{2\pi \frac{\Lambda_{21}}{m_2}}^d} \frac{1}{\sqrt{2\pi \frac{\Theta_{21}}{m_2}}^{l_2}} \exp\left(-\frac{|v - u_{21}|^2}{2\frac{\Lambda_{21}}{m_2}} - \frac{|\eta_{l_2}|^2}{2\frac{\Theta_{21}}{m_2}}\right).$$

$$(8.3)$$

The quantities in the Maxwell distributions will be defined on the following page. The quantities $\nu_{11}n_1$ and $\nu_{22}n_2$ are the collision frequencies of the particles of each species with itself, while $\nu_{12}n_2$ and $\nu_{21}n_1$ are related to interspecies collisions. To be flexible in choosing the relationship between the collision frequencies, we now assume the relationship

$$\nu_{12} = \varepsilon \nu_{21}, \quad 0 < \frac{l_1}{l_1 + l_2}\varepsilon \leq 1. \tag{8.4}$$

The restriction $\frac{l_1}{l_1+l_2}\varepsilon \leq 1$ is without loss of generality. If $\frac{l_1}{l_1+l_2}\varepsilon > 1$, exchange the notation 1 and 2 and choose $\frac{1}{\varepsilon}$ as new ε. In addition, we assume that all collision frequencies are positive.

Since rotational/vibrational and translational degrees of freedom relax at a different rate, T_k^{trans} and T_k^{rot} will first relax to partial temperatures Λ_k and Θ_k, respectively. Conservation of internal energy then requires that at each time

$$\frac{d}{2}n_k\Lambda_k = \frac{d}{2}n_k T_k^{trans} + \frac{l_k}{2}n_k T_k^{rot} - \frac{l_k}{2}n_k\Theta_k, \quad k = 1, 2. \tag{8.5}$$

Thus, Λ_k can be written as a function of Θ_k. In equilibrium we expect the two temperatures Λ_k and Θ_k to coincide, so we close the system by adding the equations

$$\partial_t M_k + v \cdot \nabla_x M_k = \frac{\nu_{kk}n_k}{Z_r^k}\frac{d + l_k}{d}(\tilde{M}_k - M_k) + \nu_{kk}n_k(M_k - f_k)$$

$$+ \nu_{kj}n_j(M_{kj} - f_k), \tag{8.6}$$

for $k, j = 1, 2$, $j \neq k$, where Z_r^k are given parameters corresponding to the different rates of decays of translational and rotational/vibrational degrees of freedom. Here M_k is given by

$$M_k(x, v, \eta_{l_k}, t) = \frac{n_k}{\sqrt{2\pi \frac{\Lambda_k}{m_k}}^d} \frac{1}{\sqrt{2\pi \frac{\Theta_k}{m_k}}^{l_k}} \exp\left(-\frac{|v - u_k|^2}{2\frac{\Lambda_k}{m_k}} - \frac{|\eta_{l_k}|^2}{2\frac{\Theta_k}{m_k}}\right), \quad k = 1, 2,$$

$$(8.7)$$

and $\tilde{M}_k$ is given by

$$\tilde{M}_k = \frac{n_k}{\sqrt{2\pi \frac{T_k}{m_k}}^{d+l_k}} \exp\left(-\frac{m_k|v - u_k|^2}{2T_k} - \frac{m_k|\eta_{l_k}|^2}{2T_k}\right), \quad k = 1, 2, \tag{8.8}$$

where T_k is the total equilibrium temperature and is given by

$$T_k := \frac{d\Lambda_k + l_k\Theta_k}{d + l_k} = \frac{dT_k^{trans} + l_kT_k^{rot}}{d + l_k}. \tag{8.9}$$

The second inequality follows from (8.5). If we multiply (8.6) by $|\eta_{l_k}|^2$, integrate with respect to v and η_{l_k} and use (8.9), we obtain

$$\partial_t(n_k\Theta_k) + \nabla_x \cdot (n_k\Theta_k u_k) = \frac{\nu_{kk}n_k}{Z_r^k}n_k(\Lambda_k - \Theta_k) + \nu_{kk}n_kn_k(\Theta_k - T_k^{rot})$$
$$+ \nu_{kj}n_jn_k(\Theta_{kj} - T_k^{rot}), \tag{8.10}$$

for $k, j = 1, 2, j \neq k$. We obtain a macroscopic equation which describes the relaxation of the temperature Θ_k towards the temperature Λ_k and the relaxation of Θ_k towards the rotational and vibrational temperature T_k^{rot} and of T_k^{rot} relaxing towards the mixture temperature Θ_{kj} in accordance with equation (8.2). Note that equation (8.10) together with mass, momentum and total energy conservation, is equivalent to (8.6). In addition, (8.2) and (8.10) are consistent. If we multiply the equations for species k of (8.2) and (8.10) by v and integrate with respect to v, we get in both cases for the right-hand side

$$\nu_{kj}n_kn_j(u_{jk} - u_j),$$

and if we compute the total internal energy of both equations, we obtain in both cases

$$\frac{1}{2}\nu_{kj}n_kn_j[d\Lambda_{jk} + l_j\Theta_{jk} - (d\Lambda_j + l_j\Theta_j)].$$

We will see this in theorem 8.2.3.

We recall that we assume that the mean-values of the momentum due to the internal degrees of freedom $\bar{\eta}_1, \bar{\eta}_2, \bar{\eta}_{12}$ and $\bar{\eta}_{21}$ are assumed to be zero. The structure of the collision terms ensures that at equilibrium or when $\nu_{kj} \to \infty$ the distribution functions become Maxwell distributions. With this choice of the Maxwell distributions M_1 and M_2 have the same densities, mean velocities and internal energies as f_1 and f_2, respectively. This guarantees the conservation of mass, momentum and energy (especially internal energy) in interactions of one species with itself. The remaining parameters $u_{12}, u_{21}, \Lambda_{12}, \Lambda_{21}, \Theta_{12}$ and Θ_{21} will be determined further down using conservation of the number of particles, total momentum and total energy, together with some symmetry considerations.

8.2 Extension to an ES-BGK model

As we know from chapter 7 that a drawback of the BGK approximation is its incapability of reproducing the correct Boltzmann hydrodynamic regime in the asymptotic continuum limit. In the polyatomic model we want to keep it as simple as possible and replace only the collision operators which represent the collisions of a species with itself by the ES-BGK collision operator. Other possible extensions are illustrated in the mono-atomic case for gas mixtures in chapter 7 and [62]. In this standard ES-BGK model, the scalar temperature T_k^{trans} is related to the distribution function f_k in the Maxwell distributions M_k and will be replaced by a linear combination of the temperature T_k^{trans} and the pressure tensor $\mathbb{P}_k$. In the polyatomic case described in this chapter the translational temperature T_k^{trans} is different from the temperature Λ_k of the Maxwell distributions M_k given by (8.7). Now, we want to extend this temperature Λ_k to a tensor Λ_k^{ten} with $\mathrm{Tr}(\Lambda_k^{ten}) = n_k \Lambda_k$ such that again we can consider a linear combination of the temperature Λ_k and the tensor Λ_k^{ten}. In the BGK case described in the previous section we determined the time evolution of Θ_k by considering equation (8.6) with the Maxwell distribution M_k given by (8.7) and the Maxwell distribution $\widetilde{M}_k$ given by (8.8) with the total equilibrium temperature T_k given by (8.9). This led to a time evolution of Θ_k given by (8.10). Λ_k is then obtained by (8.5). Now, in the ES-BGK case we determine the time evolution of Λ_k^{ten} by considering the equation

$$\partial_t \widehat{G}_k + v \cdot \nabla_x \widehat{G}_k = \frac{\nu_{kk} n_k}{Z_r^k} \frac{d + l_k}{d} (\check{G}_k - \widehat{G}_k) + \nu_{kk} n_k (G_k - f_k) + \nu_{kj} n_j (M_{kj} - f_k),$$

$$(8.11)$$

for $k = 1, 2$, with the extended Maxwell distribution $\widehat{G}_k$ given by

$$\widehat{G}_k = \frac{n_k}{\sqrt{\det(2\pi \frac{\Lambda_k^{ten}}{m_k})}} \frac{1}{\sqrt{2\pi \frac{T_k^{rot}}{m_k}}^{l_k}} \exp\left(-\frac{1}{2}(v-u_k)\cdot\left(\frac{\Lambda_k^{ten}}{m_k}\right)^{-1}\cdot(v-u_k) - \frac{m_k|\eta_{l_k}|^2}{2\Theta_k}\right),$$

$$(8.12)$$

for $k = 1, 2$, and the extended Maxwell distribution $\check{G}_k$ given by

$$\check{G}_k = \frac{n_k}{\sqrt{\det(2\pi \frac{T_k^{ten}}{m_k})}} \frac{1}{\sqrt{2\pi \frac{T_k}{m_k}}^{l_k}} \exp\left(-\frac{1}{2}(v-u_k)\cdot\left(\frac{T_k^{ten}}{m_k}\right)^{-1}\cdot(v-u_k) - \frac{1}{2}\frac{m_k|\eta_{l_k}|^2}{T_k}\right),$$

$$(8.13)$$

and the extended Maxwell distribution

$$G_k = \frac{n_k}{\sqrt{\det(2\pi \frac{\Lambda_k^{ES}}{m_k})}} \frac{1}{\sqrt{2\pi \frac{\Theta_k}{m_k}}^{l_k}} \exp\left(-\frac{1}{2}(v-u_k)\cdot\left(\frac{\Lambda_k^{ES}}{m_k}\right)^{-1}\cdot(v-u_k) - \frac{1}{2}\frac{m_k|\eta_{l_k}|^2}{\Theta_k}\right).$$

$$(8.14)$$

We define Λ_k^{ES} in the function G_k as a linear combination of Λ_k and Λ_k^{ten} given by

$$\Lambda_k^{ES} = (1 - \tilde{\mu}_k)\Lambda_k \mathbb{1}_n + \tilde{\mu}_k \frac{\Lambda_k^{ten}}{n_k}, \quad k = 1, 2,$$

with $\tilde{\mu}_k \in \mathbb{R}$, $k = 1, 2$ being free parameters which can be chosen in a way to fit physical parameters in the Navier-Stokes equations like the viscosity coefficient, analogously as in the standard ES-BGK model. The function $\tilde{G}_k$ has the total equilibrium temperature T_k and the pressure tensor of f_k on the off-diagonals, namely

$$
(T_k^{ten})_{ii} = T_k \qquad\qquad \text{for} \quad i = 1, \ldots, d,
$$

$$
(T_k^{ten})_{ij} = \frac{d}{d+l_k} \left(\frac{\mathbb{P}_k}{n_k} \right)_{ij} \qquad\qquad \text{for} \quad i, j = 1, \ldots, d, i \neq j. \tag{8.15}
$$

The factor $\frac{d}{d+l_k}$ in front of $\mathbb{P}_k$ in the definition of T_k^{ten} has the following reason. The temperature T_k given by (8.9) is a convex combination of T_k^{trans} and T_k^{rot}. Now, the off-diagonal elements of T_k^{ten} have the same structure. It is a convex combination of the pressure tensor $\mathbb{P}_k$ and the tensor corresponding to the rotational and vibrational temperature. But since the rotational effects are diagonal, we have $(T_k^{ten})_{ij} = \frac{d}{d+l_k}(\mathbb{P}_k)_{ij} + \frac{l_k}{d+l_k} 0$ for $i \neq j$.

We only extend Λ_k to a tensor and keep Θ_k as it is. This has the following reason. Since we assumed $\bar{\eta}_{lk} = 0$, the microscopic velocities related to the internal degrees of freedom are symmetric and then we do not distinguish different directions as we do in the translational degrees of freedom.

Equation (8.11) leads to a time evolution of Λ_k^{ten} given by

$$
\partial_t (n_k(\Lambda_k^{ten})_{ij}) + \nabla_x \cdot (n_k((\Lambda_k^{ten})_{ij})u_k) = \frac{\nu_{kk} n_k}{Z_r^k} \frac{d+l_k}{d} n_k((T_k^{ten})_{ij} - (\Lambda_k^{ten})_{ij})
$$

$$
+ \nu_{kk} n_k n_k((\Lambda_k^{ES})_{ij} - (\mathbb{P}_k)_{ij}) + \nu_{kj} n_j n_k(\Theta_{kj} - T_k^{rot})\delta_{ij}, \tag{8.16}
$$

for $k = 1, 2$ and $i, j = 1, ..., d$. We determine the time evolution of f_k in the ES-BGK case by

$$
\partial_t f_k + v \cdot \nabla_x f_k = \nu_{kk} n_k (G_k(f_k) - f_k) + \nu_{kj} n_j (M_{kj}(f_k, f_j) - f_k), \tag{8.17}
$$

for $k, j = 1, 2, k \neq j$. For further references we denote the relaxation operators by Q_{11}, Q_{12}, Q_{21} and Q_{22}.

Since G_k involves the term $(\Lambda_k^{ES})^{-1}$ and $\tilde{G}_k$ involves the term $(T_k^{ten})^{-1}$ we have to check if Λ_k^{ES} and T_k^{ten} are invertible.

Lemma 8.2.1. *Assume that f_k and $\tilde{G}_k$ are positive solutions to (8.17) and (8.11). Then Λ_k^{ten} and T_k^{ten} have strictly positive eigenvalues. Especially, the symmetric matrix T_k^{ten} is invertible.*

Proof. Let $y \in \mathbb{R}^d \setminus \{0\}$, then

$$
y \cdot \Lambda_k^{ten} \cdot y = \sum_{i,j=1}^{d} y_i (\Lambda_k^{ten})_{ij} y_j = \sum_{i,j=1}^{d} y_i \int (v_i - u_{k,i})(v_j - u_{k,j}) \widehat{G}_k y_j dv
$$

$$
= \int \left(\sum_{i=1}^{d} y_i (v_i - u_{k,i}) \right)^2 \widehat{G}_k dv \geq 0.
$$

The inequality is true since we assumed that $\widehat{G}_k$ is a positive solution to (8.6). If we use equation (8.9) and (8.5)

$$
\begin{aligned}
y \cdot T_k^{ten} \cdot y &= \sum_{i,j=1}^{d} y_i (T_k^{ten})_{ij} y_j = \sum_{\substack{i,j=1 \\ i \neq j}}^{d} y_i \int (v_i - u_{k,i})(v_j - u_{k,j}) f_k y_j \, dv + \sum_{i=1}^{d} y_i T_k y_i \\
&= \sum_{i,j=1}^{d} y_i \int (v_i - u_{k,i})(v_j - u_{k,j}) f_k y_j \, dv - \sum_{i=1}^{d} y_i T_k^{trans} y_i + \sum_{i=1}^{d} y_i \frac{d\Lambda_k + l_k \Theta_k}{d + l_k} y_i \\
&= \int \left(\sum_{i=1}^{d} y_i (v_i - u_{k,i}) \right)^2 f_k \, dv + \sum_{i=1}^{d} y_i T_k^{rot} y_i \geq 0,
\end{aligned}
$$

where $T_k^{rot} > 0$ because T_k^{rot} is defined via a positive integral of f_k, see the definition in (8.1). We even have strict inequality since $\{y_i(v - u_i)\}_{i=1}^{d}$ are linearly independent.

$\square$

With the previous lemma, we can prove that Λ_k^{ES} is positive. This is the next theorem. Positivity is also proven in [3] for the one species case, but for a different variant of an ES-BGK model.

Theorem 8.2.2. *Assume that $f_k > 0$ and $-\frac{1}{d-1} \leq \tilde{\mu}_k \leq 1$. Then Λ_k^{ES} has strictly positive eigenvalues. Especially Λ_k^{ES} is invertible.*

Proof. Since Λ_k^{ten} is symmetric there exists an invertible matrix S_k such that $\widetilde{\Lambda_k^{ten}} = S_k \Lambda_k^{ten} S_k^{-1}$ with a diagonal matrix $\widetilde{\Lambda_k^{ten}}$. Then $\widetilde{\Lambda_k^{ES}} := S_k \Lambda_k^{ES} S_k^{-1}$ is also diagonal since

$$
\widetilde{\Lambda_k^{ES}} = S_k \Lambda_k^{ES} S_k^{-1} = (1 - \tilde{\mu}_k) \Lambda_k \mathbb{1} + \tilde{\mu}_k \widetilde{\Lambda_k^{ten}}.
$$

Here we can see that the eigenvalues of $\widetilde{\Lambda_k^{ES}}$ are a linear combination of Λ_k and the eigenvalues of $\widetilde{\Lambda_k^{ten}}$ which coincide with the eigenvalues of Λ_k^{ten}. We denote the eigenvalues of Λ_k^{ten} by $\lambda_{k,1}, \lambda_{k,2}, \dots, \lambda_{k,d}$. Then by definition of Λ_k and Λ_k^{ten} we have

$$
d\Lambda_k = \mathbf{Tr}(\Lambda_k^{ten}) = \lambda_{k,1} + \lambda_{k,2} + \dots + \lambda_{k,d}.
$$

This means for the eigenvalues of Λ_k^{ES} denoted by $\tau_{k,i}$:

$$
\tau_{k,i} = \frac{1 - \tilde{\mu}_k}{d} \sum_{j=1}^{d} \lambda_{k,j} + \tilde{\mu}_k \lambda_{k,i} = \frac{1 + (d-1)\tilde{\mu}_k}{d} \lambda_{k,i} + \frac{1 - \tilde{\mu}_k}{d} \sum_{j=1, j \neq i}^{d} \lambda_{k,j}, \quad i = 1, 2, 3.
$$

Since $\lambda_{k,1}, \lambda_{k,2}, \dots, \lambda_{k,d}$ are strictly positive, the eigenvalues of Λ_k^{ES} are strictly positive, when $1 + (d-1)\tilde{\mu}_k$ and $1 - \tilde{\mu}_k$ are positive. Since we restricted $\tilde{\mu}_k$ to $-\frac{1}{d-1} \leq \tilde{\mu}_k \leq 1$, Λ_k^{ES} is strictly positive.

$\square$

8.2.1 Conservation properties

Conservation of the number of particles and total momentum of the BGK model for mixtures described in section 8.1 are shown in the same way as in the case of mono-atomic molecules. In the extension described in section 8.2 these conservation properties are still satisfied since G_1 and G_2 have the same density, mean velocity and internal energy as f_1 and f_2, respectively. Conservation of the number of particles and of total momentum are guaranteed by the following choice of the mixture parameters:

If we assume that

$$n_{12} = n_1 \quad \text{and} \quad n_{21} = n_2, \tag{8.18}$$

we have conservation of the number of particles, see theorem 2.1.1. If we further assume that u_{12} is a linear combination of u_1 and u_2

$$u_{12} = \delta u_1 + (1 - \delta)u_2, \quad \delta \in \mathbb{R}, \tag{8.19}$$

then we have conservation of total momentum provided that

$$u_{21} = u_2 - \frac{m_1}{m_2}\varepsilon(1 - \delta)(u_2 - u_1), \tag{8.20}$$

see theorem 2.1.2.

In the case of total energy we have a difference for the polyatomic case compared to the mono-atomic one. So we explicitly consider this in the following theorem.

Theorem 8.2.3 (Conservation of total energy). *Assume* (8.4), *conditions* (8.18), (8.19) *and* (8.20) *and assume that* Λ_{12} *and* Θ_{12} *are of the following form*

$$\Lambda_{12} = \alpha\Lambda_1 + (1 - \alpha)\Lambda_2 + \gamma|u_1 - u_2|^2, \quad 0 \le \alpha \le 1, \gamma \ge 0.$$
$$\Theta_{12} = \frac{l_1\Theta_1 + l_2\Theta_2}{l_1 + l_2}. \tag{8.21}$$

Then we have conservation of total energy

$$\int \frac{m_1}{2}(|v|^2 + |\eta_{l_1}|^2)(Q_{11}(f_1, f_1) + Q_{12}(f_1, f_2))dvd\eta_{l_1}$$
$$+ \int \frac{m_2}{2}(|v|^2 + |\eta_{l_2}|^2)(Q_{22}(f_2, f_2) + Q_{21}(f_2, f_1))dvd\eta_{l_2} = 0,$$

provided that

$$\Lambda_{21} + \frac{l_2}{d}\Theta_{21} = \left[\frac{1}{d}\varepsilon m_1(1 - \delta)\left(\frac{m_1}{m_2}\varepsilon(\delta - 1) + \delta + 1\right) - \varepsilon\gamma\right]|u_1 - u_2|^2$$
$$+\varepsilon(1 - \alpha)\Lambda_1 + (1 - \varepsilon(1 - \alpha))\Lambda_2 + \frac{1}{d}\varepsilon\frac{l_1 l_2}{l_1 + l_2}\Theta_1 + \frac{1}{d}\left(l_2 - \varepsilon\frac{l_1 l_2}{l_1 + l_2}\right)\Theta_2. \tag{8.22}$$

Proof. Using the energy exchange of species 1 and equation (8.5), we obtain

$$
F_{E_{1,2}} := \int \frac{m_1}{2}(|v|^2 + |\eta_{l_1}|^2)\nu_{11}n_1(G_1 - f_1)dvd\eta_{l_1}
$$
$$
+ \int \frac{m_1}{2}(|v|^2 + |\eta_{l_1}|^2)\nu_{12}n_2(M_{12} - f_1)dvd\eta_{l_1}
$$
$$
= \varepsilon\nu_{21}\frac{1}{2}n_2n_1m_1(|u_{12}|^2 - |u_1|^2) + \frac{d}{2}\varepsilon\nu_{21}n_1n_2(\Lambda_{12} - T_1^{trans})
$$
$$
+ \frac{l_1}{2}\varepsilon\nu_{21}n_1n_2(\Theta_{12} - T_1^{rot})
$$
$$
= \varepsilon\nu_{21}\frac{1}{2}n_2n_1m_1(|u_{12}|^2 - |u_1|^2) + \frac{d}{2}\varepsilon\nu_{21}n_1n_2(\Lambda_{12} - \Lambda_1)
$$
$$
+ \frac{l_1}{2}\varepsilon\nu_{21}n_1n_2(\Theta_{12} - \Theta_1).
$$

Next, we will insert the definitions of u_{12}, Λ_{12} and Θ_{12} given by (8.19) and (8.21). Analogously the energy exchange of species 2 towards 1 is

$$
F_{E_{2,1}} = \nu_{21}\frac{1}{2}n_2n_1m_2(|u_{21}|^2 - |u_2|^2) + \frac{d}{2}\nu_{21}n_1n_2(\Lambda_{21} - \Lambda_2) + \frac{l_2}{2}\nu_{21}n_1n_2(\Theta_{21} - \Theta_2).
$$

Substitute u_{21} with (8.20) and $\Lambda_{21} + \frac{l_2}{d}\Theta_{21}$ from (8.22). This permits to rewrite the energy exchange as

$$
F_{E_{1,2}} = \varepsilon\nu_{21}\frac{1}{2}n_2n_1m_1 \left[(\delta^2 - 1)|u_1|^2 + (1 - \delta)^2|u_2|^2 + 2\delta(1 - \delta)u_1 \cdot u_2\right]
$$
$$
+ \frac{1}{2}\varepsilon\nu_{21}n_1n_2 \left[(1 - \alpha)d(\Lambda_2 - \Lambda_1) + \frac{l_1 l_2}{l_1 + l_2}(\Theta_2 - \Theta_1) + \gamma d|u_1 - u_2|^2\right],
\tag{8.23}
$$

$$
F_{E_{2,1}} = \frac{1}{2}\nu_{21}m_2n_1n_2 \left[\left(\left(1 - \frac{m_1}{m_2}\varepsilon(1 - \delta)\right)^2 - 1\right) |u_2|^2 + \left(\frac{m_1}{m_2}\varepsilon(\delta - 1)\right)^2 |u_1|^2\right.
$$
$$
+ 2\left(1 - \frac{m_1}{m_2}\varepsilon(1 - \delta)\right)\frac{m_1}{m_2}\varepsilon(1 - \delta)u_1 \cdot u_2\right] + \frac{1}{2}\nu_{21}n_1n_2\left[\varepsilon(1 - \alpha)d(\Lambda_1 - \Lambda_2)\right.
$$
$$
\left. + \varepsilon\frac{l_1 l_2}{l_1 + l_2}(\Theta_1 - \Theta_2) + \left(\varepsilon m_1(1 - \delta)\left(\frac{m_1}{m_2}\varepsilon(\delta - 1) + \delta + 1\right) - \varepsilon\gamma d\right)|u_1 - u_2|^2\right].
\tag{8.24}
$$

Adding these two terms, we see that the total energy is conserved. $\qquad\square$

Remark 8.2.1. Since we assumed $\bar{\eta}_{l_1} = \bar{\eta}_{l_2} = 0$, we expect no exchange terms of the form $\bar{\eta}_{l_2} - \bar{\eta}_{l_1}$ in the momentum equation or a corresponding internal energy exchange in the energy equation. Furthermore, for this reason, we did not add a term of the form $|\bar{\eta}_{l_1} - \bar{\eta}_{l_2}|^2$ in the definitions of the mixture temperatures (8.21).

Remark 8.2.2. The energy flux between the two species is zero if and only if $u_1 = u_2$, $\Lambda_1 = \Lambda_2$, $\Theta_1 = \Theta_2$ provided that $\alpha, \delta < 1$ and $\gamma > 0$.

From conservation of total energy we get only one condition on $\Lambda_{21} + \frac{l_2}{d}\Theta_{21}$ given by (8.22), but not an explicit formula for Λ_{21} and Θ_{21}. In order to keep the model symmetric we again separate the temperatures corresponding to the translational part and the one corresponding to the rotational and vibrational part and choose

$$\Lambda_{21} = \varepsilon(1-\alpha)\Lambda_1 + (1 - \varepsilon(1-\alpha))\Lambda_2$$
$$+ \left[\frac{1}{d}\varepsilon m_1(1-\delta)\left(\frac{m_1}{m_2}\varepsilon(\delta-1) + \delta + 1\right) - \varepsilon\gamma\right]|u_1 - u_2|^2, \tag{8.25}$$

$$\Theta_{21} = \left(1 - \varepsilon\frac{l_1}{l_1 + l_2}\right)\Theta_2 + \varepsilon\frac{l_1}{l_1 + l_2}\Theta_1. \tag{8.26}$$

Remark 8.2.3. If $l_1 = l_2$, we have $\Theta_{12} = \frac{1}{2}(\Theta_1 + \Theta_2)$. We then find $\Theta_{21} = \Theta_{12}$ if the two species have the same interspecies collision frequency ($\varepsilon = 1$).

Remark 8.2.4. The fact that we only consider the two species case is just for simplicity. We can also extend the model to more than two species, because we assume that we only have binary interactions. So if we consider collision terms given by

$$\nu_{ii}n_i(G_i - f_i) + \sum_{j\neq i}^{N} \nu_{ij}n_j(G_{ij} - f_i), \quad i = 1, ..., N,$$

we expect that we have conservation of total momentum and total energy in every interaction of species i with species j. This means we require

$$\int \begin{pmatrix} v \\ v^2 \end{pmatrix} \nu_{ij}n_j(G_{ij} - f_i)dv + \int \begin{pmatrix} v \\ v^2 \end{pmatrix} \nu_{ji}n_i(G_{ji} - f_j)dv = 0,$$

for every $i, j = 1, ...N$, $i \neq j$ and so it reduces to the two species case.

8.2.2 Positivity of the temperatures

Theorem 8.2.4. *Assume that $f_1(x, v, \eta_{l_1}, t), f_2(x, v, \eta_{l_2}, t) > 0$. Then all temperatures $\Lambda_1, \Lambda_2, \Theta_1, \Theta_2, \Lambda_{12}, \Theta_{12}$ given by (8.21), and $\Lambda_{21}, \Theta_{21}$ determined by (8.25), (8.26) are positive provided that*

$$0 \leq \gamma \leq \frac{m_1}{d}(1-\delta)\left[\left(1 + \frac{m_1}{m_2}\varepsilon\right)\delta + 1 - \frac{m_1}{m_2}\varepsilon\right]. \tag{8.27}$$

Proof. The temperatures $\Lambda_1, \Lambda_2, \Theta_1, \Theta_2, \Lambda_{12}, \Theta_{12}$ and Θ_{21} are positive by definition because they are integrals of positive functions. Thus, the only thing is to check the temperature Λ_{21} in (8.25). This is done in theorem 2.1.4 for $d = 3$, so we skip the proof here. The resulting condition is given by (8.27). $\qquad\square$

Remark 8.2.5. Since $\gamma \geq 0$ the right-hand side of the inequality in (8.27) must be non-negative. This condition is equivalent to

$$\frac{\frac{m_1}{m_2}\varepsilon - 1}{1 + \frac{m_1}{m_2}\varepsilon} \leq \delta \leq 1. \tag{8.28}$$

8.2.3 The structure of equilibrium

Theorem 8.2.5 (Equilibrium). *Assume $f_1, f_2 > 0$ and the conditions (8.18), (8.19), (8.20), (8.21) and (8.22), $\delta \neq 1, \alpha \neq 1, l_1, l_2 \neq 0$, the positivity of all temperatures. Then, if f_1 and f_2 are independent of x and t; f_1 and f_2 are Maxwell distributions with equal mean velocities $u_1 = u_2 = u_{12} = u_{21}$ and temperatures $T_1^{rot} = T_2^{rot} = T_1^{trans} = T_2^{trans} = \Lambda_1 = \Lambda_2 = \Theta_1 = \Theta_2 = \Theta_{12} = \Theta_{21} = \Lambda_{12} = \Lambda_{21}$.*

Proof. Equilibrium means that $f_1, f_2, \Lambda_1, \Lambda_2, \Theta_1, \Theta_2$ are independent of x and t. Thus, in the situation of equilibrium the right-hand side of the equations (8.17) and (8.11) have to be zero. From this condition on (8.17) we can deduce

$$(\nu_{11}n_1 + \nu_{12}n_2)f_1 = \nu_{11}n_1 G_1 + \nu_{12}n_2 M_{12}, \tag{8.29}$$

$$(\nu_{22}n_2 + \nu_{21}n_1)f_2 = \nu_{22}n_2 G_2 + \nu_{21}n_1 M_{21}. \tag{8.30}$$

Since the right-hand side of (8.17) and the right-hand side of (8.11) have to be zero, the difference of the right-hand side of (8.17) and the right-hand side of (8.11) has to be equal to zero. If we compute the translational temperature of this difference, we obtain

$$\Lambda_1^{ten} = T_1^{ten}, \tag{8.31}$$

$$\Lambda_2^{ten} = T_2^{ten}. \tag{8.32}$$

Especially, from the diagonal part of (8.31) and (8.32) we can deduce

$$\Lambda_1 = \Theta_1, \tag{8.33}$$

$$\Lambda_2 = \Theta_2. \tag{8.34}$$

When we consider the moment of the velocity of (8.29), we get

$$(\nu_{11}n_1 + \nu_{12}n_2)u_1 = \nu_{11}n_1 u_1 + \nu_{12}n_2 u_{12},$$

which is equivalent to

$$u_1 = u_2, \tag{8.35}$$

for $\delta \neq 1$. When we consider the moments of the translational and the rotational/vibrational temperatures of (8.29) and (8.30), we get

$$(\nu_{11}n_1 + \nu_{12}n_2)T_1^{trans} = (\nu_{11}n_1 + \nu_{12}n_2\alpha)\Lambda_1 + \nu_{12}n_2(1-\alpha)\Lambda_2, \tag{8.36}$$

$$(\nu_{11}n_1 + \nu_{12}n_2)T_1^{rot} = (\nu_{11}n_1 + \nu_{12}n_2\frac{l_1}{l_1+l_2})\Lambda_1 + \nu_{12}n_2\frac{l_2}{l_1+l_2}\Lambda_2, \tag{8.37}$$

$$(\nu_{22}n_2 + \nu_{21}n_1)T_2^{trans} = \nu_{22}n_2\Lambda_2 + \nu_{21}n_1\Lambda_{21}, \tag{8.38}$$

$$(\nu_{22}n_2 + \nu_{21}n_1)T_2^{rot} = \nu_{22}n_2\Lambda_2 + \nu_{21}n_1\Theta_{21}. \tag{8.39}$$

To arrive at these equations, we used the definitions of the mixture velocities and temperatures (8.18), (8.19), (8.20), (8.21) and equations (8.33), (8.34) and (8.35).

Using (8.33), (8.34) and (8.35), the temperatures of the mixture Maxwell distributions (8.21) and (8.25), (8.26) simplify to

$$\Lambda_{12} = \alpha\Lambda_1 + (1-\alpha)\Lambda_2, \qquad\qquad \Theta_{12} = \frac{l_1}{l_1+l_2}\Lambda_1 + \frac{l_2}{l_1+l_2}\Lambda_2, \qquad (8.40)$$

$$\Lambda_{21} = \varepsilon(1-\alpha)\Lambda_1 + (1-\varepsilon(1-\alpha))\Lambda_2, \quad \Theta_{21} = \varepsilon\frac{l_1}{l_1+l_2}\Lambda_1 + \left(1-\varepsilon\frac{l_1}{l_1+l_2}\right)\Lambda_2.$$
$$(8.41)$$

Analogue, equations (8.5) simplify to

$$\frac{d+l_1}{2}\Lambda_1 = \frac{d}{2}T_1^{trans} + \frac{l_1}{2}T_1^{rot}, \qquad\qquad (8.42)$$

$$\frac{d+l_2}{2}\Lambda_2 = \frac{d}{2}T_2^{trans} + \frac{l_2}{2}T_2^{rot}. \qquad\qquad (8.43)$$

Inserting (8.36) and (8.37) in (8.42), we obtain

$$\frac{d}{2}\left(\frac{\nu_{11}n_1 + \nu_{12}n_2\alpha}{\nu_{11}n_1 + \nu_{12}n_2}\Lambda_1 + \frac{\nu_{12}n_2(1-\alpha)}{\nu_{11}n_1 + \nu_{12}n_2}\Lambda_2\right)$$
$$+ \frac{l_1}{2}\left(\frac{\nu_{11}n_1 + \nu_{12}n_2\frac{l_1}{l_1+l_2}}{\nu_{11}n_1 + \nu_{12}n_2}\Lambda_1 + \frac{\nu_{12}n_2\frac{l_2}{l_1+l_2}}{\nu_{11}n_1 + \nu_{12}n_2}\Lambda_2\right) = \frac{d+l_1}{2}\Lambda_1,$$

which, provided $d\alpha + l_1\frac{l_1}{l_1+l_2} \neq d+l_1$, is equivalent to

$$\Lambda_1 = \Lambda_2. \qquad\qquad (8.44)$$

This condition is equivalent to $d(1-\alpha) + \frac{l_1 l_2}{l_1+l_2} \neq 0$ which is satisfied since $\alpha \neq 1, l_1, l_2 \neq 0$. With (8.44) we can deduce from (8.36) and (8.37) that

$$T_1^{trans} = \Lambda_1 \quad \text{and} \quad T_1^{rot} = \Lambda_1. \qquad\qquad (8.45)$$

Condition (8.41) together with (8.44) leads to

$$\Lambda_{21} = \Theta_{21} = \Lambda_1. \qquad\qquad (8.46)$$

Inserting (8.44) and (8.46) in (8.38) and (8.39) leads to

$$T_2^{trans} = T_2^{rot} = \Lambda_1.$$

If we compute the pressure tensor of (8.29) using that all temperatures are equal to Λ_1 we obtain

$$(\nu_{11}n_1 + \nu_{12}n_2)\frac{\mathbb{P}_1}{n_1} = \nu_{11}n_1(1-\tilde{\mu}_1)\Lambda_1\mathbf{1} + \nu_{11}n_1\tilde{\mu}_1\Lambda_1^{ten} + \nu_{12}n_2\Lambda_1\mathbf{1}.$$

Using (8.15), (8.31) and (8.45), we have that

$$\frac{d}{d+l_k}\frac{\mathbb{P}_1}{n_1} + \frac{l_k}{d+l_k}\Lambda_1^{rot}\mathbf{1}_d = \Lambda_1^{ten} = T_1^{ten}$$

and therefore

$$(\nu_{11}n_1(1 - \tilde{\mu}_1\frac{d}{d + l_k}) + \nu_{12}n_2)\frac{\mathbb{P}_1}{n_1} = (\nu_{11}n_1(1 - \tilde{\mu}_1\frac{d}{d + l_k}) + \nu_{12}n_2)\Lambda_1\mathbf{1},$$

which shows that the pressure tensor of f_1 is diagonal since $\tilde{\mu}_1 \leq 1$. Similar for $\frac{\mathbb{P}_2}{n_2}$ using (8.30), (8.32) and (8.45).

So all in all, in equilibrium we get that f_1 and f_2 are Maxwell distributions with equal mean velocities $u_1 = u_2 = u_{12} = u_{21}$ and temperatures $T_1^{rot} = T_2^{rot} = T_1^{trans} = T_2^{trans} = \Lambda_1 = \Lambda_2 = \Theta_1 = \Theta_2 = \Theta_{12} = \Theta_{21} = \Lambda_{12} = \Lambda_{21}$. $\qquad\square$

Definition 8.2.1. If f_1 and f_2 are Maxwell distributions with equal mean velocities $u_1 = u_2 = u_{12} = u_{21}$ and temperatures $T_1^{rot} = T_2^{rot} = T_1^{trans} = T_2^{trans} = \Lambda_1 = \Lambda_2 = \Theta_1 = \Theta_2 = \Theta_{12} = \Theta_{21} = \Lambda_{12} = \Lambda_{21}$, then we say that f_1 and f_2 are in local equilibrium.

8.2.4 H-Theorem

In this section we will prove that our model admits an entropy with an entropy inequality. For this, we have to prove an inequality on the term $\int \ln f_k (G_k - f_k) dv d\eta_{l_k}$ coupled with $\ln \hat{G}_k$ times the right-hand side of equation (8.11) and an inequality on $\nu_{12}n_2 \int (M_{12} - f_1) \ln f_1 dv d\eta_{l_1} + \nu_{21}n_1 \int (M_{21} - f_2) \ln f_2 dv d\eta_{l_2}$ coupled with $\ln \tilde{G}$ times the right-hand side of equation (8.11). We prove the first one in section 8.2.4 and the second one in section 8.2.4.

H-theorem for the one species relaxation term

Remark 8.2.6. From the definition of the moments of $f_k, k = 1, 2$ in (8.1) and the definitions of the extended Maxwell distributions $G_k, k = 1, 2$ in (8.14), we see that the pressure tensors and the temperatures do not coincide. Now, we consider extended Maxwell distributions $\bar{G}_k, k = 1, 2$ which have the same pressure tensor and temperatures as $f_k, k = 1, 2$. Then from the case of one species ES-BGK model we know that

$$\int \bar{G}_k \ln \bar{G}_k dv d\eta_{l_k} \leq \int f_k \ln f_k dv d\eta_{l_k},$$

for $k = 1, 2$, see equations (20) and (21) in [3] in the mono-atomic case. The polyatomic case is analogous to the mono-atomic case.

Lemma 8.2.6. Assume that $f_1, f_2 > 0$. As in remark 8.2.6 let $\bar{G}_k$ be the extended Maxwell distributions with the same pressure tensor and temperatures as $f_k, k = 1, 2$ and $\tilde{G}$ the extended Maxwell distribution defined by (8.12). Then we have

$$\int \tilde{G}_k \ln \tilde{G}_k dv d\eta_{l_k} \leq \int \bar{G}_k \ln \bar{G}_k dv d\eta_{l_k}, \quad k = 1, 2,$$

$$\int \hat{G}_k \ln \hat{G}_k dv d\eta_{l_k} \geq \int G_k \ln G_k dv \eta_{l_k}, \quad k = 1, 2,$$

$$\int G_k \ln G_k dv d\eta_{l_k} \geq \int M_k \ln M_k dv \eta_{l_k}, \quad k = 1, 2.$$

Proof. The proof of the second inequality is analogously to the proof in the mono-atomic case of equation (21) in [3]. So we only prove the first and the third one. Using that $\ln M_k = \ln(\frac{n_k}{\sqrt{2\pi\frac{\Lambda_k}{m_k}}^d}\frac{1}{\sqrt{2\pi\frac{\Theta_k}{m_k}}^{l_k}}) - \frac{|v-u_k|^2}{2\frac{\Lambda_k}{m_k}} - \frac{|\eta_{l_k}|^2}{2\frac{\Theta_k}{m_k}}$,

$$\ln \bar{G}_k = \ln(\frac{n_k}{\sqrt{\det(2\pi\frac{\mathbb{P}_k}{m_k})}}\frac{1}{\sqrt{2\pi\frac{T_k^{rot}}{m_k}}^{l_k}}) - \frac{1}{2}(v-u_k)\cdot\left(\frac{\mathbb{P}_k}{m_k}\right)^{-1}\cdot(v-u_k) - \frac{|\eta_{l_k}|^2}{2\frac{T_k^{rot}}{m_k}},$$

$$\ln \widetilde{G}_k = \ln(\frac{n_k}{\sqrt{\det(2\pi\frac{T_k^{ten}}{m_k})}}\frac{1}{\sqrt{2\pi\frac{T_k}{m_k}}^{l_k}}) - \frac{1}{2}(v-u_k)\cdot\left(\frac{T_k^{ten}}{m_k}\right)^{-1}\cdot(v-u_k) - \frac{|\eta_{l_k}|^2}{2\frac{T_k}{m_k}},$$

and $\ln G_k = \ln(\frac{n_k}{\sqrt{\det(2\pi\frac{\Lambda_k^{ES}}{m_k})}}\frac{1}{\sqrt{2\pi\frac{\Theta_k}{m_k}}^{l_k}}) - \frac{1}{2}m_k(v-u_k)\cdot(\Lambda_k^{ES})^{-1}\cdot(v-u_k) - \frac{|\eta_{l_k}|^2}{2\frac{\Theta_k}{m_k}}$,

we compute the integrals and obtain that the required inequalities are equivalent to

$$\ln(\frac{n_k}{\sqrt{\det(2\pi\frac{T_k^{ten}}{m_k})}}\frac{1}{\sqrt{2\pi\frac{T_k}{m_k}}^{l_k}}) \leq \ln(\frac{n_k}{\sqrt{\det(2\pi\frac{\mathbb{P}_k}{m_k})}}\frac{1}{\sqrt{2\pi\frac{T_k^{rot}}{m_k}}^{l_k}}),$$

$$\ln(\frac{n_k}{\sqrt{\det(2\pi\frac{\Lambda_k^{ES}}{m_k})}}) \geq \ln(\frac{n_k}{\sqrt{2\pi\frac{\Lambda_k}{m_k}}^d}).$$

This is equivalent to the conditions

$$\ln\det(T_k^{ten}) + l_k\ln T_k \geq \ln\det\mathbb{P}_k + l_k\ln T_k^{rot},$$
$$(\Lambda_k)^d \geq \det(\Lambda_k^{ES}). \tag{8.47}$$

We first look at the first inequality. If we insert the expression for T_k given by (8.9) and use the concavity of ln, we obtain

$$\ln\det(T_k^{ten}) + l_k\frac{l_k}{d+l_k}\ln T_k^{rot} + l_k\frac{d}{d+l_k}\ln T_k^{trans} \geq \ln\det\mathbb{P}_k + l_k\ln T_k^{rot}. \tag{8.48}$$

Now we use the Brunn-Minkowsky inequality (see section 7.5). Since we can write T_k^{ten} as

$$T_k^{ten} = \frac{d}{d+l_k}\mathbb{P}_k + \frac{l_k}{d+l_k}T_k^{rot}\mathbf{1}_d,$$

we can apply the Brunn-Minkowsky inequality on (8.48) and obtain

$$\frac{d}{d+l_k}\ln\det\mathbb{P}_k + d\frac{l_k}{d+l_k}\ln T_k^{rot} + l_k\frac{l_k}{d+l_k}\ln T_k^{rot} + l_k\frac{d}{d+l_k}\ln T_k^{trans}$$
$$\geq \ln\det\mathbb{P}_k + l_k\ln T_k^{rot}.$$

So it remains to show that

$$(T_k^{trans})^d \geq \det\mathbb{P}_k.$$

This inequality has the same structure as the second inequality in (8.47). So we only prove the second inequality in (8.47). We observe that $\text{Tr}(\Lambda_k^{ES}) = d\Lambda_k$, so we have to show

$$\left(\frac{\text{Tr}(\Lambda_k^{ES})}{d}\right)^d \geq \det(\Lambda_k^{ES}).$$

Let $\lambda_1, \ldots, \lambda_d$ be the eigenvalues of the symmetric positive matrix Λ_k^{ES}, then this inequality is equivalent to

$$\left(\frac{\lambda_1 + \cdots + \lambda_d}{d}\right)^d \geq \lambda_1 \cdots \lambda_d.$$

This is true since it is the inequality of arithmetic and geometric means, see theorem A.1.5 in the appendix. $\qquad\square$

Lemma 8.2.7 (H-theorem for one species). *Assume $f_1, f_2 > 0$. Then*

$$\int \ln f_k (G_k - f_k) dv d\eta_{l_k} + \int \ln \widehat{G}_k (\widetilde{G}_k - \widehat{G}_k) dv d\eta_{l_k} \leq 0, \quad k = 1, 2,$$

with equality if and only if $M_k = f_k$ and $\Lambda_k = \Theta_k = T_k^{rot} = T_k^{trans}$.

Proof. Apply lemma 1.3.12 on both terms of

$$S_k(f_k) := \int \ln f_k (G_k - f_k) dv d\eta_{l_k} + \int \ln \widehat{G}_k (\widetilde{G}_k - \widehat{G}_k) dv d\eta_{l_k}.$$

Then we obtain

$$S_k(f_k) \leq \int G_k \ln G_k dv d\eta_{l_k} - \int f_k \ln f_k dv d\eta_{l_k} - \int G_k dv d\eta_{l_k} + \int f_k dv d\eta_{l_k}$$
$$+ \left[\int \widetilde{G}_k \ln \widetilde{G}_k dv d\eta_{l_k} - \int \widehat{G}_k \ln \widehat{G}_k dv d\eta_{l_k} - \int \widetilde{G}_k dv d\eta_{l_k} + \int \widehat{G}_k dv d\eta_{l_k}\right],$$

with equality if and only if $f_k = G_k$ and $G_k = \widetilde{G}_k$ from which we can deduce $f_k = M_k$ by computing macroscopic quantities of $f_k = G_k$ and $G_k = \widetilde{G}_k$. Since $f_k, G_k, \widehat{G}_k$ and $\widetilde{G}_k$ have the same density, we obtain

$$S(f_k) \leq \int G_k \ln G_k dv d\eta_{l_k} - \int f_k \ln f_k dv d\eta_{l_k} + \left[\int \widetilde{G}_k \ln \widetilde{G}_k dv d\eta_{l_k} - \int \widehat{G}_k \ln \widehat{G}_k dv d\eta_{l_k}\right]. \qquad (8.49)$$

According to the second part of lemma 8.2.6, we obtain

$$S(f_k) \leq \int \widetilde{G}_k \ln \widetilde{G}_k dv d\eta_{l_k} - \int f_k \ln f_k dv d\eta_{l_k}.$$

Here we have equality if and only if $G_k = \widetilde{G}_k$, which means $\Lambda_k = \Theta_k$. Now, using the first part of lemma 8.2.6 and remark 8.2.6, we can estimate $\int \widetilde{G}_k \ln \widetilde{G}_k dv d\eta_{l_k}$ by $\int f_k \ln f_k dv d\eta_{l_k}$. So, all in all, we obtain $S_k(f_k) \leq 0$ with equality if and only if $f_k = M_k$ and $\Lambda_k = \Theta_k = T_k^{rot} = T_k^{trans}$.

$\qquad\square$

H-theorem for the mixture of polyatomic molecules

Lemma 8.2.8. *Assume $f_1, f_2 > 0$, the relationship between the collision frequencies (8.4), the conditions for the interspecies Maxwell distributions (8.18), (8.19), (8.20), (8.21) and (8.22) and the positivity of all temperatures, then*

$$\varepsilon \tfrac{d}{2} \ln \Lambda_{12} + \varepsilon \tfrac{l_1}{2} \ln \Theta_{12} + \tfrac{d}{2} \ln \Lambda_{21} + \tfrac{l_2}{2} \ln \Theta_{21} \geq \tfrac{d}{2} \varepsilon \ln \Lambda_1 + \tfrac{d}{2} \ln \Lambda_2 + \tfrac{l_1}{2} \varepsilon \ln \Theta_1 + \tfrac{l_2}{2} \ln \Theta_2. \qquad (8.50)$$

Proof. First we consider the part $E_1 := \tfrac{d}{2} \ln \Lambda_{12} + \tfrac{l_1}{2} \ln \Theta_{12}$. We insert the definitions of Λ_{12} and Θ_{12} into E_1 and use the monotonicity of $\ln$ to drop the velocity term. Then we obtain

$$E_1 \geq \frac{d}{2} \ln(\alpha \Lambda_1 + (1-\alpha)\Lambda_2) + \frac{l_1}{2} \ln(\frac{l_1}{l_1 + l_2} \Theta_1 + \frac{l_2}{l_1 + l_2} \Theta_2).$$

Now we use that $\ln$ is concave and get

$$E_1 \geq \frac{d}{2} \alpha \ln \Lambda_1 + \frac{d}{2}(1-\alpha) \ln \Lambda_2 + \frac{l_1}{2} \frac{l_1}{l_1 + l_2} \ln \Theta_1 + \frac{l_1}{2} \frac{l_2}{l_1 + l_2} \ln \Theta_2. \qquad (8.51)$$

Doing the same with the second part $E_2 := \tfrac{d}{2} \ln \Lambda_{21} + \tfrac{l_2}{2} \ln \Theta_{21}$ using that $\frac{l_1}{l_1 + l_2} \varepsilon \leq 1$, we obtain

$$E_2 \geq \tfrac{d}{2} \varepsilon(1-\alpha) \ln \Lambda_1 + \tfrac{d}{2}(1-\varepsilon(1-\alpha)) \ln \Lambda_2 + \tfrac{l_2}{2} \varepsilon \tfrac{l_1}{l_1 + l_2} \ln \Theta_1 + \tfrac{l_2}{2}(1-\varepsilon \tfrac{l_1}{l_1 + l_2}) \ln \Theta_2. \qquad (8.52)$$

Multiplying (8.51) by ε and adding (8.52), we get

$$\varepsilon E_1 + E_2 \geq \frac{d}{2} \varepsilon \ln \Lambda_1 + \frac{d}{2} \ln \Lambda_2 + \frac{l_1}{2} \varepsilon \ln \Theta_1 + \frac{l_2}{2} \ln \Theta_2,$$

which is the required inequality. □

Lemma 8.2.9. *Assume $f_1, f_2 > 0$. Assume the relationship between the collision frequencies (8.4), the conditions for the interspecies Maxwell distributions (8.18), (8.19), (8.20), (8.21) and (8.22) and the positivity of all temperatures. Then*

$$\nu_{12} n_2 \int M_{12} \ln M_{12} dv d\eta_{l_1} + \nu_{21} n_1 \int M_{21} \ln M_{21} dv d\eta_{l_2}$$

$$\leq \nu_{12} n_2 \int M_1 \ln M_1 dv d\eta_{l_1} + \nu_{21} n_1 \int M_2 \ln M_2 dv d\eta_{l_2}.$$

Proof. Using that $\ln M_{12} = \ln \left(\frac{n_2}{\sqrt{2\pi \frac{\Lambda_{12}}{m_1}}^d} \frac{1}{\sqrt{2\pi \frac{\Theta_{12}}{m_1}}^{l_1}} \right) - \frac{|v - u_{12}|^2}{2 \frac{\Lambda_{12}}{m_1}} - \frac{|\eta_{l_1}|^2}{2 \frac{\Theta_{12}}{m_1}}$,

$$\ln M_{21} = \ln \left(\frac{n_1}{\sqrt{2\pi \frac{\Lambda_{21}}{m_2}}^d} \frac{1}{\sqrt{2\pi \frac{\Theta_{21}}{m_2}}^{l_2}} \right) - \frac{|v - u_{21}|^2}{2 \frac{\Lambda_{21}}{m_2}} - \frac{|\eta_{l_2}|^2}{2 \frac{\Theta_{21}}{m_2}},$$

$$\ln M_k = \ln \left(\frac{n_k}{\sqrt{2\pi \frac{\Lambda_k}{m_k}}^d} \frac{1}{\sqrt{2\pi \frac{\Theta_k}{m_k}}^{l_k}} \right) - \frac{|v - u_k|^2}{2\frac{\Lambda_k}{m_k}} - \frac{|\eta_{l_k}|^2}{2\frac{\Theta_k}{m_k}}, \ k = 1, 2,$$ we compute the integrals

and obtain that the required inequalities are equivalent to

$$\varepsilon \ln \left(\frac{n_1}{\sqrt{2\pi \frac{\Lambda_{12}}{m_1}}^d} \frac{1}{\sqrt{2\pi \frac{\Theta_{12}}{m_1}}^{l_1}} \right) + \ln \left(\frac{n_2}{\sqrt{2\pi \frac{\Lambda_{21}}{m_2}}^d} \frac{1}{\sqrt{2\pi \frac{\Theta_{21}}{m_2}}^{l_2}} \right)$$

$$\leq \varepsilon \ln \left(\frac{n_1}{\sqrt{2\pi \frac{\Lambda_1}{m_1}}^d} \frac{1}{\sqrt{2\pi \frac{\Theta_1}{m_1}}^{l_1}} \right) + \ln \left(\frac{n_2}{\sqrt{2\pi \frac{\Lambda_2}{m_2}}^d} \frac{1}{\sqrt{2\pi \frac{\Theta_2}{m_2}}^{l_2}} \right),$$

which is equivalent to the condition proven in lemma 8.2.8. $\qquad \square$

Theorem 8.2.10 (H-theorem for mixture). *Assume $f_1, f_2 > 0$. Assume $\nu_{11}n_1 \geq \nu_{12}n_2$, $\nu_{22}n_2 \geq \nu_{21}n_1$, $\alpha, \delta \neq 1, l_1, l_2 \neq 0$. Assume the relationship between the collision frequencies (8.4), the conditions for the interspecies Maxwell distributions (8.18), (8.19), (8.20), (8.21) and (8.22) and the positivity of all temperatures, then*

$$\sum_{k=1}^{2} [\nu_{kk}n_k \int (G_k - f_k) \ln f_k dv d\eta_{l_k} + \nu_{kk}n_k \int (\widetilde{G}_k - \widehat{G}_k) \ln \widehat{G}_k dv d\eta_{l_k}]$$

$$+ \nu_{11}n_1 \int (\widetilde{G}_1 - \widehat{G}_1) \ln \widehat{G}_1 dv d\eta_{l_1} + \nu_{22}n_2 \int (\widetilde{G}_2 - \widehat{G}_2) \ln \widehat{G}_2 dv d\eta_{l_2}$$

$$+ \nu_{12}n_2 \int (M_{12} - f_1) \ln f_1 dv d\eta_{l_1} + \nu_{21}n_1 \int (M_{21} - f_2) \ln f_2 dv d\eta_{l_2} \leq 0,$$

with equality if and only if f_1 and f_2 are in local equilibrium (see definition 8.2.1).

Remark 8.2.7. The inequality in the H-theorem is still true if $l_1 = 0$ or $l_2 = 0$ which means that one species is mono-atomic. In this case only the equalities with Θ_1 and Θ_2, respectively in the local equilibrium vanish.

Proof. The fact that $\nu_{kk}n_k \int (G_k - f_k) \ln f_k dv d\eta_{l_k} + \nu_{kk}n_k \int (\widetilde{G}_k - \widehat{G}_k) \ln \widehat{G}_k dv d\eta_{l_k} \leq 0, k = 1, 2$ is shown in lemma 8.2.7. In both cases we have equality if and only if $f_1 = G_1$ with $\Lambda_1 = \Theta_1 = T_1^{trans} = T_1^{rot}$ and $f_2 = G_2$ with $\Lambda_2 = \Theta_2 = T_2^{trans} = T_2^{rot}$. Let us define

$$S(f_1, f_2) := \nu_{11}n_1 \int (\widetilde{G}_1 - \widehat{G}_1) \ln \widehat{G}_1 dv d\eta_{l_1} + \nu_{22}n_2 \int (\widetilde{G}_2 - \widehat{G}_2) \ln \widehat{G}_2 dv d\eta_{l_2}$$

$$+ \nu_{12}n_2 \int (M_{12} - f_1) \ln f_1 dv d\eta_{l_1} + \nu_{21}n_1 \int (M_{21} - f_2) \ln f_2 dv d\eta_{l_2}.$$

The task is to prove that $S(f_1, f_2) \leq 0$. Consider now $S(f_1, f_2)$ and apply the inequality from lemma 1.3.12 to each of the terms in S.

$$S \leq \nu_{11}n_1\left[\int \widetilde{G}_1 \ln \widetilde{G}_1 dvd\eta_{l_1} - \int \widehat{G}_1 \ln \widehat{G}_1 dvd\eta_{l_1} + \int \widehat{G}_1 dvd\eta_{l_1} - \int \widetilde{G}_1 dvd\eta_{l_1}\right]$$

$$+\nu_{12}n_2\left[\int M_{12} \ln M_{12} dvd\eta_{l_1} - \int f_1 \ln f_1 dvd\eta_{l_1} + \int f_1 dv\eta - \int M_{12} dvd\eta_{l_1}\right]$$

$$+\nu_{21}n_1\left[\int M_{21} \ln M_{21} dvd\eta_{l_2} - \int f_2 \ln f_2 dvd\eta_{l_2} + \int f_2 dvd\eta - \int M_{21} dvd\eta_{l_2}\right]$$

$$+\nu_{22}n_2\left[\int \widetilde{G}_2 \ln \widetilde{G}_2 dvd\eta_{l_2} - \int \widehat{G}_2 \ln \widehat{G}_2 dvd\eta_{l_2} + \int \widehat{G}_2 dvd\eta_{l_2} - \int \widetilde{G}_2 dvd\eta_{l_2}\right],$$

with equality if and only if $f_1 = M_{12}$, $f_2 = M_{21}$, $\widetilde{G}_1 = \widehat{G}_1$ and $\widetilde{G}_2 = \widehat{G}_2$. Combining this with the condition for equality of the single collision term $f_1 = G_1$ with $\Lambda_1 = \Theta_1 = T_1^{trans} = T_1^{rot}$ and $f_2 = G_2$ with $\Lambda_2 = \Theta_2 = T_2^{trans} = T_2^{rot}$, we get that we have equality if and only if we are in local equilibrium (see definition 8.2.1). Since $\widehat{G}_1, \widetilde{G}_1, f_1$ and M_{12} have the same density and $\widehat{G}_2, \widetilde{G}_2, M_{21}$ and f_2 have the same density, too, the right-hand side reduces to

$$S \leq \nu_{11}n_1\left[\int \widetilde{G}_1 \ln \widetilde{G}_1 dvd\eta_{l_1} - \int \widehat{G}_1 \ln \widehat{G}_1 dvd\eta_{l_1}\right]$$

$$+\nu_{12}n_2\left[\int M_{12} \ln M_{12} dvd\eta_{l_1} - \int f_1 \ln f_1 dvd\eta_{l_1}\right]$$

$$+\nu_{21}n_1\left[\int M_{21} \ln M_{21} dvd\eta_{l_2} - \int f_2 \ln f_2 dvd\eta_{l_2}\right]$$

$$+\nu_{22}n_2\left[\int \widetilde{G}_2 \ln \widetilde{G}_2 dvd\eta_{l_2} - \int \widehat{G}_2 \ln \widehat{G}_2 dvd\eta_{l_2}\right].$$

According to the second part of lemma 8.2.6, we obtain

$$S \leq \nu_{11}n_1\left[\int \widetilde{G}_1 \ln \widetilde{G}_1 dvd\eta_{l_1} - \int G_1 \ln G_1 dvd\eta_{l_1}\right]$$

$$+\nu_{12}n_2\left[\int M_{12} \ln M_{12} dvd\eta_{l_1} - \int f_1 \ln f_1 dvd\eta_{l_1}\right]$$

$$+\nu_{21}n_1\left[\int M_{21} \ln M_{21} dvd\eta_{l_2} - \int f_2 \ln f_2 dvd\eta_{l_2}\right]$$

$$+\nu_{22}n_2\left[\int \widetilde{G}_2 \ln \widetilde{G}_2 dvd\eta_{l_2} - \int G_2 \ln G_2 dvd\eta_{l_2}\right].$$

According to lemma 8.2.9, the last part of lemma 8.2.6 and the assumption that $\nu_{kk}n_k \geq \nu_{kj}n_j$, $k,j = 1,2, k \neq j$, we get

$$S \leq \nu_{11}n_1\Big[\int \widetilde{G}_1 \ln \widetilde{G}_1 dvd\eta_{l_1} - \int G_1 \ln G_1 dvd\eta_{l_1}\Big]$$

$$+ \nu_{12}n_2\Big[\int G_1 \ln G_1 dvd\eta_{l_1} - \int f_1 \ln f_1 dvd\eta_{l_1}\Big]$$

$$+ \nu_{21}n_1\Big[\int G_2 \ln G_2 dvd\eta_{l_2} - \int f_2 \ln f_2 dvd\eta_{l_2}\Big]$$

$$+ \nu_{22}n_2\Big[\int \widetilde{G}_2 \ln \widetilde{G}_2 dvd\eta_{l_2} - \int G_2 \ln G_2 dvd\eta_{l_2}\Big]$$

$$\leq \nu_{12}n_2\Big[\int \widetilde{G}_1 \ln \widetilde{G}_1 dvd\eta_{l_1} - \int f_1 \ln f_1 dvd\eta_{l_1}\Big]$$

$$+ \nu_{21}n_1\Big[\int \widetilde{G}_2 \ln \widetilde{G}_2 dvd\eta_{l_2} - \int f_2 \ln f_2 dvd\eta_{l_2}\Big],$$

which leads to $S \leq 0$ using the first part of lemma 8.2.6 and remark 8.2.6.

$\square$

Define $\frac{1}{z_k} := \frac{1}{Z_k^r}\frac{d+l_k}{d}$, $k = 1,2$ and the total entropy

$$H(f_1, f_2) = \int (f_1 \ln f_1 + 2z_1 \widehat{G}_1 \ln \widehat{G}_1)dvd\eta_{l_1} + \int (f_2 \ln f_2 + 2z_2 \widehat{G}_2 \ln \widehat{G}_2)dvd\eta_{l_2}.$$

We can compute

$$\partial_t H(f_1, f_2) + \nabla_x \cdot \int (f_1 \ln f_1 + 2z_1 \widehat{G}_1 \ln \widehat{G}_1)vdvd\eta_{l_1}$$

$$+\nabla_x \cdot \int (f_2 \ln f_2 + 2z_2 \widehat{G}_2 \ln \widehat{G}_2)vdvd\eta_{l_2} = S(f_1, f_2) + R(f_1, f_2),$$

by multiplying the BGK equation for species 1 by $\ln f_1$, the BGK equation for the species 2 by $\ln f_2$, the equations (8.11) by $2z_k \ln G_k$ and add the integrals with respect to v and η_{l_1} and η_{l_2}, respectively. The remaining term $R(f_1, f_2)$ can be bounded by zero from below by an explicit computation assuming that Λ_k and Θ_k are bounded from below and above and assume that $T_k^{rot} \geq \widetilde{C}\Theta_k$ for an appropriate $\widetilde{C}$ and z_k small enough. The additional estimate $T_k^{rot} \geq \widetilde{C}\Theta_k$ helps to indicate how to choose the initial data of the artificial temperature Θ_k.

Corollary 8.2.11 (Entropy inequality for mixtures). *Assume $f_1, f_2 > 0$, Λ_k and Θ_k are bounded from below and above and $T_k^{rot} \geq \widetilde{C}\Theta_k$ for an appropriate $\widetilde{C}$ and z_k small enough. Assume relationship (8.4), the conditions (8.18), (8.19), (8.20), (8.21) and (8.22) and the positivity of all temperatures (8.27), then we have the following entropy inequality*

$$\partial_t (H(f_1, f_2)) + \nabla_x \cdot \left(\int v(f_1 \ln f_1 + 2z_1 \widehat{G}_1 \ln \widehat{G}_1 + f_2 \ln f_2 + 2z_2 \widehat{G}_2 \ln \widehat{G}_2)dvd\eta_{l_2}\right) \leq 0$$

with equality if and only if f_1 and f_2 are in local equilibrium (see definition 8.2.1).

Remark 8.2.8. By computing the integrals $\int \widehat{G}_k \ln \widehat{G}_k dv d\eta_{l_k}$ for $k = 1, 2$, and $\int v \widehat{G}_k \ln \widehat{G}_k dv d\eta_{l_k}$, we see that

$$\partial_t \left[\int \widehat{G}_1 \ln \widehat{G}_1 dv d\eta_{l_1} + \int \widehat{G}_2 \ln \widehat{G}_2 dv d\eta_{l_2} \right] + \nabla_x \cdot \left[\int v(\widehat{G}_1 \ln \widehat{G}_1 + \widehat{G}_2 \ln \widehat{G}_2) dv d\eta_{l_2} \right] \leq 0,$$

is equivalent to

$$\partial_t (\det(\Lambda_1^{ten})\Theta_1^{l_1} + \det(\Lambda_2^{ten})\Theta_2^{l_2}) + \nabla_x \cdot ((\det(\Lambda_1^{ten})\Theta_1^{l_1} + \det(\Lambda_2^{ten})\Theta_2^{l_2})u_k) \leq 0.$$

Hence, we could also consider the entropy

$$H(f_1, f_2) = \sum_{k=1}^{2} \int f_k \ln f_k dv d\eta_{l_k} + z_1 \det(\Lambda_1^{ten})\Theta_1^{l_1} + z_2 \det(\Lambda_2^{ten})\Theta_2^{l_2}.$$

8.3 Comparison with the ES-BGK model of Andries, Le Tallec, Perlat and Perthame for one species of polyatomic molecules

We will now consider a different ES-BGK model for a single species ES-BGK model of polyatomic molecules. In [2], a distribution function $f(t, x, v, I)$ depending on the position $x \in \mathbb{R}^3$, the velocity $v \in \mathbb{R}^3$ and internal energy $\varepsilon(I) = I^{\frac{2}{\delta}}, I \in \mathbb{R}^+$ at time t is considered. δ denotes the number of degrees of freedom in internal energy. In [2], it is assumed that the mass of the particles is equal to 1. In the following, we assume additionally that $k_B = 1$ in this model. The density ρ and mean velocity u are defined as n and u in the model described in the previous section but now the integration is with respect to v and I instead of v and η. The energy is defined as

$$E(x, t) = \int \int (\frac{1}{2}|v|^2 + I^{\frac{\delta}{2}})f dv dI = \frac{1}{2}\rho|u|^2 + \rho e.$$

The specific internal energy can be divided into

$$e_{tr} = \frac{1}{\rho} \int \int \frac{1}{2}|v - u|^2 f dv dI,$$

$$e_{int} = \frac{1}{\rho} \int \int I^{\frac{2}{\delta}} f dv dI,$$

and associate with this the corresponding temperatures

$$e = e_{tr} + e_{int} = \frac{3 + \delta}{2}T_{equ},$$

$$e_{tr} = \frac{3}{2}T_{tr},$$

$$e_{int} = \frac{\delta}{2} T_{int},$$

and define $T_{rel} = \theta T_{equ} + (1 - \theta) T_{int}$. In [2] the following generalized Gaussian for the single species ES-BGK model is considered

$$\widetilde{G}[f] = \frac{\rho \Lambda_\delta}{\sqrt{\det(2\pi\mathcal{T})}} \frac{1}{T_{rel}^{\frac{\delta}{2}}} \exp\left(-\frac{1}{2}(v - u) \cdot \mathcal{T}^{-1} \cdot (v - u) + \frac{I^{\frac{\delta}{2}}}{T_{rel}}\right),$$

with the tensor $\mathcal{T} = (1 - \theta)((1 - \tilde{\mu})T_{tr}\mathbb{1} + \tilde{\mu}\Theta) + \theta T_{equ}\mathbb{1}$ where only the translational part is replaced by a tensor. Θ denotes the pressure tensor, Λ_δ is a constant ensuring that the integral of $\widetilde{G}[f]$ with respect to v and I is equal to the density ρ. The convex combination in θ takes into account that T_{tr} and T_{int} relax towards the common value T_{equ}. In the space-homogeneous case we see that we get the following macroscopic equations

$$\partial_t T_{tr} = C(T_{tr}(1 - \theta) + \theta T_{equ} - T_{tr}) = C\theta(T_{equ} - T_{tr}),$$
$$\partial_t T_{int} = C\theta(T_{equ} - T_{int}),$$

with some coefficient C. These macroscopic equations describe a relaxation of T_{tr} and T_{int} towards T_{equ}.

In this chapter, we took [14] as basis to extend it to mixtures. The main differences of the model in [2] and the model in [14] are the following. The model in [2] has one variable $I \in \mathbb{R}^+$ for all degrees of freedom in internal energy and the model in [14] has one variable $\eta \in \mathbb{R}^M$ to each degree of freedom in internal energy. Moreover, the relaxation of the translational and rotational temperatures to a common value is done in [2] by introducing a relaxation temperature T_{rel} and in the model [14] it is done by the additional relaxation equation (8.11).

8.4   Applications

In this section, we consider two applications of the polyatomic ES-BGK model presented in section 8.2. First, we apply the Chu reduction on our model in order to reduce the complexity of this model for numerical purposes. Second, we consider a gas mixture of one species of mono-atomic and one species of diatomic molecules.

8.4.1   Chu reduction

In the polyatomic setting the distribution functions f_k depend on $2d + l_k + 1$ independent variables. This makes the problem extremely complex from a computational point of view, due to its high dimensionality. However, it is possible to reduce the number of dimensions of the distribution function with Chu's reduction, proposed in [27]. In the standard BGK model, Chu's reduction can be applied whenever the distribution function f_k depends only on $r < d$ degrees of freedom in space. Then it is possible to rewrite the kinetic equation using only r degrees of freedom, also in the microscopic velocity space. In order to reduce the complexity of the variables for rotational and vibrational energy degrees of freedom $\mu_1, ..., \mu_{l_k}$ we apply the Chu reduction in the polyatomic case. We will apply the reduction to aggregate the internal energy degrees of freedom. Let us consider the case in which we want to reduce the M rotational and vibrational degrees of freedom, while the system has d translational degrees of freedom. It is possible to apply the Chu reduction since $\eta_1, ..., \eta_{l_k}$ do not appear in the transport operators on (8.2). We consider the system of equations

$$\partial_t f_1 + v \cdot \nabla_x f_1 = \nu_{11} n_1 (G_1 - f_1) + \nu_{12} n_2 (M_{12} - f_1),$$
$$\partial_t f_2 + v \cdot \nabla_x f_2 = \nu_{22} n_2 (G_2 - f_2) + \nu_{21} n_1 (M_{21} - f_2).$$

Consider the reduced functions

$$g_1 = \int f_1 d\eta_{l_1}, \quad g_2 = \int f_2 d\eta_{l_2}.$$

Then they satisfy the equations

$$\partial_t g_1 + v \cdot \nabla_x g_1 = \nu_{11} n_1 (\widetilde{G}_1 - g_1) + \nu_{12} n_2 (\widetilde{M}_{12} - g_1),$$
$$\partial_t g_2 + v \cdot \nabla_x g_2 = \nu_{22} n_2 (\widetilde{G}_2 - g_2) + \nu_{21} n_1 (\widetilde{M}_{21} - g_2),$$

where $\widetilde{G}_1, \widetilde{G}_2, \widetilde{M}_{12}$ and $\widetilde{M}_{21}$ are given by

$$\widetilde{G}_1 = \int G_1 d\eta_{l_1}, \quad \widetilde{M}_{12} = \int M_{12} d\eta_{l_1},$$

$$\widetilde{G}_2 = \int G_2 d\eta_{l_2}, \quad \widetilde{M}_{21} = \int M_{21} d\eta_{l_2}.$$

It is possible to compute the densities

$$n_1 = \int \int f_1 d\eta_{l_1} dv = \int g_1 dv,$$

$$n_2 = \int \int f_2 d\eta_{l_2} dv = \int g_2 dv,$$

the velocities

$$u_1 = \int \int v f_1 d\eta_{l_1} dv = \int v g_1 dv,$$

$$u_2 = \int \int v f_2 d\eta_{l_2} dv = \int v g_2 dv,$$

the temperatures

$$\Lambda_1 = \frac{1}{n_1} \int \int |v - u_1|^2 f_1 d\eta_{l_1} dv, = \frac{1}{n_1} \int |v - u_1|^2 g_1 dv$$

$$\Lambda_2 = \frac{1}{n_2} \int |v - u_2|^2 g_2 dv,$$

$$\Theta_1 = \frac{1}{n_1} \int \int |\eta_{l_1}|^2 f_1 d\eta_{l_1} dv = \frac{1}{n_1} \int |\eta_{l_1}|^2 h_1 dv,$$

$$\Theta_2 = \frac{1}{n_2} \int |\eta_{l_2}|^2 h_2 dv,$$

if we define the reduced functions

$$h_1 = \int |\eta_{l_1}|^2 f_1 d\eta_{l_1}, \quad h_2 = \int |\eta_{l_2}|^2 f_2 d\eta_{l_2},$$

which solve the equations

$$\partial_t h_1 + v \cdot \nabla_x h_1 = \nu_{11} n_1 (\widetilde{G}_1 - h_1) + \nu_{12} n_2 (\widetilde{M}_{12} - h_1),$$

$$\partial_t h_2 + v \cdot \nabla_x h_2 = \nu_{22} n_2 (\widetilde{G}_2 - h_2) + \nu_{21} n_1 (\widetilde{M}_{21} - h_2),$$

where $\widetilde{G}_1, \widetilde{G}_2, \widetilde{M}_{12}$ and $\widetilde{M}_{21}$ are given by

$$\widetilde{G}_1 = \int |\eta_{l_1}|^2 G_1 d\eta_{l_1}, \quad \widetilde{M}_{12} = \int |\eta_{l_1}|^2 M_{12} d\eta,$$

$$\widetilde{G}_2 = \int |\eta_{l_2}|^2 G_2 d\eta_{l_2}, \quad \widetilde{M}_{21} = \int |\eta_{l_2}|^2 M_{21} d\eta_{l_2}.$$

If we compute $\widetilde{G}_k, \widetilde{M}_{12}, \widetilde{M}_{21}, \widetilde{G}_k, \widetilde{M}_{12}, \widetilde{M}_{21}$ for $k = 1, 2$, we get

$$\widetilde{G}_k(x, v, t) = \frac{n_k}{\sqrt{\det(2\pi \frac{\Lambda_k^{ES}}{m_k})}} \exp(-m_k(v - u_k)(\Lambda_k^{ES})^{-1} \cdot (v - u_k)), \quad k = 1, 2,$$

$$\widetilde{M}_{12}(x, v, t) = \frac{n_1}{\sqrt{2\pi \frac{\Lambda_{12}}{m_1}}^d} \exp\left(-\frac{|v - u_{12}|^2}{2 \frac{\Lambda_{12}}{m_1}}\right),$$

$$\widetilde{M}_{21}(x, v, t) = \frac{n_2}{\sqrt{2\pi \frac{\Lambda_{21}}{m_2}}^d} \exp\left(-\frac{|v - u_{21}|^2}{2 \frac{\Lambda_{21}}{m_2}}\right),$$

$$\widetilde{\widetilde{G}}_k(x,v,t) = \frac{n_k}{\sqrt{\det(2\pi\frac{\Lambda_k^{ES}}{m_k})}} \exp(-m_k(v-u_k)(\Lambda_k^{ES})^{-1}\cdot(v-u_k))\Theta_k, \quad k=1,2,$$

$$\widetilde{\widetilde{M}}_{12}(x,v,t) = \frac{n_1}{\sqrt{2\pi\frac{\Lambda_{12}}{m_1}}^d} \exp\left(-\frac{|v-u_{12}|^2}{2\frac{\Lambda_{12}}{m_1}}\right)\Theta_{12},$$

$$\widetilde{\widetilde{M}}_{21}(x,v,t) = \frac{n_2}{\sqrt{2\pi\frac{\Lambda_{21}}{m_2}}^d} \exp\left(-\frac{|v-u_{21}|^2}{2\frac{\Lambda_{21}}{m_2}}\right)\Theta_{21}.$$

We are able to compute all the six Maxwell distributions because we can compute all moments by the previous computation.

8.4.2 A mixture consisting of a mono-atomic and a diatomic gas

We consider now the special case of two species, one species is mono-atomic and has only translational degrees of freedom $l_1 = 0$, the other one is diatomic and has in addition two rotational degrees of freedom $l_2 = 2$ and both have a general number of degrees of freedom in translations $d \in \mathbb{N}$. In this case the total number of rotational degrees of freedom is $M = l_1 + l_2 = 2$ since in sum we have two possible rotations. Our variable for the rotational energy degrees of freedom are $\eta \in \mathbb{R}^2$, $\eta_{l_1} = (0,0)^T$, $\eta_{l_2} = \eta$, since η_{l_k} coincides with η in the components corresponding to the rotational degrees of freedom of species k and is zero in the other components. So our distribution function $f_1(x,v,t)$ of species 1 depends on x,v and t and our distribution function $f_2(x,v,\eta,t)$ of species 2 depends on x,v,η and t. The moments of f_1 are given by

$$\int f_1(v) \begin{pmatrix} 1 \\ v \\ m_1|v-u_1|^2 \\ m_1(v-u_1)\otimes(v-u_1) \end{pmatrix} dv =: \begin{pmatrix} n_1 \\ n_1 u_1 \\ dn_1 T_1^{trans} \\ \mathbb{P}_1 \end{pmatrix},$$

and the moments of species 2 are given by

$$\int f_2(v,\eta) \begin{pmatrix} 1 \\ v \\ \eta \\ m_2|v-u_2|^2 \\ m_2|\eta|^2 \\ m_2(v-u_2)\otimes(v-u_2) \end{pmatrix} dvd\eta =: \begin{pmatrix} n_2 \\ n_2 u_2 \\ 0 \\ dn_2 T_2^{trans} \\ l_2 n_2 T_k^{rot} \\ \mathbb{P}_2 \end{pmatrix}.$$

The third equality is an assumption. We could also define a general $\bar{\eta} := \int f_2(v,\eta)\eta dvd\eta$. Our model reduces to

$$\partial_t f_1 + v\cdot\nabla_x f_1 = \nu_{11}n_1(G_1(f_1)-f_1) + \nu_{12}n_2(M_{12}(f_1,f_2)-f_1),$$
$$\partial_t f_2 + v\cdot\nabla_x f_2 = \nu_{22}n_2(G_2(f_2)-f_2) + \nu_{21}n_1(M_{21}(f_1,f_2)-f_2),$$

with the modified Maxwell distributions

$$G_1(x, v, t) = \frac{n_1}{\sqrt{\det(2\pi\frac{\Lambda_1^{ES}}{m_1})}} \exp\left(-\frac{1}{2}(v - u_1) \cdot \left(\frac{\Lambda_1^{ES}}{m_1}\right)^{-1} \cdot (v - u_1)\right),$$

$$G_2(x, v, \eta, t) = \frac{n_2}{\sqrt{\det(2\pi\frac{\Lambda_2^{ES}}{m_2})}} \frac{1}{\sqrt{2\pi\frac{\Theta_2}{m_2}}^{l_2}} \exp\left(-\frac{1}{2}(v - u_2) \cdot \left(\frac{\Lambda_2^{ES}}{m_2}\right)^{-1} \cdot (v - u_2) - \frac{1}{2}\frac{m_2|\eta|^2}{\Theta_2}\right),$$

$$M_{12}(x, v, t) = \frac{n_{12}}{\sqrt{2\pi\frac{\Lambda_{12}}{m_1}}^d} \exp\left(-\frac{|v - u_{12}|^2}{2\frac{\Lambda_{12}}{m_1}}\right),$$

$$M_{21}(x, v, \eta, t) = \frac{n_{21}}{\sqrt{2\pi\frac{\Lambda_{21}}{m_2}}^d} \frac{1}{\sqrt{2\pi\frac{\Theta_{21}}{m_2}}^{l_2}} \exp\left(-\frac{|v - u_{21}|^2}{2\frac{\Lambda_{21}}{m_2}} - \frac{|\eta|^2}{2\frac{\Theta_{21}}{m_2}}\right),$$

where

$$\Lambda_1^{ES} = (1 - \tilde{\mu}_1)T_1^{trans}\mathbb{1}_n + \tilde{\mu}_1\frac{\mathbb{P}_1}{n_1},$$

$$\Lambda_2^{ES} = (1 - \tilde{\mu}_2)\Lambda_2\mathbb{1}_n + \tilde{\mu}_2\frac{\Lambda_2^{ten}}{n_2},$$

with $\tilde{\mu}_k \in \mathbb{R}$, $k = 1, 2$. For Λ_2^{ten} we use the additional relaxation equation

$$\partial_t\widehat{G}_2 + v \cdot \nabla_x\widehat{G}_2 = \frac{\nu_{22}n_2}{Z_r^2}\frac{d+2}{d}(\check{G}_2 - \widehat{G}_2) + \nu_{22}n_2(G_2 - f_2) + \nu_{21}n_1(M_{21} - f_2). \tag{8.53}$$

Here $\widehat{G}_2$ is given by

$$\widehat{G}_2 = \frac{n_2}{\sqrt{\det(2\pi\Lambda_2^{ten})}} \exp\left(-\frac{1}{2}(v - u_2) \cdot \left(\frac{\Lambda_2^{ten}}{m_2}\right)^{-1} \cdot (v - u_2) - \frac{m_2|\eta|^2}{2T_2^{rot}}\right), \tag{8.54}$$

and $\check{G}_2$ is given by

$$\check{G}_2 = \frac{n_2}{\sqrt{\det(2\pi\frac{T_2^{ten}}{m_2})}} \frac{1}{\sqrt{2\pi\frac{T_2}{m_2}}^2} \exp\left(-\frac{1}{2}(v - u_2) \cdot \left(\frac{T_2^{ten}}{m_2}\right)^{-1} \cdot (v - u_2) - \frac{1}{2}\frac{m_2|\eta|^2}{T_2}\right),$$

where the components of T_2^{ten} are defined in the following way

$$(T_2^{ten})_{ii} = T_2 := \frac{d}{d+2}\Lambda_2 + \frac{2}{d+2}\Theta_2 \qquad \text{for} \quad i = 1, \dots, d,$$

$$(T_2^{ten})_{ij} = \frac{d}{d+2}(\mathbb{P}_2)_{ij} \qquad \text{for} \quad i, j = 1, \dots, d, i \neq j. \tag{8.55}$$

We couple this with conservation of internal energy of species 2

$$\frac{d}{2}n_2\Lambda_2 = \frac{d}{2}n_2T_2^{trans} + \frac{l_2}{2}n_2T_2^{rot} - \frac{l_2}{2}n_2\Theta_2.$$

If we multiply (8.53) by $|\eta|^2$, integrate with respect to v and η and use conservation of mass, we obtain the following macroscopic equation

$$\partial_t(\Lambda_2^{ten}) + u_2 \cdot \nabla_x(\Lambda_2^{ten}) = \frac{\nu_{22}n_2}{Z_r^2}\frac{d+2}{d}(T_2^{ten} - \Lambda_2^{ten}) + \nu_{22}n_2(\Lambda_2^{ES} - \mathbb{P}_2)$$

$$+ \nu_{21}n_1(\Theta_{12} - T_2^{rot}).$$

If we assume that

$$n_{12} = n_1 \quad \text{and} \quad n_{21} = n_2,$$
$$u_{12} = \delta u_1 + (1-\delta)u_2, \quad \delta \in \mathbb{R},$$

and

$$\Lambda_{12} = \alpha T_1^{trans} + (1-\alpha)\Lambda_2 + \gamma|u_1 - u_2|^2, \quad 0 \le \alpha \le 1, \gamma \ge 0,$$

we have conservation of mass, total momentum and total energy provided that

$$u_{21} = u_2 - \frac{m_1}{m_2}\varepsilon(1-\delta)(u_2 - u_1),$$

$$\Lambda_{21} + \frac{l_2}{d}\Theta_{21} = \left[\frac{1}{d}\varepsilon m_1(1-\delta)\left(\frac{m_1}{m_2}\varepsilon(\delta-1) + \delta + 1\right) - \varepsilon\gamma\right]|u_1 - u_2|^2$$

$$+ \varepsilon(1-\alpha)T_1^{trans} + (1 - \varepsilon(1-\alpha))\Lambda_2 + \frac{l_2}{d}\Theta_2.$$

We take into account the symmetry of the both temperatures and choose

$$\Lambda_{21} = \varepsilon(1-\alpha)\Lambda_1 + (1 - \varepsilon(1-\alpha))\Lambda_2$$
$$+ \left[\frac{1}{d}\varepsilon m_1(1-\delta)\left(\frac{m_1}{m_2}\varepsilon(\delta-1) + \delta + 1\right) - \varepsilon\gamma\right]|u_1 - u_2|^2,$$
$$\Theta_{21} = \Theta_2.$$

8.5 A kinetic model leading to an equation of state different from the ideal gas law

Let $x \in \mathbb{R}^3$ and $v \in \mathbb{R}^3$ be the phase space variables and $t \ge 0$ the time. Let l the number of rotational and vibrational degrees of freedom of the gas. Further, $\eta \in \mathbb{R}^l$ is the variable for the internal energy degrees of freedom.

We consider a distribution function $f(x, v, \eta, t) > 0$. Furthermore, we relate the distribution function to macroscopic quantities by mean-values of f as follows

$$\int f(v, \eta) \begin{pmatrix} 1 \\ v \\ \eta \\ m|v - u|^2 \\ m|\eta - \bar{\eta}_l|^2 \end{pmatrix} dv d\eta =: \begin{pmatrix} n \\ n\,u \\ n\,\bar{\eta}_l \\ 3\,n\,T^{trans} \\ l\,n\,T^{rot} \end{pmatrix}, \tag{8.56}$$

where n is the number density, u the mean velocity and T^{trans} the temperature of the translation, T^{rot} the temperature of the rotation and vibration. Note that we write T^{trans} and T^{rot} instead of $k_B T^{trans}$ and $k_B T^{rot}$, where k_B is Boltzmann's constant.

We consider the following model

$$\partial_t f + v \cdot \nabla_x f = \nu n (M - f), \tag{8.57}$$

with the Maxwell distribution

$$M(x, v, \eta, t) = \frac{n}{\sqrt{2\pi\frac{T}{m}}^3} \frac{1}{\sqrt{2\pi\frac{T}{m}}^l} \exp\left(-\frac{|v-u|^2}{2\frac{T}{m}} - \frac{|\eta-w|^2}{2\frac{T}{m}}\right), \tag{8.58}$$

where T is the total equilibrium temperature and is given by

$$T := \frac{3T^{trans} + lT^{rot}}{d + l}. \tag{8.59}$$

Now, assume that we assume that $\bar{\eta}_l$ is fixed and equal to a vector w in $\mathbb{R}^l$ such that $|w|^2 = 2\frac{p_\infty}{mn}$ for a given constant p_∞ in the Maxwell distribution in (8.3). Since $|w|^2$ represents the kinetic energy in the rotation and vibration, p_∞ may be related to the moment of inertia in the case of rotations or the Hook'sches law in the case of vibrations. In this case, we will obtain an equation of state given by

$$p = nT + \text{const.}$$

This is shown in the following. The additional constant takes into account an attractive force between the particles which is neglected in the case of an ideal gas.

Theorem 8.5.1 (Macroscopic equations). *Assume $f \in L^\infty(dvd\eta)$ decays fast enough to zero in the v and η variables and is a solution to (8.57) in the sense of distributions. If in addition f coincides with the Maxwell distribution M given by (8.58), it satisfies the following local macroscopic conservation laws.*

$$\partial_t n + \nabla_x (nu) = 0$$
$$\partial_t (mnu) + \nabla_x (nT) + \nabla_x \cdot (mnu \otimes u) = 0,$$
$$\partial_t \left(\frac{m}{2}n|u|^2 + \frac{3+l}{2}nT\right) + \nabla_x \cdot \left(\left(\frac{5+l}{2}nT + p_\infty\right)u\right) + \nabla_x \cdot \left(\frac{m}{2}n|u|^2 u\right) = 0.$$

Proof. If we integrate the equation (8.57) with respect to v and η and use $f = M$, we get:

$$\int \partial_t M dv d\eta + \int v \cdot \nabla_x M dv d\eta = 0.$$

If we formally exchange integration and derivatives, we obtain

$$\partial_t \int M dv d\eta + \nabla_x \cdot \int v M dv d\eta = 0.$$

This is equivalent to

$$\partial_t n + \nabla_x \cdot (nu) = 0,$$

since we have

$$\int M \, dv \, d\eta = \int \frac{n}{\sqrt{2\pi \frac{T}{m}}^3} \exp\left(-\frac{|v-u|^2}{2\frac{T}{m}}\right) dv \int \frac{1}{\sqrt{2\pi \frac{T}{m}}^l} \exp\left(-\frac{|\eta-w|^2}{2\frac{T}{m}}\right) d\eta$$

$$= n$$

and

$$\int M v \, dv \, d\eta = \int v \frac{n}{\sqrt{2\pi \frac{T}{m}}^3} \exp\left(-\frac{|v-u|^2}{2\frac{T}{m}}\right) dv \int \frac{1}{\sqrt{2\pi \frac{T}{m}}^l} \exp\left(-\frac{|\eta-w|^2}{2\frac{T}{m}}\right) d\eta$$

$$= nu.$$

Multiplying the equation (8.57) by mv, integrating it with respect to v and η and using $f = M$, leads to

$$m \int v \partial_t M \, dv \, d\eta + m \int v \, v \cdot \nabla_x M \, dv \, d\eta = 0.$$

We formally exchange derivative and integration and obtain

$$m \, \partial_t (nu) + \nabla_x \cdot \int mv \otimes v M \, dv \, d\eta = 0.$$

We can compute

$$\int v \otimes v M \, dv \, d\eta = \int v \otimes v \frac{n}{\sqrt{2\pi \frac{T}{m}}^3} \exp\left(-\frac{|v-u|^2}{2\frac{T}{m}}\right) dv \int \frac{1}{\sqrt{2\pi \frac{T}{m}}^l} \exp\left(-\frac{|\eta-w|^2}{2\frac{T}{m}}\right) d\eta$$

$$= nu \otimes u + n\frac{T}{m},$$

so the second term turns into

$$\nabla_x (nT) + \nabla_x \cdot (mnu \otimes u).$$

So all in all, we get

$$\partial_t (mnu) + \nabla_x (nT) + \nabla_x \cdot (mnu \otimes u) = 0.$$

Multiplying the equation (8.57) by $\frac{m}{2}(|v|^2 + |\eta|^2)$, integrating it with respect to v and η and using $f = M$, leads to

$$\frac{m}{2} \int (|v|^2 + |\eta|^2) \partial_t M \, dv \, d\eta + \frac{m}{2} \int (|v|^2 + |\eta|^2) v \cdot \nabla_x M \, dv \, d\eta = 0.$$

We formally exchange derivative and integration and obtain

$$\partial_t \left(\frac{m}{2} n |u|^2 + \frac{3+l}{2} nT \right) + \nabla_x \cdot \int mv(|v|^2 + |\eta|^2) M \, dv d\eta = 0,$$

since we have

$$m \int (|v|^2 + |\eta|^2) M \, dv d\eta = \frac{mn|u|^2}{2} + \frac{3}{2} nT + \frac{mn|w|^2}{2} + \frac{l}{2} nT$$

$$= \frac{mn|u|^2}{2} + \frac{3+l}{2} nT + p_\infty,$$

where p_∞ is a constant, so its time derivative vanishes. Last, we compute

$$m \int v(|v|^2 + |\eta|^2) M \, dv d\eta = \int |v|^2 v M \, dv d\eta + \int |\eta|^2 v M \, dv d\eta$$

$$= \left(\frac{mn}{2} |u|^2 + \frac{5}{2} nT \right) u$$

$$+ \int v \frac{n}{\sqrt{2\pi \frac{T}{m}}^3} \exp \left(-\frac{|v-u|^2}{2\frac{T}{m}} \right) dv \int |\eta|^2 \frac{1}{\sqrt{2\pi \frac{T}{m}}^l} \exp \left(-\frac{|\eta - w|^2}{2\frac{T}{m}} \right) d\eta$$

$$= \left(\frac{m}{2} n |u|^2 + \frac{5}{2} nT \right) u + \left(\frac{l}{2} nT + p_\infty \right) u$$

and obtain

$$\partial_t \left(\frac{m}{2} n |u|^2 + \frac{3+l}{2} nT \right) + \nabla_x \cdot \left[\left(\frac{mn}{2} |u|^2 + \frac{5+l}{2} nT + p_\infty \right) u \right] = 0.$$

$\square$

8.6 Existence, uniqueness and positivity of solutions

In this section we want to give some remarks to the existence, uniqueness and positivity of mild solutions in the case of polyatomic molecules.

8.6.1 Existence and uniqueness for the single polyatomic BGK model

First of all, we want to consider the polyatomic BGK model in the case of one species. So we consider the case when $k = 1$ in section 8.1. That means we have only one species and in the following we omit the index k. The proof of existence and uniqueness of non-negative solutions for polyatomic molecules is analogous to the proof of existence and uniqueness for mono-atomic molecules. So we just sketch the main differences in the proof compared to the mono-atomic case. We start with the definition of mild solutions in the polyatomic BGK case for one species. We assume that the collision frequency ν is equal to a constant $\tilde{\nu}$ as in the existence and uniqueness proof in the mono-atomic case and denote the position space by Λ_{poly} in order to avoid confusion with the rotational and vibrational temperature Λ.

We observe that equation (8.6) for one species has not the same structure as equation (8.2) for one species. In equation (8.2), we only have the Maxwell distribution M on the right-hand side after absorbing f into the left-hand side with the trick described in section 4.1.1. This leads to the mild formulation as in definition 4.2.1 for one species, where we have only M in the integral. Whereas, in equation (8.6), we also have an f on the right-hand side which we can not absorb into the left-hand side. What we do is the following. We define $g = (1 - z_r)M - f$. Then g satisfies

$$\partial_t g + v \cdot \nabla_x g = z_r \tilde{\nu}[(1 - z_r)\tilde{M} - g],$$

and then we consider the following mild formulation.

Definition 8.6.1. We call f, M with $(1 + |v|^2 + |\eta|^2)f, M \in L^1(dv), f, M \geq 0$ a mild solution to (8.2), (8.6) under the conditions for the collision frequencies (4.11) if and only if f satisfies

$$f(x, v, \eta, t) = e^{-\alpha(t)} f^0(x - tv, v, \eta) + e^{-\alpha(t)} \int_0^t \tilde{\nu} M(x + (s - t)v, v, \eta, s)e^{\alpha(s)}ds,$$

$$g(x, v, \eta, t) = e^{-\tilde{\alpha}(t)} g^0(x - vt, v, \eta) + e^{-\tilde{\alpha}(t)} \int_0^t \tilde{\nu} z_r (1 - z_r)\widetilde{M}(x + (s - t)v, v, \eta, s)e^{\tilde{\alpha}(s)}ds,$$

where α is given by

$$\alpha(x, v, t) = \tilde{\nu}t \quad \text{and} \quad \tilde{\alpha} = z_r \tilde{\nu}t$$

and

$$M = \frac{g + f}{1 - z_r}.$$

In order to prove existence and uniqueness, we have the following inequalities for $n, u, \bar{\eta}$ and T given by (8.9).

Remark 8.6.1. In the polyatomic case for one species we define $\xi = (v, \eta), \bar{\xi} = (u, \bar{\eta})$ and $N_q(f) = \sup_\xi |\xi|^q f$. Then we have the following estimates which correspond to the estimates $(i.1)$ in theorem 4.3.1, $(ii.1)$ in theorem 4.3.2, $(iii.1)$ in theorem 4.3.4 and $(iv.1)$ in consequences 4.3.5.

(i.1)* $\frac{n}{(T)^{\frac{d+l}{2}}} \leq CN_0(f),$

(ii.1)* $n(T + |u|^2 + |\bar{\eta}|^2)^{\frac{q-d-l}{2}} \leq C_q N_q(f),$ for $q > d + l + 2,$

(iii.1)* $\frac{n|\bar{\xi}|^{d+l+q}}{[(T+|\bar{\xi}|^2)T]^{\frac{d+l}{2}}} \leq C_q N_q(f),$ for $q > 1.$

(iv.1)* $\sup_\xi |\xi|^q \widetilde{M}(f) \leq C_q N_q(f)$ for $q > N + 2$ or $q = 0.$

These estimates can be proven analogously to the estimates $(i.1)$, $(ii.1)$, $(iii.1)$ and $(iv.1)$ in the mono-atomic case using ξ instead of v.

Concerning the integration we also replace v by ξ and consider the L^1-norm with the weight $(1+|\xi|^2)d\xi dx = (1+|v|^2+|\eta|^2)dvd\eta dx$ in the following. Furthermore, in the mono-atomic case we proved existence and uniqueness of mild solutions under the assumptions 4.3.1. In the polyatomic case we also assume assumptions 1 to 7 in 4.3.1 with the new L^1-norm and we make the following new assumptions.

Assumption 8.6.1. In the polyatomic case we make the following additional assumptions:

1.-7. We assume 1 to 7 in 4.3.1 with the L^1-norm with the weight $(1+|\xi|^2)$.

8. Additional initial values Θ^0, Λ^0 which satisfy condition (8.5) and are integrable with respect to the space variable $x \in \Lambda_{poly}$.

9. Assume that the relaxation parameter in front of $(\widetilde{M} - M)$ in (8.6) is constant and non-negative.

10. Assume $(1 - z_r) \geq c > 0$ for a $c \in \mathbb{R}$.

With this additional assumptions we obtain the following theorem in the poly-atomic case for one species

Theorem 8.6.1. *Under the assumptions 8.6.1 and the definitions (8.1) and (8.9), there exists a unique non-negative mild solution $f \in C(\mathbb{R}^+; L^1((1 + |v|^2)dvdx))$ of the initial value problem (8.2) for one species. Moreover, for all $t > 0$ the following bounds hold:*

$$|u(t)|, T(t), N_q(f)(t) \leq A(t) < \infty,$$
$$n(t) \geq C_0 e^{-t} > 0,$$
$$T(t) \geq B(t) > 0,$$

for $k = 1, 2$ and some constants $A(t), B(t)$.

Proof. The proof is analogue to the proof of theorem 4.3.6. We describe only the main differences.

Step 1: As in the mono-atomic case we conclude that since f is a mild solution according to the definition 8.6.1. Then we have

$$N_q(f) \leq A_0 + \int_0^t C \sup_\xi |\xi|^q M(x,\xi,s)ds.$$

Then we use that M is also part of the solution according to definition 8.6.1. So we take g from definition 8.6.1 and estimate g from below by cM. Then we use the mild formulation for g given by the definition 8.6.1, and obtain the following inequality for M

$$M(x,v,\eta,t) \leq \frac{1}{c}[e^{-\tilde{\alpha}(t)}g^0(x - vt, v, \eta)$$
$$+ e^{-\tilde{\alpha}(t)}\int_0^t \tilde{\nu}z_r(1 - z_r)\widetilde{M}(x + (s - t)v, v, \eta, s)e^{\tilde{\alpha}(s)}ds].$$

We put this inequality into the previous equation. Then we can use $(iv.1)^*$ and conclude as in step 1 of the mono-atomic case

$$N_q(f)(t) \le A_0 e^{C_q t},$$

for $q > N + 2$ or $q = 0$.

Step 2-4: Step 2-4 can be proven analogously to the mono-atomic case.

Step 5: Here we prove Lipschitz continuity of the operator $f \mapsto \widetilde{M}(f)$ when f is restricted to the set

$$\Omega = \{f \in L^1(\Lambda \times \mathbb{R}^d \times \mathbb{R}^l; (1+|\xi|^2) d\xi dx) | f \ge 0, N_q(f) < A, \min(n, T, n, \Theta, \Lambda, T_{rot}, T_{trans}) > C\}. \tag{8.60}$$

The proof of the Lipschitz continuity of $\widetilde{M}[f]$ differs from the Lipschitz continuity of M in the mono-atomic case only in the fact that we will get terms with the global equilibrium temperature T^Θ instead of the single temperature $T^{trans,\theta}$. Nevertheless, since we have estimates of the equilibrium temperature T in remark 8.6.1 the proof is very similar to the mono-atomic case.

Step 6: We consider the sequence $\{(f^k, \Theta^k, \Lambda^k)\}$ given by mild solutions to

$$\partial_t f^k + v \cdot \nabla_x f^k = \tilde{\nu}(M^{k-1|k-2} - f^k),$$

$$\partial_t g^{k-1|k-2} + v \cdot \nabla_x g^{k-1|k-2} = z_r \tilde{\nu}[(1-z_r)\widetilde{M}^{k-2} - g^{k-1|k-2}],$$

$$\frac{d}{2} n \Lambda^{k-1} + \frac{l}{2} n \Theta^{k-2} = \frac{d}{2} n T_{k-2}^{trans} + \frac{l}{2} n T_{k-2}^{rot},$$

$$f^2 = f(0), \quad \Theta^0 = \Theta(0), \quad k \ge 2,$$

where the meaning of the notation $k - 1|k - 2$ is the following. We take the value of Θ^{k-2} but all the other functions have the index $k - 1$. For fixed k we obtain inhomogeneous transport equations from which we know existence of unique mild solutions in the periodic setting.

In the proof of the Cauchy sequence we obtain analogously as in the mono-atomic case the following estimate

$$\|f^n - f^{n-1}\|_{L^1((1+|\xi|^2) d\xi dx)}$$

$$\le e^{-Ct} \int_0^t e^{Cs} [\|M^{n-1|n-2}(s) - M^{n-2|n-3}(s)\|_{L^1((1+|\xi|^2) d\xi dx)} ds,$$

from the first equation of (8.2) and

$$\|M^{n-1|n-2} - M^{n-2|n-3}\|_{L^1(1+|\xi|^2) d\xi dx)}$$

$$\le e^{-Ct} \int_0^t e^{Cs} [\|\widetilde{M}^{n-2}(s) - \widetilde{M}^{n-3}(s)\|_{L^1((1+|\xi|^2) d\xi dx)} ds,$$

from the second equation of (8.2). In the second estimate we use Lipschitz continuity of $\widetilde{M}$ and combine both estimates. The rest is analogously to the mono-atomic case. $\qquad \square$

Remark 8.6.2 (Existence and uniqueness for the mixture BGK model for polyatomic molecules). The proof of existence and uniqueness for the single polyatomic BGK model can be extended for mixtures in an analogous way as in the case of monoatomic BGK model.

8.6.2 Positivity of solutions

Theorem 8.6.2. *Let (f_1, f_2) be a mild solution to (8.2) - (8.22) under the modified assumptions for existence and uniqueness described in the previous section with positive initial data. Then the solution is positive meaning $f_1, f_2 > 0$.*

Proof. The proof is exactly the same as in the case of the BGK model for mixtures, see section 4.4. $\square$

Chapter 9

Convergence to dissipative incompressible Euler equations for two species

In chapter 7 we derived formally a hydrodynamic limit which will lead to the compressible Navier-Stokes equations for two species. But this derivation was formal, we did not prove convergence. To prove convergence to hydrodynamic limits from kinetic equations is a recent area of research. In the case of one species Saint-Raymond proved the convergence to hydrodynamic limits of the BGK model for one species to the incompressible Euler equations in [77] and the incompressible Navier-Stokes equations in [78]. As far as we know, there is no result in the case of gas mixtures. In this chapter we extend the limit to the incompressible Euler equations for one species in [77] to the model described in chapter 2 in the case of gas mixtures.

9.1 The BGK approximation

We repeat the model from chapter 2 for the convenience of the reader. For simplicity in the following we consider a mixture composed of two different species. Thus, our kinetic model has two distribution functions $f_1(x, v, t) > 0$ and $f_2(x, v, t) > 0$ where $x \in \mathbb{R}^3$ and $v \in \mathbb{R}^3$ are the phase space variables and $t \geq 0$ the time. For any $f_1, f_2 : \Lambda \times \mathbb{R}^3 \times \mathbb{R}_0^+ \to \mathbb{R}$, $\Lambda \subset \mathbb{R}^3$ with $(1 + |v|^2)f_1, (1 + |v|^2)f_2 \in L^1(dv), f_1, f_2 \geq 0$ we relate the distribution functions to macroscopic quantities by mean-values of f_k, $k = 1, 2$

$$\int f_k(v) \begin{pmatrix} 1 \\ v \\ m_k |v - u_k|^2 \end{pmatrix} dv =: \begin{pmatrix} n_k \\ n_k u_k \\ 3 n_k T_k \end{pmatrix}, \quad k = 1, 2, \tag{9.1}$$

where n_k is the number density, u_k the mean velocity and T_k the temperature of species k, $k = 1, 2$. Note that in this chapter we shall write T_k instead of $k_B T_k$, where k_B is Boltzmann's constant.

We are interested in a BGK approximation of the interaction terms. Then the model can be written as:

$$\begin{aligned}
\partial_t f_1 + v \cdot \nabla_x f_1 &= \nu_{11} n_1 (M_1 - f_1) + \nu_{12} n_2 (M_{12} - f_1), \\
\partial_t f_2 + v \cdot \nabla_x f_2 &= \nu_{22} n_2 (M_2 - f_2) + \nu_{21} n_1 (M_{21} - f_2), \\
f_1(t = 0) &= f_1^0, \\
f_2(t = 0) &= f_2^0,
\end{aligned} \tag{9.2}$$

with the Maxwell distributions

$$M_k(x, v, t) = \frac{n_1}{\sqrt{2\pi \frac{T_k}{m_k}}^3} \exp\left(-\frac{|v - u_k|^2}{2\frac{T_k}{m_k}}\right), \quad k = 1, 2$$

$$M_{12}(x, v, t) = \frac{n_{12}}{\sqrt{2\pi \frac{T_{12}}{m_1}}^3} \exp\left(-\frac{|v - u_{12}|^2}{2\frac{T_{12}}{m_1}}\right), \tag{9.3}$$

$$M_{21}(x, v, t) = \frac{n_{21}}{\sqrt{2\pi \frac{T_{21}}{m_2}}^3} \exp\left(-\frac{|v - u_{21}|^2}{2\frac{T_{21}}{m_2}}\right),$$

where $\nu_{11}n_1$ and $\nu_{22}n_2$ are the collision frequencies of the particles of each species with itself, while $\nu_{12}n_2$ and $\nu_{21}n_1$ are related to interspecies collisions. To be flexible in choosing the relationship between the collision frequencies, we now assume the relationship

$$\nu_{12} = \varepsilon\nu_{21}, \quad 0 < \varepsilon \leq 1. \tag{9.4}$$

The restriction on ε is without loss of generality. If $\varepsilon > 1$, exchange the notation 1 and 2 and choose $\frac{1}{\varepsilon}$ as new ε. In addition, we assume that all collision frequencies are positive. In order to ensure existence and uniqueness of solutions we assume the following restrictions on our collision frequencies

$$\nu_{kj}(x, t) = \frac{\tilde{\nu}_{kj}}{n_k(x, t) + n_j(x, t)}, \quad k, j = 1, 2 \tag{9.5}$$

with constants $\tilde{\nu}_{11}, \tilde{\nu}_{12}, \tilde{\nu}_{21}, \tilde{\nu}_{22} > 0$.

With this choice of the Maxwell distributions M_1 and M_2 with the same densities, velocities and internal energies as f_1 and f_2, respectively, we guarantee the conservation of mass, momentum and energy in interactions of one species with itself. The remaining parameters $n_{12}, n_{21}, u_{12}, u_{21}, T_{12}$ and T_{21} will be determined using conservation of total momentum and energy, together with some symmetry considerations.

If we assume that

$$n_{12} = n_1 \quad \text{and} \quad n_{21} = n_2, \tag{9.6}$$

we have conservation of the number of particles. If we further assume that u_{12} is a linear combination of u_1 and u_2

$$u_{12} = \delta u_1 + (1 - \delta)u_2, \quad \delta \in \mathbb{R}, \tag{9.7}$$

then we have conservation of total momentum provided that

$$u_{21} = u_2 - \frac{m_1}{m_2}\varepsilon(1 - \delta)(u_2 - u_1), \tag{9.8}$$

and T_{12} and T_{21} given by (2.10) and (2.11).

9.2 From the BGK equation for gas mixtures to incompressible dissipative Euler

First, we need some definitions concerning the equilibrium distribution and the entropy.

Definition 9.2.1. We denote a Maxwell distribution with mass m_k, zero mean velocity and temperature 1 by

$$\bar{M}_k(v) = \frac{1}{\sqrt{2\pi \frac{1}{m_k}}^3} \exp\left(-\frac{m_k |v|^2}{2}\right), \quad k = 1, 2.$$

For simplicity, we assume that both $\bar{M}_1$ and $\bar{M}_2$ have density 1, but it also would be possible to choose different constant densities here.

Definition 9.2.2 (Relative entropy). Let f, f' be two distribution functions. Then, we define the relative entropy of f and f' by

$$H_\varepsilon(f|f') = \frac{1}{\varepsilon^2} \int \int [f \ln \frac{f}{f'} - f + f'] dx dv \quad \text{for} \quad \varepsilon > 0,$$

and the local relative entropy of f and f' by

$$\bar{H}_\varepsilon(f|f') = \frac{1}{\varepsilon^2} \int [f \ln \frac{f}{f'} - f + f'] dv \quad \text{for} \quad \varepsilon > 0.$$

Note that this is a more general definition of the relative entropy as in section 6.2.1. This definition coincides with the definition in section 6.2.1 if the two distribution functions have the same density and if we choose $\varepsilon = 1$.

In the following we denote by M_f the Maxwell distributions with the same moments as the moments of a distribution function f.

Theorem 9.2.1. Let $q > 1$, $T > 0$ and $u_0^1, u_0^2 \in L^2 \cap L^\infty(\mathbb{R}^3)$ be two vector fields with $\nabla_x \cdot u_0^1 = \nabla_x \cdot u_0^2 = 0$. Let $(g_\varepsilon^{1,0}), (g_\varepsilon^{2,0})$ be two families of functions bounded in $L^2(\mathbb{R}^3 \times \mathbb{R}^3, dx \bar{M}_1(v) dv)$ and $L^2(\mathbb{R}^3 \times \mathbb{R}^3, dx \bar{M}_2(v) dv)$, respectively, such that

- $1 + \varepsilon g_\varepsilon^{1,0} \geq 0, 1 + \varepsilon g_\varepsilon^{2,0} \geq 0$ for every $\varepsilon > 0$, $(x, v) \in \mathbb{R}^3 \times \mathbb{R}^3$,

- $H_\varepsilon(\bar{M}_1(1 + \varepsilon g_\varepsilon^{1,0})|M_\varepsilon^{1,0}) \to 0, H_\varepsilon(\bar{M}_2(1 + \varepsilon g_\varepsilon^{2,0})|M_\varepsilon^{2,0}) \to 0$ for $\varepsilon \to 0$,

where $M_\varepsilon^{1,0}, M_\varepsilon^{2,0}$ are local Maxwell distributions with mass m_k, density 1, mean velocity εu_0^1 and εu_0^2, respectively, and temperature 1. Let $(f_1^\varepsilon, f_2^\varepsilon)$ be a solution to

$$\begin{cases} \varepsilon \partial_t f_1^\varepsilon + v \cdot \nabla_x f_1^\varepsilon &= \frac{1}{\varepsilon^q}(M_{f_1^\varepsilon} - f_1^\varepsilon) + \kappa_1(\varepsilon)(M_{12}^\varepsilon - f_1^\varepsilon), \\ \varepsilon \partial_t f_2^\varepsilon + v \cdot \nabla_x f_2^\varepsilon &= \frac{1}{\varepsilon^q}(M_{f_2^\varepsilon} - f_2^\varepsilon) + \kappa_2(\varepsilon)(M_{21}^\varepsilon - f_2^\varepsilon), \\ f_1^\varepsilon(0, x, v) &= \bar{M}_1(1 + \varepsilon g_\varepsilon^{1,0}(x, v)), \\ f_2^\varepsilon(0, x, v) &= \bar{M}_2(1 + \varepsilon g_\varepsilon^{2,0}(x, v)). \end{cases}$$

Then, in the limit $\varepsilon \to 0$, the part of $\frac{1}{\varepsilon} \int f_1^\varepsilon v dv$ and $\frac{1}{\varepsilon} \int f_2^\varepsilon v dv$ which is divergence-free, converge in $C^0([0,T], \mathcal{D}'(\mathbb{R}^3))$ to a solution of the dissipative incompressible Euler equations for mixtures. We call two functions $u_1, u_2 \in L^\infty(\mathbb{R}^+; L^2(\mathbb{R}^3)) \cap C^0(\mathbb{R}^+, w - L^2(\mathbb{R}^3))$ with divergence zero a solution to the dissipative incompressible Euler equations if for all test functions $w_1, w_2 \in C_c^\infty([0,T] \times \mathbb{R}^3)$ such that $\nabla_x \cdot w_1 = \nabla_x \cdot w_2 = 0$ and for all $t \in [0,T]$ we have the inequalities

$$\int |w_1 - u_1|^2(x,t)dx \leq \int |w_1 - u_1|(x,0)dx e^{\int_0^t C||X(w_1)(s)||_\infty ds}$$

$$-2 \int_0^t e^{\int_\tau^t C||X(w_1)(s)||_\infty ds} \int E(w_1) \cdot (u_1 - w_1)dxd\tau$$

$$-2 \int_o^t e^{\int_0^\tau C||X(w_1)(s)||_\infty ds} \int C(u_2 - u_1) \cdot w_1 dxd\tau,$$

$$\int |w_2 - u_2|^2(x,t)dx \leq \int |w_2 - u_2|(x,0)dx e^{\int_0^t C||X(w_2)(s)||_\infty ds}$$

$$-2 \int_0^t e^{\int_\tau^t C||X(w_2)(s)||_\infty ds} \int E(w_2) \cdot (u_2 - w_2)dxd\tau$$

$$-2 \int_0^t e^{\int_0^\tau 2||X(w_2)(s)||_\infty ds} \int C(u_1 - u_2) \cdot w_1 dxd\tau,$$

if $\kappa(\varepsilon)(1 - \delta(\varepsilon)) = \frac{1}{\varepsilon} + O(\varepsilon)$, and

$$\int |w_1 - u_1|^2(x,t)dx \leq \int |w_1 - u_1|(x,0)dx e^{\int_0^t C||X(w_1)(s)||_\infty ds}$$

$$-2 \int_0^t e^{\int_\tau^t C||X(w_1)(s)||_\infty ds} \int E(w_1) \cdot (u_1 - w_1)dxd\tau,$$

$$\int |w_2 - u_2|^2(x,t)dx \leq \int |w_2 - u_2|(x,0)dx e^{\int_0^t C||X(w_2)(s)||_\infty ds}$$

$$-2 \int_0^t e^{\int_\tau^t C||X(w_2)(s)||_\infty ds} \int E(w_2) \cdot (u_2 - w_2)dxd\tau,$$

if $\kappa(\varepsilon)(1 - \delta(\varepsilon)) = O(\varepsilon)$, where the matrix $X(w_1)$ is given by

$$X_{ij}(w_1) = \frac{1}{2}(\partial_{x_i} w_{1,j} + \partial_{x_j} w_{1,i}),$$

the dissipation is given by

$$E(w_k) = \partial_t w_k + P(w_k \cdot \nabla_x w_k), \quad k = 1,2 \tag{9.9}$$

with the Leray projection P on the divergence-free part (see appendix A.3 for the existence), and corresponding equations for the second species. δ is the parameter in (9.7). We allow a dependence of δ on ε here.

The notation $w - L^2(\mathbb{R}^3)$ is explained in the appendix A.3. The non-negativity of the terms $1 + \varepsilon g_\varepsilon^{1,0}$, $1 + \varepsilon g_\varepsilon^{2,0}$ in the assumptions is to ensure non-negativity of the initial data which leads to non-negative solutions for all later times.

The considered system of equations in this theorem is a non-dimensionalized version of the model presented in chapter 2. It is derived in the case of electrons and ions in section 6.1.4. The meaning of the system obtained in the limit will be motivated in the next section.

9.3 Motivation of the dissipative incompressible Euler equations

We want to start with a motivation of the inequality in theorem 9.2.1 in the case of one species. This motivation is also given in [8]. Let $w(x, t)$ be a divergence-free test function and E given by $E(w) = \partial_t w + P(w \cdot \nabla_x w)$. Then for any smooth divergence-free solution $u(x, t)$ of the Euler equations one has

$$\partial_t u + \nabla_x \cdot (u \otimes u) + \nabla_x p = 0,$$
$$\partial_t w + \nabla_x \cdot (w \otimes w) + \nabla_x p = E(w).$$

If we subtract the second equation from the first one and multiply the result by $(u - w)$, we obtain

$$\frac{d|u - w|^2}{dt} + 2(u - w) \cdot S(w) \cdot (u - w) = 2E(w) \cdot (u - w),$$

with $S(w) = \frac{1}{2}(\nabla_x w + (\nabla_x w)^T)$. Duhamel's formula leads to

$$|u - w|^2 \le e^{\int_0^t 2\|S(w)(\tau)\|_\infty ds}|u(0) - w(0)|^2$$
$$+ 2\int_0^t e^{\int_s^t 2\|S(w)(s)\|_\infty d\tau} E(w)(s) \cdot (u - w)(s)ds. \tag{9.10}$$

Definition 9.3.1. We call $u \in L^\infty(\mathbb{R}^+; L^2(\mathbb{R}^3)) \cap C^0(\mathbb{R}^+; w - L^2(\mathbb{R}^3))$ with divergence zero a dissipative solution of the incompressible Euler equations if (9.10) holds for any smooth divergence-free vector field $w \in C_c^\infty([0, T] \times \mathbb{R}^3)$ for almost all $t \in [0, T]$.

We have the following properties of dissipative solutions:

Theorem 9.3.1.

i) Any classical solution u of the incompressible Euler equations is a dissipative solution.

ii) Every dissipative solution satisfies the energy inequality

$$|u(t)|^2 \le |u(0)|^2.$$

iii) If w is a classical solution to the incompressible Euler equations and u is a dissipative solution of the incompressible Euler equations, then

$$|u(t) - w(t)|^2 \leq e^{\int_0^t 2||S(w)(s)||_\infty ds} |u(0) - w(0)|^2.$$

Proof. i) is a direct consequence by the construction. We get ii) if we consider $w = 0$ as classical solution. iii) follows from the fact that if w is a classical solution then $E(w) = 0$. $\square$

The meaning of the theorem is the following. Property *i)* shows a relationship between classical and dissipative solutions. Property *ii)* motivates the notation of dissipative solutions. The kinetic energy of such a solution is dissipative. Property *iii)* states that if there exists a classical solution to the Euler equations then every dissipative solution with the same initial data coincides with this classical solution.

Now, we consider the two species case. Consider the functions u_1, w_1, u_2, w_2 : $\mathbb{R}^3 \times [0, T] \to \mathbb{R}$ which satisfy the equations

$$\partial_t u_1 + \nabla_x \cdot (u_1 \otimes u_1) + \nabla_x p_1 = C_1(u_2 - u_1),$$

$$\partial_t w_1 + \nabla_x \cdot (w_1 \otimes w_1) + \nabla_x p_1 = C_1(w_2 - w_1) + E(w_1) - C_1(w_2 - w_1) + Y_1(u_1, u_2, w_1),$$

$$\partial_t u_2 + \nabla_x \cdot (u_2 \otimes u_2) + \nabla_x p_2 = C_2(u_1 - u_2),$$

$$\partial_t w_2 + \nabla_x \cdot (w_2 \otimes w_2) + \nabla_x p_2 = C_2(w_1 - w_2) + E(w_2) - C_2(w_1 - w_2) + Y_2(u_2, u_1, w_2),$$

(9.11)

where Y_1 is a vector orthogonal to w_1 and not orthogonal to u_1. C_1 is given by $\frac{Y_1 \cdot u_1}{(u_1 - u_2) \cdot u_1}$. If w_1 is parallel to u_1, set $Y_1 = 0$ and $C_1 = 1$. If $u_1 = u_2$ or $u_1 = 0$ set $C_1 = 0$.

Similar for the second species. Y_2 is a vector orthogonal to w_2 and not orthogonal to u_2. C_2 is given by $\frac{Y_2 \cdot u_2}{(u_2 - u_1) \cdot u_2}$. If w_2 is parallel to u_2, set $Y_2 = 0$ and $C_2 = 1$. If $u_1 = u_2$ or $u_2 = 0$ set $C_2 = 0$. We want to derive similar inequalities as in the one species case. We show it only for species 1, the other species is analogue to the derivation of the first one.

If we multiply the first two equations of equations (9.11) by $u_1 - w_1$ and subtract the second equation from the first one, we obtain

$$\partial_t |u_1 - w_1|^2 + 2(u_1 - w_1) \cdot X(w_1) \cdot (u_1 - w_1)$$
$$= 2C_1(u_2 - u_1) \cdot (u_1 - w_1) - 2C_1(w_2 - w_1) \cdot (u_1 - w_1) - 2E(w_1) \cdot (u_1 - w_1)$$
$$+ 2C_1(w_2 - w_1) \cdot (u_1 - w_1) - 2Y_1(u_1, u_2, w_1) \cdot (u_1 - w_1).$$

Due to the properties of Y_1, we have $Y_1 \cdot w_1 = 0$ and $Y_1 \cdot u_1 = C_1(u_1 - u_2) \cdot u_1$ and obtain

$$\partial_t |u_1 - w_1|^2 + 2(u_1 - w_1) \cdot X(w_1) \cdot (u_1 - w_1)$$
$$= -2C_1(u_2 - u_1) \cdot w_1 - 2C_1(w_2 - w_1) \cdot (u_1 - w_1)$$
$$- 2E(w_1) \cdot (u_1 - w_1) + 2C_1(w_2 - w_1) \cdot (u_1 - w_1).$$

According to Duhamel's formula, we obtain the following inequality

$$
\int |w_1 - u_1|^2(x,t)dx \leq \int |w_1 - u_1|(x,0)dx e^{\int_0^t 2||X(w_1)(s)||_\infty ds}
$$

$$
- 2 \int_0^t e^{\int_\tau^t 2||X(w_1)(s)||_\infty ds} \int E(w_1) \cdot (u_1 - w_1)dxd\tau \quad (9.12)
$$

$$
- 2 \int_0^t e^{\int_\tau^t 2||X(w_1)(s)||_\infty ds} \int C_1(u_2 - u_1) \cdot w_1 dxd\tau.
$$

If we do the same for the second species, we obtain

$$
\int |w_2 - u_2|^2(x,t)dx \leq \int |w_2 - u_2|(x,0)dx e^{\int_0^t 2||X(w_2)(s)||_\infty ds}
$$

$$
- 2 \int_0^t e^{\int_\tau^t 2||X(w_2)(s)||_\infty ds} \int E(w_2) \cdot (u_2 - w_2)dxd\tau \quad (9.13)
$$

$$
- 2 \int_0^t e^{\int_\tau^t 2||X(w_2)(s)||_\infty ds} \int C_2(u_1 - u_2) \cdot w_2 dxd\tau.
$$

For the second case in theorem 9.2.1 consider the equations

$$
\partial_t u_k + \nabla \cdot (u_k \otimes u_k) + \nabla p_k = 0,
$$
$$
\partial_t w_k + \nabla \cdot (w_k \otimes w_k) + \nabla p_k = E(w_k), \quad k = 1, 2,
$$

and derive inequalities in the same way as in the previous case. We extend the notion of dissipative solutions analogously as in definition 9.3.1. We observe that the different choice of the dependence on ε is reflected in the appearance of an exchange term between the two species. In the first case, we obtain equations where we have an exchange term of momentum. In the second case, there is no exchange term of the momentum. Then, in the case of mixtures we observe the following properties of dissipative solutions.

Theorem 9.3.2.

i) *Any classical solution satisfies the derived inequality for dissipative incompressible Euler.*

ii) *We have*
$$
|u_k(t)|^2 \leq |u_k(0)|^2, \quad k = 1, 2,
$$
which describes the dissipation of the kinetic energy.

iii) *Let w_1, w_2 be classical solutions of the incompressible Euler equations with exchange terms $C_1(u_1 - u_2)$ and $C_2(u_2 - u_1)$ with $C_1 = C_2$ for conservation of momentum. Then $u_1 = w_1$ and $u_2 = w_2$ if the initial values of u_1 and w_1 and the initial values of u_2 and w_2 coincide.*

Proof. i) is again true by construction. If we take $w_1 = w_2 = 0$, we obtain ii). We prove iii) in the following way. If w_1 and w_2 are classical solutions to the incompressible Euler equations with opposite exchange terms ($C_1 = C_2$), then

$$E(w_1) - C_1(w_2 - w_1) + Y_1(u_1, u_2, w_1) = 0,$$
$$E(w_2) - C_1(w_1 - w_2) + Y_2(u_2, u_1, w_2) = 0. \tag{9.14}$$

From the inequalities (9.12) and (9.13), we obtain by using (9.14) and the properties of Y_1 and Y_2.

$$\int |w_1 - u_1|^2(x, t)dx \leq \int |w_1 - u_1|(x, 0)dx e^{\int_0^t 2||X(w_1)(s)||_\infty ds}$$

$$-2\int_0^t e^{\int_\tau^t 2||X(w_1)(s)||_\infty ds} \int C_1(u_1 - u_2 + w_2 - w_1) \cdot (u_1 - w_1)dxd\tau.$$

$$\int |w_2 - u_2|^2(x, t)dx \leq \int |w_2 - u_2|(x, 0)dx e^{\int_0^t 2||X(w_2)(s)||_\infty ds}$$

$$-2\int_0^t e^{\int_\tau^t 2||X(w_2)(s)||_\infty ds} \int C_1(u_2 - u_1 + w_1 - w_2) \cdot (u_2 - w_2)dxd\tau.$$

We estimate $e^{\int_\tau^t 2||X(w_1)(s)||_\infty ds}$, $e^{\int_\tau^t 2||X(w_2)(s)||_\infty ds}$ from below by 1 if the integrand is positive and by $\max\{e^{\int_\tau^t 2||X(w_1)(s)||_\infty ds}, e^{\int_\tau^t 2||X(w_2)(s)||_\infty ds}\}$ from above if the integrand is negative. The two integrands have the same sign, since we have

$$(u_2 - u_1 + w_1 - w_2) \cdot (u_2 - w_2) = |u_2 - w_2|^2 + (w_1 - u_1) \cdot (u_2 - w_2),$$
$$(u_1 - u_2 + w_2 - w_1) \cdot (u_1 - w_1) = |u_1 - w_1|^2 + (w_2 - u_2) \cdot (u_1 - w_1).$$

So we denote the common estimate of $e^{\int_\tau^t 2||X(w_1)(s)||_\infty ds}$ and $e^{\int_\tau^t 2||X(w_2)(s)||_\infty ds}$ by C^{com} and add both inequalities

$$\int |w_1 - u_1|^2(x, t)dx + \int |w_2 - u_2|^2(x, t)dx$$

$$\leq \int |w_1 - u_1|(x, 0)dx e^{\int_0^t 2||X(w_1)(s)||_\infty ds} + \int |w_2 - u_2|(x, 0)dx e^{\int_0^t 2||X(w_2)(s)||_\infty ds}$$

$$- 2\int_0^t C^{com}(u_1 - u_2 + w_2 - w_1) \cdot (u_1 - w_1 - u_2 + w_2)dxd\tau$$

$$= \int |w_1 - u_1|(x, 0)dx e^{\int_0^t 2||X(w_1)(s)||_\infty ds} + \int |w_2 - u_2|(x, 0)dx e^{\int_0^t 2||X(w_2)(s)||_\infty ds}$$

$$- 2\int_0^t C^{com}|u_1 - u_2 + w_2 - w_1|^2 dxd\tau.$$

This is equivalent to

$$\int |w_1 - u_1|^2(x, t)dx + \int |w_2 - u_2|^2(x, t)dx + 2\int_0^t \int C^{com}|u_2 - u_1 + w_2 - w_1|^2 dxd\tau$$

$$\leq \int |w_1 - u_1|(x, 0)dx e^{\int_0^t 2||X(w_1)(s)||_\infty ds} + \int |w_2 - u_2|(x, 0)dx e^{\int_0^t 2||X(w_2)(s)||_\infty ds}.$$

Now, if u_1 and u_2 have the same initial data as w_1 and w_2, respectively, this leads to

$$\int |w_1 - u_1|^2(x,t)dx + \int |w_2 - u_2|^2(x,t)dx + 2\int_0^t \int C^{com}|u_2 - u_1 + w_2 - w_1|^2 dxd\tau \leq 0,$$

which leads to $w_1 = u_1$ and $w_2 = u_2$. So for the same initial data, the solutions coincide. $\qquad\square$

9.4 Estimates on the entropy and entropy dissipation for gas mixtures

Lemma 9.4.1. *Define* $D_\varepsilon(f_k^\varepsilon) := \frac{1}{\varepsilon^{q+3}} \int \int (M_{f_k^\varepsilon} - f_k^\varepsilon) \ln\left(\frac{f_\varepsilon^k}{M_k}\right) dxdv$. *With the requirements of theorem 9.2.1,* $(H_\varepsilon(f_1^\varepsilon|\bar{M}_k))_\varepsilon$, $(H_\varepsilon(f_2^\varepsilon|\bar{M}_k))_\varepsilon$, $(D_\varepsilon(f_1^\varepsilon))_\varepsilon$, $(D_\varepsilon(f_2^\varepsilon))_\varepsilon$ *are uniformly bounded in* $L^\infty(\mathbb{R}^+)$ *and* $L^1(\mathbb{R}^+)$, *respectively. Especially, in the limit* $\varepsilon \to 0$, $(g_\varepsilon^1)_\varepsilon$, $(g_\varepsilon^2)_\varepsilon$ *defined by* $f_1^\varepsilon = \bar{M}_1(1 + \varepsilon g_\varepsilon^1)$, $f_2^\varepsilon = \bar{M}_2(1 + \varepsilon g_\varepsilon^2)$ *are relatively compact in* $w - L^1_{loc}(dtdx; w - L^1((1+|v|^2)dv))$.

An explanation of the notation $w - L^1_{loc}(dtdx; w - L^1((1+|v|^2)dv))$ can be found in appendix A.3.

Proof. For the time derivative of the entropy, we get in the weak sense

$$\partial_t \bar{H}(f_1^\varepsilon|\bar{M}_1) = \partial_t \left[\frac{1}{\varepsilon^2} \int (f_1^\varepsilon \ln\left(\frac{f_1^\varepsilon}{\bar{M}_1}\right) - f_1^\varepsilon + \bar{M}_1)dv \right]$$

$$= \frac{1}{\varepsilon^2} \int (\partial_t f_1^\varepsilon \ln\left(\frac{f_1^\varepsilon}{\bar{M}_1}\right) + \partial_t f_1^\varepsilon - \partial_t f_1^\varepsilon + \partial_t \bar{M}_1)dv$$

$$= \frac{1}{\varepsilon^2} \int \partial_t f_1^\varepsilon \ln\left(\frac{f_1^\varepsilon}{\bar{M}_1}\right) dv$$

$$= -\frac{1}{\varepsilon^3} \int v \cdot \nabla_x f_1^\varepsilon \ln\left(\frac{f_1^\varepsilon}{\bar{M}_1}\right) dv + \frac{1}{\varepsilon^{q+3}} \int (M_{f_1^\varepsilon} - f_1^\varepsilon) \ln\left(\frac{f_1^\varepsilon}{\bar{M}_1}\right) dv$$

$$+ \kappa_1(\varepsilon)\frac{1}{\varepsilon^3} \int (M_{12}^\varepsilon - f_1^\varepsilon) \ln\left(\frac{f_1^\varepsilon}{\bar{M}_1}\right) dv$$

$$=: -\frac{1}{\varepsilon^3} \int v \cdot \nabla_x f_1^\varepsilon \ln\left(\frac{f_1^\varepsilon}{\bar{M}_1}\right) dv + \bar{D}_{1,\varepsilon}(f_1^\varepsilon) + \bar{D}_{12,\varepsilon}(f_1^\varepsilon, f_2^\varepsilon)$$

$$=: -\frac{1}{\varepsilon^3} \int v \cdot \nabla_x f_1^\varepsilon \ln\left(\frac{f_1^\varepsilon}{\bar{M}_1}\right) dv + \bar{D}_{1,\varepsilon}^{ges}(f_1^\varepsilon, f_2^\varepsilon).$$

Now we integrate this equation with respect to x and t. Using appropriate boundary conditions in x, we see with integration by parts and Gauß's theorem that the first term on the right-hand side vanishes and we obtain

$$H_\varepsilon(f_\varepsilon^1|\bar{M}_1) + \int_0^t D_{1,\varepsilon}^{ges}(f_\varepsilon^1, f_\varepsilon^2)ds \leq H_\varepsilon(\bar{M}_1(1 + \varepsilon g_\varepsilon^{1,0})|\bar{M}_1),$$

with $D_{1,\varepsilon}^{ges}(f_\varepsilon^1, f_\varepsilon^2) = \int \bar{D}_{1,\varepsilon}^{ges}(f_\varepsilon^1, f_\varepsilon^2)dx$. In the same way, we can also show this inequality for the second species

$$H_\varepsilon(f_\varepsilon^2|\bar{M}_2) + \int_0^l D_{2,\varepsilon}^{ges}(f_\varepsilon^1, f_\varepsilon^2)ds \leq H_\varepsilon(\bar{M}_2(1 + \varepsilon g_\varepsilon^{2,0})|\bar{M}_2).$$

Taking the sum of both, we obtain

$$H_\varepsilon(f_\varepsilon^1|\bar{M}_1) + H_\varepsilon(f_\varepsilon^2|\bar{M}_2) + \int_0^t (D_{1,\varepsilon}^{ges}(f_\varepsilon^1, f_\varepsilon^2) + D_{2,\varepsilon}^{ges}(f_\varepsilon^1, f_\varepsilon^2))ds \tag{9.15}$$
$$\leq H_\varepsilon(\bar{M}_1(1 + \varepsilon g_\varepsilon^{1,0})|\bar{M}_1) + H_\varepsilon(\bar{M}_2(1 + \varepsilon g_\varepsilon^{2,0})|\bar{M}_2).$$

According to the proof of lemma 2.1 in [77] $H_\varepsilon(\bar{M}_1(1 + \varepsilon g_\varepsilon^{1,0})|\bar{M}_1)$ and $H_\varepsilon(\bar{M}_2(1 + \varepsilon g_\varepsilon^{2,0})|\bar{M}_2)$ are bounded by a term just depending on $H_\varepsilon(\bar{M}_1(1 + \varepsilon g_\varepsilon^{1,0})|M_\varepsilon^{1,0})$ and $H_\varepsilon(\bar{M}_2(1 + \varepsilon g_\varepsilon^{2,0})|M_\varepsilon^{1,0})$, respectively. Therefore they are uniformly bounded in $L^\infty(\mathbb{R}^+)$ due to the assumptions on the initial data. By definition, we have $H_\varepsilon(f_1^\varepsilon|\bar{M}_1) \geq 0$ and $H_\varepsilon(f_2^\varepsilon|\bar{M}_2) \geq 0$. Due to the H-theorem for mixtures, we also know that $D_{1,\varepsilon}(f_1^\varepsilon, f_2^\varepsilon) \geq 0$, $D_{2,\varepsilon}(f_1^\varepsilon, f_2^\varepsilon) \geq 0$ and $(D_{12,\varepsilon}(f_1^\varepsilon, f_2^\varepsilon) + D_{21,\varepsilon}(f_1^\varepsilon, f_2^\varepsilon)) \geq 0$. Here, the entropy dissipations without a bar are defined in the same way as $D_{1,\varepsilon}^{ges}$ via integration with respect to x over the entropy dissipation with a bar. Therefore, we can deduce from (9.15), that $H_\varepsilon(f_1^\varepsilon|\bar{M}_1)$ and $H_\varepsilon(f_2^\varepsilon|\bar{M}_2)$ are uniformly bounded in $L^\infty(\mathbb{R}^+)$, and $D_{1,\varepsilon}(f_1^\varepsilon, f_2^\varepsilon)$ and $D_{2,\varepsilon}(f_1^\varepsilon, f_2^\varepsilon)$ are uniformly bounded in $L^1(\mathbb{R}^+)$. So we proved the first statement of the theorem. The proof of the second statement is the same as in [77], lemma 2.1. The idea is to prove that $(g_\varepsilon^1(1 + |v|^2))_\varepsilon$, $(g_\varepsilon^2(1 + |v|^2))_\varepsilon$ are equi-integrable and conclude that it is relatively compact in the weak topology with the theorem of Dunford-Pettis. For the definition of equi-integrability and the theorem of Dunford-Pettis see appendix A.4. The equi-integrability is shown in [77] in the following way. Let h^* be the Legendre transformation of $h : z \mapsto (1 + z)\ln(1 + z) - z$, see appendix A.2. Then one obtains with Young's inequality from appendix A.2

$$\frac{1}{4}(1 + |v|^2)|g_\varepsilon^k| \leq \frac{c}{\varepsilon^2}h^*\left(\frac{\varepsilon}{c}\frac{1}{4}(1 + |v|^2)\right) + \frac{c}{\varepsilon^2}h(\varepsilon g_\varepsilon^k), \quad c \in \mathbb{R}, k = 1, 2.$$

From the Taylor series one can see that $h^* : p \mapsto e^p - p - 1$ has the property that $h^*(\lambda p) \leq \lambda^2 h^*(p)$ for λ small enough, see appendix A.2. This leads to

$$\frac{1}{4}(1 + |v|^2)|g_\varepsilon^k| \leq \frac{1}{\alpha}h^*\left(\frac{1}{4}(1 + |v|^2)\right) + \frac{\alpha}{\varepsilon^2}h(\varepsilon g_\varepsilon).$$

One can show that the integral of $\frac{h(\varepsilon g_\varepsilon)}{\varepsilon^2}$ with respect to (x, v) is bounded by a constant, see the proof of lemma 2.1 in [77]. Then, if we multiply the inequality by $\bar{M}_k$, integrate with respect to Ω such that $\int \int \int \mathbf{1}_\Omega \bar{M}_k dv dx dt \leq \bar{\delta}^2$. Then choose $c = \tilde{\delta}$, one obtains

$$\int \int_\Omega \int (1 + |v|^2)|g_\varepsilon^k|\bar{M}_k dv dx dt \leq K\tilde{\delta},$$

for all $\varepsilon < \tilde{\delta}$. $\qquad\square$

Remark 9.4.1. The sum of the entropies is also uniformly bounded. Therefore it exists a convergent subsequence for the sum. Therefore we can find a common sequence going to zero for both single entropies.

Lemma 9.4.2. *Let w_1, w_2 be two vector fields on $\mathbb{R}^3 \times [0, T]$ with $\nabla_x \cdot w_1 = \nabla_x \cdot w_2 = 0$ and $M_\varepsilon^{w_1}, M_\varepsilon^{w_2}$ Maxwell distributions with mass m_k, density 1, velocity εw_1 and εw_2, respectively, and temperature 1. With the requirements of theorem 9.2.1, there exist $\pi_1, \pi_2 : \mathbb{R}^3 \times [0, T] \to \mathbb{R}$ such that*

$$H_\varepsilon(f_1^\varepsilon | M_\varepsilon^{w_1})(t) - H_\varepsilon(\bar{M}_1(1 + \varepsilon g_\varepsilon^{1,0}) | M_\varepsilon^{w_1,0})$$

$$\leq -\frac{1}{\varepsilon^2} \int_0^t \int \int (v - \varepsilon w_1) \cdot \nabla_x w_1 \, f_1^\varepsilon(x, v, s) dx dv ds$$

$$-\frac{1}{\varepsilon} \int_0^t \int \int (E(w_1) + \nabla_x \pi_1)(s, x) \cdot (v - \varepsilon w_1(x, s)) f_1^\varepsilon(x, v, s) dx dv ds$$

$$- \kappa(\varepsilon) \int_0^t \int \rho_{1,\varepsilon}(1 - \delta)(u_2^\varepsilon - u_1^\varepsilon) \cdot w_1 dx ds,$$

$$H_\varepsilon(f_2^\varepsilon | M_\varepsilon^{w_2})(t) - H_\varepsilon(\bar{M}_2(1 + \varepsilon g_\varepsilon^{2,0}) | M_\varepsilon^{w_2,0})$$

$$\leq -\frac{1}{\varepsilon^2} \int_0^t \int \int (v - \varepsilon w_2) \cdot \nabla_x w_2 \, f_2^\varepsilon(s, x, v) dx dv ds$$

$$-\frac{1}{\varepsilon} \int_0^t \int \int (E(w_2) + \nabla_x \pi_2)(x, s) \cdot (v - \varepsilon w_1(x, s)) f_2^\varepsilon(x, v, s) dx dv ds$$

$$- \kappa(\varepsilon) \int_0^t \int \rho_{2,\varepsilon}(1 - \delta)(u_1^\varepsilon - u_2^\varepsilon) \cdot w_2 dx ds.$$

Proof. We prove only the estimate for species 1, the other one is the same exchanging the indices 1 and 2.
The H-theorem leads to a decrease of the total entropy

$$H_\varepsilon(f_1^\varepsilon | \bar{M}_1) + H_\varepsilon(f_2^\varepsilon | \bar{M}_2) \leq H_\varepsilon^1(\bar{M}_1(1 + \varepsilon g_\varepsilon^{1,0}) | \bar{M}_1) + H_\varepsilon^2(\bar{M}_2(1 + \varepsilon g_\varepsilon^{2,0}) | \bar{M}_2),$$

or, since $H_\varepsilon(f_1^\varepsilon | \bar{M}_1), H_\varepsilon(f_2^\varepsilon | \bar{M}_2) \geq 0$,

$$H_\varepsilon(f_1^\varepsilon | \bar{M}_1) \leq H_\varepsilon^1(\bar{M}_1(1 + \varepsilon g_\varepsilon^{1,0}) | \bar{M}_1) + H_\varepsilon^2(\bar{M}_2(1 + \varepsilon g_\varepsilon^{2,0}) | \bar{M}_2).$$

A calculation shows

$$H_\varepsilon(f_1^\varepsilon | M_\varepsilon^{w_1}) = H_\varepsilon(f_1^\varepsilon | \bar{M}_1) + \frac{1}{\varepsilon^2} \int \int f_1^\varepsilon \ln \left(\frac{\bar{M}_1}{M_\varepsilon^{w_1}} \right) dx dv + \frac{1}{\varepsilon^2} \int \int (M_\varepsilon^{w_1} - \bar{M}_1) dx dv$$

$$= H_\varepsilon(f_1^\varepsilon | \bar{M}_1) + \frac{1}{\varepsilon^2} \int \int f_1^\varepsilon \ln \left(\frac{\bar{M}_1}{M_\varepsilon^{w_1}} \right) dx dv,$$

since $M_\varepsilon^{w_1}$ and $\bar{M}_1$ have the same density. We can compute the second term on the right-hand side and obtain

$$\int \int f_1^\varepsilon \ln \left(\frac{\bar{M}_1}{M_\varepsilon^{w_1}} \right) dx dv = \frac{1}{2} \int \int f_1^\varepsilon (\varepsilon^2 w_1^2 - 2\varepsilon v \cdot w_1) dx dv.$$

Then after differentiating and integrating in time, we use

$$H_\varepsilon(f_1^\varepsilon|\bar{M}_1) \leq H_\varepsilon(f_1^\varepsilon(0)|\bar{M}_1(0))$$

and get

$$H_\varepsilon(f_1^\varepsilon|M_\varepsilon^{w_1}) - H_\varepsilon(\bar{M}_1(1 + \varepsilon g_\varepsilon^{1,0})|M_\varepsilon^{w_1}(0)) \leq \frac{1}{2\varepsilon^2}\int_0^t \frac{d}{ds}\int\int f_1^\varepsilon(\varepsilon^2 w_1^2 - 2v\cdot\varepsilon w_1)dxdvds$$

$$= \frac{1}{2\varepsilon^2}\int_0^t\int\int \partial_s f_1^\varepsilon(\varepsilon^2 w_1^2 - 2v\cdot\varepsilon w_1)dxdvds + \frac{1}{2\varepsilon^2}\int_0^t\int\int f_1^\varepsilon \partial_s(\varepsilon^2 w_1^2 - 2v\cdot\varepsilon w_1)dxdvds$$

$$= \frac{1}{2\varepsilon}\int_0^t\int\int(-\frac{v}{\varepsilon}\cdot\nabla_x f_1^\varepsilon + \frac{1}{\varepsilon^{q+1}}(M_{f_1^\varepsilon} - f_1^\varepsilon) + \kappa_1(\varepsilon)(M_{12}^\varepsilon - f_1^\varepsilon))(\varepsilon^2 w_1^2 - 2v\cdot\varepsilon w_1)dxdvds$$

$$+\frac{1}{2\varepsilon}\int_0^t\int\int f_1^\varepsilon(\varepsilon w_1 - v)\cdot\partial_t w_1 dxdvds$$

$$= -\frac{1}{2\varepsilon^2}\int_0^t\int\int v\cdot\nabla_x f_1^\varepsilon(\varepsilon^2 w_1^2 - 2v\cdot\varepsilon w_1)dxdvds + \frac{1}{2\varepsilon}\int_0^t\int\int f_1^\varepsilon(\varepsilon w_1 - v)\cdot\partial_t w_1 dxdvds$$

$$-\frac{\kappa_1(\varepsilon)}{2\varepsilon}\int_0^t\int\int(M_{12}^\varepsilon - f_1^\varepsilon)2v\cdot\varepsilon w_1 dxdvds,$$

since f_1^ε, $M_{f_1^\varepsilon}$ and M_{12}^ε have the same density. In the third term we carry out the integration with respect to v and use the expressions (9.1) and (9.7). So, all in all, we obtain

$$H_\varepsilon(f_\varepsilon^1|M_\varepsilon^{w_1}) - H_\varepsilon(\bar{M}_k(1 + \varepsilon g_\varepsilon^{1,0})|M_\varepsilon^{w_1}(0))$$

$$\leq -\frac{1}{\varepsilon^2}\int_0^t\int\int(v - \varepsilon w_1)\cdot\nabla_x w_1 \, f_\varepsilon^1 dxdvds$$

$$+\frac{1}{\varepsilon}\int_0^t\int\int f_\varepsilon^1(\varepsilon w_1 - v)\cdot(\partial_t w_1 + w_1\cdot\nabla_x w_1)dxdvds$$

$$-\kappa_1(\varepsilon)\int_0^t\int\rho_{1,\varepsilon}(1 - \delta)(u_2^\varepsilon - u_1^\varepsilon)\cdot w_1 dxds.$$

According to the Leray projection (see appendix A.3), it exists a π_1 such that

$$w_1\cdot\nabla_x w_1 - P(w_1\cdot\nabla_x w_1) = \nabla_x \pi_1.$$

By using the definition of E given by (9.9) we obtain the result. □

Lemma 9.4.3. *Let* w_1, w_2 *be smooth vector fields on* $\mathbb{R}^3\times[0,T]$ *with* $\nabla_x\cdot w_1 = \nabla_x\cdot w_2 = 0$ *and* $M_\varepsilon^{w_1}, M_\varepsilon^{w_2}$ *Maxwell distributions with mass* m_k, *density 1, velocity* εw_1 *and* εw_2, *respectively and temperature 1. With the requirements of theorem 9.2.1, we have*

$$\left|\frac{1}{\varepsilon^2}\int\int(v - \varepsilon w_1)\cdot\nabla_x w_1 \, f_1^\varepsilon(x,v,t)dxdv\right|$$

$$\leq C||X(w_1)(t)||_{L^\infty(\mathbb{R}^3)}H_\varepsilon(f_1^\varepsilon|M_\varepsilon^{w_1})(t) + \varepsilon^{\frac{q-1}{2}}||X(w_1)(t)||_{L^\infty(\mathbb{R}^3)}D_\varepsilon^1(f_1^\varepsilon)(t)$$

$$+ C\varepsilon^{\frac{q-1}{2}}||X(w_1)(t)||_{L^1(\mathbb{R}^3)},$$

$$\left| \frac{1}{\varepsilon^2} \int \int (v - \varepsilon w_2) \cdot \nabla_x w_2 \, f_1^\varepsilon(x, v, t) dx dv \right|$$

$$\leq C \|X(w_2)(t)\|_{L^\infty(\mathbb{R}^3)} H_\varepsilon(f_1^\varepsilon | M_\varepsilon^{w_2})(t) + \varepsilon^{\frac{q-1}{2}} \|X(w_2)(t)\|_{L^\infty(\mathbb{R}^3)} D_\varepsilon^1(f_1^\varepsilon)(t)$$

$$+ C \varepsilon^{\frac{q-1}{2}} \|X(w_2)(t)\|_{L^1(\mathbb{R}^3)}.$$

Proof. By a computation one can relate the term $(v - \varepsilon w_k) \cdot w_k$, $k = 1, 2$ with $X(w_k)$. This is done in the end of the proof of lemma 3.1 in [77]. Then the estimate is proven in [77], lemma 4.1. $\qquad\square$

Lemma 9.4.4. *Let w_1, w_2 be smooth vector fields on $\mathbb{R}^3 \times [0, T]$ with $\nabla_x \cdot w_1 = \nabla_x \cdot w_2 = 0$. With the requirements of theorem 9.2.1, there exist $(\varepsilon_n)_n$ with $\varepsilon_n \to 0$ and $u_1, u_2 \in L^\infty(\mathbb{R}^+, L^2(\mathbb{R}^3))$ with $\nabla_x \cdot u_1 = \nabla_x \cdot u_2 = 0$ such that*

$$\frac{1}{\varepsilon_n} \int \int [\partial_t w_1 + w_1 \cdot \nabla_x w_1](x, t) \cdot (v - \varepsilon_n w_1(t, x)) f_1^{\varepsilon_n}(x, v, t) dx dv,$$

and

$$\frac{1}{\varepsilon_n} \int \int [\partial_t w_2 + w_2 \cdot \nabla_x w_2](x, t) \cdot (v - \varepsilon_n w_2(x, t)) f_2^{\varepsilon_n}(x, v, t) dx dv,$$

converge in $L^1_{loc}(\mathbb{R}^+)$ weakly to

$$\int [\partial_t w_1 + w_1 \cdot \nabla_x w_1](x, t) \cdot [u_1 - w_1](x, t) dx,$$

respective

$$\int [\partial_t w_2 + w_2 \cdot \nabla_x w_2](x, t) \cdot [u_2 - w_2](x, t) dx.$$

Proof. The proof is given in [77], lemma 4.2. $\qquad\square$

9.5 Proof of theorem 9.2.1 for mixtures

Let w_1 be a smooth vector field on $\mathbb{R}^3 \times [0, T]$ with $\nabla_x \cdot w_1 = 0$. According to the proof of lemma 9.4.2, we have

$$H_\varepsilon(f_1^\varepsilon | M_\varepsilon^{w_1})(t) = H_\varepsilon(f_1^\varepsilon | \bar{M}_1) + \frac{1}{2\varepsilon^2} \int \int f_1^\varepsilon(\varepsilon^2 w_1^2 - 2\varepsilon w_1 \cdot v) dx dv.$$

Using the ansatz $f_1^\varepsilon = \bar{M}_1(1 + \varepsilon g_\varepsilon^1)$, we obtain

$$H_\varepsilon(f_1^\varepsilon | M_\varepsilon^{w_1})(t) = H_\varepsilon(f_1^\varepsilon | \bar{M}_1) + \frac{1}{2} \int w_1^2(x, t) dx$$

$$+ \frac{1}{2} \int \int \bar{M}_1 g_\varepsilon^1(x, v, t) w_1 \cdot (\varepsilon w_1(x, t) - 2v) dv dx.$$
(9.16)

According to lemma 9.4.1 $H_\varepsilon(f_1^\varepsilon|\bar{M}_1)$ is uniformly bounded in $L^\infty([0,T])$ and therefore with (9.16) we get that $H_\varepsilon(f_1^\varepsilon|M_\varepsilon^{w_1})$ is also uniformly bounded in $L^\infty([0,T])$. Since $L^1([0,T])$ is separable, we can apply the theorem of Banach-Alaoglu (see appendix A.4) and obtain that there exists a sequence $\varepsilon_n \to 0$ and a function $H^{w_1} \in L^\infty([0,T])$ such that

$$H_{\varepsilon_n}(f_1^{\varepsilon_n}|M_{\varepsilon_n}^{w_1}) \to H^{w_1} \quad \text{in} \quad L^\infty([0,T]) \quad \text{weak*}.$$

According to lemma 9.4.2, we have

$$H_\varepsilon(f_1^\varepsilon|M_\varepsilon^{w_1})(t) - H_\varepsilon(f_1^\varepsilon|M_\varepsilon^{w_1})(0)$$
$$\leq -\frac{1}{\varepsilon^2} \int_0^t \int \int (v - \varepsilon w_1) \cdot \nabla_x w_1 \, f_1^\varepsilon(x,v,s) dx dv ds$$
$$- \frac{1}{\varepsilon} \int_0^t \int \int (E(w_1) + \nabla_x \pi)(x,s) \cdot (v - \varepsilon w_1(x,s)) f_1^\varepsilon(x,v,s) dx dv ds$$
$$- \kappa(\varepsilon) \int_0^t \int \rho_{1,\varepsilon}(1-\delta)(u_2^\varepsilon - u_1^\varepsilon) \cdot w_1 dx.$$

According to lemma 9.4.3, we get

$$H_\varepsilon(f_1^\varepsilon|M_\varepsilon^{w_1})(t) - H_\varepsilon(f_1^\varepsilon|M_\varepsilon^{w_1})(0) \leq C \int_0^t \|X(w_1)(s)\|_{L^\infty(\mathbb{R}^3)} H_\varepsilon(f_1^\varepsilon|M_\varepsilon^{w_1})(s) ds$$
$$+ \varepsilon^{\frac{q-1}{2}} \int_0^t [\|X(w_1)(t)\|_{L^\infty(\mathbb{R}^3)} D_\varepsilon^1(f_1^\varepsilon)(s) + C\|X(w_1)(s)\|_{L^1(\mathbb{R}^3)}] ds$$
$$- \frac{1}{\varepsilon} \int_0^t \int \int (E(w_1) + \nabla_x \pi)(x,s) \cdot (v - \varepsilon w_1(x,s)) f_1^\varepsilon(x,v,s) dx dv ds$$
$$- \kappa(\varepsilon) \int_0^t \int \rho_{1,\varepsilon}(1-\delta)(u_2^\varepsilon - u_1^\varepsilon) \cdot w_1 dx ds.$$

According to lemma 9.4.4, it exists a subsequence $\varepsilon_n \to 0$ such that in the limit

$$H^{w_1}(t) - H^{w_1}(0) \leq C \int_0^t \|X(w_1)(s)\|_{L^\infty(\mathbb{R}^3)} H^{w_1}(s) ds$$
$$- \int_0^t \int [\partial_t w_1 + w_1 \cdot \nabla_x w_1](t,x) \cdot [u_1 - w_1](x,s) dx ds$$
$$- \begin{cases} 0 & \text{if } \kappa(\varepsilon)(1-\delta(\varepsilon)) = O(\varepsilon) \\ \int_0^t \int \rho_1(u_2 - u_1) \cdot w_1 dx ds & \text{if } \kappa(\varepsilon)(1-\delta(\varepsilon)) = \frac{1}{\varepsilon} + O(\varepsilon). \end{cases}$$

Analogue to the proof of theorem 1 in [77], we conclude using Gronwall's inequality and relating the difference of the entropy to the difference of the velocities $|u_1(x,t) - w_1(x,t)|$. See the proof of theorem 1 in [77] for details.

$$\int |u_1(x,t) - w_1(x,t)|^2 dx \leq \int |u_0^1(x) - w_1(x,0)|^2 dx e^{\int_0^t C||X(w_1)(\tau)||_{L^\infty(\mathbb{R}^3)} d\tau}$$

$$- \int_0^t e^{\int_s^t C||X(w_1)(\tau)||_{L^\infty(\mathbb{R}^3)} d\tau} \int E(w_1)(x,s) \cdot [u_1 - w_1](x,s) dx ds$$

$$- \int_0^t e^{\int_s^t C||X(w_1)(\tau)||_{L^\infty(\mathbb{R}^3)} d\tau} \begin{cases} 0 \\ \int \rho_1(u_2 - u_1) \cdot w_1 dx \end{cases} \quad ds,$$

depending on δ and κ. If we choose

$$\kappa(\varepsilon)(1 - \delta(\varepsilon)) = \frac{1}{\varepsilon} + O(\varepsilon),$$

then in the limit we obtain an exchange term for the velocities. If

$$\kappa(\varepsilon)(1 - \delta(\varepsilon)) = O(\varepsilon),$$

there is no exchange term. Similar for species 2.

Appendix A

A.1 Elementary equalities and inequalities

We start with an equality for computing a double cross product. It can be found in section 5.2 in Fischer [42].

Theorem A.1.1 (Grassmann's identity). *Let a, b, c be three vectors in $\mathbb{R}^3$. Then we have*

$$a \times (b \times c) = (a \cdot c)b - (a \cdot b)c.$$

We continue with some inequality used in this thesis. The first inequality is Hölder's inequality. The proof can be found for example in theorem 4.4 in volume 2 of [33].

Theorem A.1.2 (Hölder's inequality). *Assume $1 \leq p, q \leq \infty$, $\frac{1}{p} + \frac{1}{q} = 1$ and let U be an open subset of $\mathbb{R}^n$. Let $u_\infty : U \to \mathbb{R}^+$ be a strictly positive probability density on U. Then if $u \in L^p(u_\infty dx)$, $v \in L^q(u_\infty dx)$, we have*

$$\int_U |uv| u_\infty dx \leq ||u||_{L^p(u_\infty dx)} ||v||_{L^q(u_\infty dx)}.$$

The next inequality is the inequality of Gronwall. The proof can be found for example in theorem 6.21 in Eck, Garcke, Knabner [38].

Theorem A.1.3 (Gronwall's inequality). *Let $a > 0$, $h, \Phi \in C^0([0, a], \mathbb{R})$. Let $h \geq 0, \beta \in \mathbb{R}$ and assume*

$$\Phi(t) \leq \beta + \int_0^t \Phi(s)h(s)ds,$$

for all $t \in [0, a]$. Then we have

$$\Phi(t) \leq \beta e^{\int_0^t h(s)ds},$$

for all $t \in [0, a]$.

The next inequality is an inequality which relates the norm of a probability density to an entropy.

Theorem A.1.4 (Ciszar-Kullback inequality). *Let $\Omega \subset \mathbb{R}^d$ and $u_\infty : \Omega \to \mathbb{R}^+$ be a strictly positive probability density on Ω. Assume that $\phi : \mathbb{R}^+ \to \mathbb{R}$ is a smooth and strictly convex function with $\phi''(1) \geq c > 0$ for all $0 \leq x \leq 1$. For the relative entropy functional*

$$H_\phi[u] := \int_\Omega \phi\left(\frac{u}{u_\infty}\right) u_\infty dx,$$

the following inequality holds

$$\|u - u_\infty\|_{L^1(\Omega)} \le C(H_\phi[u] - H_\phi[u_\infty])^{\frac{1}{2}},$$

for all probability densities $u \in \mathcal{P}(\Omega)$.

Proof. The proof is given in Matthes, [66]. For the convenience of the reader we want to repeat it here. Since ϕ is assumed to be smooth, we can write the Taylor expansion of ϕ as

$$\phi(s) = \phi(1) + \phi'(1)(s - 1) + \frac{\phi''(\tau)}{2}(1 - s)^2,$$

for $s > 0$ and τ between s and 1. Since ϕ is convex with $\phi''(x) \ge 2c > 0$ for all $x \in [0, 1]$, we obtain

$$\phi(s) - \phi(1) \ge \phi'(1)(s - 1) + c(1 - s)^2 \mathbf{1}_{\{s<1\}},$$

for all $s > 0$. Since $u_\infty(x)dx$ defines a probability measure on Ω, we have

$$
\begin{aligned}
H_\phi[u] - H_\phi[u_\infty] &= \int_\Omega \left(\phi\left(\frac{u}{u_\infty}\right) - \phi(1) \right) u_\infty dx \\
&\ge \phi'(1) \int_\Omega (u - u_\infty)dx + c \int_{u<u_\infty} \left(\frac{u}{u_\infty} - 1 \right)^2 u_\infty dx.
\end{aligned}
\tag{A.1}
$$

The integral $\int_\Omega (u - u_\infty)dx$ vanishes because udx and $u_\infty dx$ are both probability measures meaning the integrals of both functions are 1. So (A.1) reduces to

$$H_\phi[u] - H_\phi[u_\infty] \ge c \int_{u<u_\infty} \left(\frac{u}{u_\infty} - 1 \right)^2 u_\infty dx. \tag{A.2}$$

Another consequence for probability measures is that

$$
\begin{aligned}
\|u - u_\infty\|_{L^1(\Omega)} = 2 \int_{u<u_\infty} |u - u_\infty| dx &= 2 \int_{u<u_\infty} u_\infty \left| \frac{u}{u_\infty} - 1 \right| dx \\
&\le 2 \left(\int_{u<u_\infty} \left(\frac{u}{u_\infty} - 1 \right)^2 u_\infty dx \right)^{\frac{1}{2}}.
\end{aligned}
\tag{A.3}
$$

The last inequaltity uses the Hölder inequality for the functions $\sqrt{u_\infty}$ and $\sqrt{u_\infty} \left| \frac{u}{u_\infty} - 1 \right|$. A combination of (A.2) and (A.3) leads to the result. □

The next inequality is the inequality of the arithmetic and geometric means. It can be found for example in example 11.30 in volume 1 of Denk, Racke[33].

Theorem A.1.5 (Inequality of arithmetic and geometric means). *For any list of n non-negative real numbers $x_1, x_2, \ldots, x_n$, we have*

$$\frac{x_1 + x_2 + \cdots + x_n}{n} \ge \sqrt[n]{x_1 \cdot x_2 \cdots \cdots x_n},$$

and with equality if and only if $x_1 = x_2 = \cdots = x_n$.

A.2 Transforms

In this section we want to repeat the most important things on Fourier and Laplace transform. We start with the definition of the Fourier transform.

Definition A.2.1. For any $f \in L^1(\mathbb{R}^n)$, we call the function $\mathcal{F}f : \mathbb{R}^n \to \mathbb{C}$ with

$$\mathcal{F}f(\omega) := \frac{1}{\sqrt{2\pi}^n} \int_{\mathbb{R}^n} f(x)e^{-i\omega \cdot x}dx, \quad \omega \in \mathbb{R}^n,$$

the Fourier transform of f.

This definition and the following property of the Fourier transform are given in definition 23.2 and remark 23.4 in volume 2 of Denk, Racke [33].

Theorem A.2.1 (Plancherel). *We have*

$$\langle f, g \rangle_{L^2(\mathbb{R}^n)} = \langle \mathcal{F}f, \mathcal{F}g \rangle_{L^2(\mathbb{R}^n)}$$

for all $f, g \in L^2(\mathbb{R}^n)$. Especially, we have

$$||\mathcal{F}f||_2 = ||f||_2$$

for all $f \in L^2(\mathbb{R}^n)$.

Now, we give an overview on the Laplace transform. This overview is also given in chapter 8 in Blatter [19]. We consider functions $f : \mathbb{R} \to \mathbb{C}$ with the properties

i) $f(t) = 0$ if $t < 0$.

ii) It exists $\alpha \in \mathbb{R}$ and $c > 0$ such that

$$|f(t)| \leq Ce^{\alpha t} \quad \text{for} \quad t \in \mathbb{R}.$$

iii) f is integrable.

We denote the space of functions f with this properties by E and introduce the variable $s = x + iy$.

Definition A.2.2. We define the Laplace transform $\mathcal{L}f$ for a function $f \in E$ in the right half-plane by

$$\mathcal{L}f(s) := \int_0^\infty f(t)e^{-st}dt \quad \text{for} \quad \text{Re}(s) > \alpha_f.$$

Lemma A.2.2. *Let $f \in E$, then $\mathcal{L}f$ is well-defined in $P := \{s \,|\, \text{Re}(s) > \alpha_f\}$ and we have*

$$\lim_{\text{Re}(s) \to \infty} \mathcal{L}f(s) = 0.$$

Proof. The proof is given in [19]. For the convenience of the reader, we want to repeat it here. Consider a point $s_0 = x_0 + iy_0 \in P$. Since $x_0 > \alpha_f$, there exist $\alpha < x_0$ and C such that

$$|f(t)| \leq Ce^{\alpha t}.$$

Therefore

$$|f(t)e^{-s_0 t}| = |f(t)|e^{-x_0 t} \leq Ce^{(\alpha - x_0)t}.$$

Since $\alpha - x_0 < 0$, the function $f(t)e^{-s_0 t}$ is exponentially decreasing and the existence of $\mathcal{L}f(s_0)$ is ensured and we have

$$|\mathcal{L}f(s_0)| \leq \int_0^\infty Ce^{(\alpha - x_0)t}dt = \frac{C}{x_0 - \alpha},$$

and therefore

$$\lim_{\mathrm{Re}(s) \to \infty} \mathcal{L}f(s) = 0.$$

$\square$

The last transform we want to mention is the Legendre transform. We present the construction of the Legendre transform presented in chapter 14 C in Arnold [5] and some properties presented in section 3.1 in Saint-Raymond [78].

Definition A.2.3 (Legendre transform). Let $h(z)$ be a strictly convex twice differentiable function. Then the Legendre transform of h is a new function h^* of a new variable p which is constructed in the following way. Let p be a given number. Consider the straight line $y = px$ in $x - y$−space. We take the point at which the curve belonging to the graph of h has the largest distance from the straight line in the vertical direction. The vertical distance is given by

$$H(p, z) = pz - h(z).$$

For each p the function H has a maximum with respect to z at $z(p)$ since h is convex. The point $x(p)$ is defined by the extremal condition

$$\frac{\partial H}{\partial z}(p, z) = 0,$$

which is equivalent to $h'(z) = p$. Since h is convex, the point $z(p)$ is unique. Now define

$$h^*(p) = H(p, z(p)).$$

We call this function h^* Legendre transform.

Lemma A.2.3 (Young's inequality). *Let $h^*(p)$ the Legendre transform of a convex twice differentiable function $h(z)$, then we have*

$$pz \leq h^*(p) + h(z)$$

for all p and z.

Proof. Since $h^*(p)$ is defined as the maximum of the function $H(p, z) = pz - h(z)$ in $z(p)$, we have

$$pz - h(z) \le h^*(p)$$

for all p and z. ◻

Example A.2.1. Consider the function $h(z) = (1+z)\ln(1+z) - z$. Then the Legendre transform of h^* is given by $h^*(p) = e^p - p - 1$. Furthermore, h and h^* in this example satisfy

$$h^*(\lambda p) \le \lambda^2 h^*(p) \quad \text{for all} \quad p \ge 0, \lambda \in [0, 1], \tag{A.4}$$

$$h(|z|) \le h(z) \quad \text{for all} \quad z > -1, \tag{A.5}$$

$$p|z| \le \lambda h^*(p) + \frac{1}{\lambda} h(z) \quad \text{for all} \quad p \ge 0, z \ge -1, \lambda \in (0, 1]. \tag{A.6}$$

Proof. This example is presented in [78]. Consider the function

$$H(p, z) = pz - h(z) = pz - (z+1)\ln(1+z) + z.$$

Then the maximum is determined by

$$\frac{\partial H}{\partial z}(p, z) = p - \ln(1+z) = 0,$$

which is equivalent to $z = e^p - 1$. Therefore h^* is given by

$$h^*(p) = H(p, z(p)) = e^p - p - 1.$$

Then (A.4) follows from the Taylor expansion of h^* in the following way

$$h^*(\lambda p) = e^{\lambda p} - \lambda p - 1 = \sum_{k=2}^{\infty} \frac{(\lambda p)^k}{k!} \le \lambda^2 \sum_{k=2}^{\infty} \frac{p^k}{k!} = \lambda^2 h^*(p)$$

for $\lambda \in (0, 1]$. The inequality (A.5) follows from the Taylor expansion of h for $z > -1$. The inequality (A.6) can be deduced using (A.4), (A.5) and Young's inequality

$$\lambda h^*(p) + \frac{1}{\lambda} h(z) \ge \frac{1}{\lambda} h^*(\lambda p) + \frac{1}{\lambda} h(|z|) \ge \frac{1}{\lambda} \lambda p |z|.$$

◻

A.3 Functional spaces

In this section we present some definitions from the area of functional analysis. We define some functional spaces, also presented in the appendix A of Bardos, Golse and Levermore [7], and the projection onto divergence-free functions, also presented in Majda, Bertozzi and Ogawa [65].

Definition A.3.1. Let E be a normed linear space with norm $||\cdot||_E$ and dual space E^*. With $w - E$ we denote the space E equipped with its weak topology which is the coarsest topology on E for which each of the linear forms $u \mapsto \langle w; u \rangle_{E^*, E}$ for $w \in E^*$ is continuous.

Definition A.3.2. Let X be a locally compact topological space and E a normed linear space. Then $C^0(X; w - E)$ denotes the space of continuous functions from X to $w - E$ which is the set of functions u for which $x \mapsto \langle w; u(x) \rangle_{E^*, E}$ is in $C^0(X)$ for each $w \in E^*$.

Definition A.3.3. Let X be a locally compact topological space and E a normed linear space. Then $L^1_{loc}(X; w - E)$ denotes the space of L^1_{loc} functions from X to $w - E$ which is the set of functions u for which $x \mapsto \langle w; u(x) \rangle_{E^*, E}$ is in $L^1_{loc}(X)$ for each $w \in E^*$.

Theorem A.3.1 (Leray projection). *Every vector field v in the Sobolev space $H^m(\mathbb{R}^N; \mathbb{R}^d)$, $m \in \mathbb{N}_0$ has the unique orthogonal projection*

$$v = w + \nabla \phi,$$

with $w, \nabla \phi \in H^m(\mathbb{R}^N; \mathbb{R}^d)$. The operator P which projects on the divergence-free functions $Pv = w$ is called the Leray projection.

Proof. The proof is similar to the proof of the existence of the Hodge decomposition in $L^2(\mathbb{R}^N) \cap C^\infty(\mathbb{R}^N)$. One can deduce that ϕ has to satisfy a Poisson equation, construct the unique solution and compute w as $w = v - \nabla \phi$, see Proposition 1.16 and lemma 3.6 in [65]. $\qquad\square$

A.4 Compactness arguments in the weak and weak* topology

In this section, we present the theorem of Banach-Alaoglu, the notion of equi-integrability and the theorem of Dunford-Pettis. This is taken from Bell [10].

Theorem A.4.1 (Banach-Alaoglu). *Let E be a normed space and E^* its dual space with the operator norm. Then the set $\bar{B}_{E^*} = \{T \in E^*, ||T|| \leq 1\}$ is compact in the weak* topology. Furthermore, if E is a separable Banach space, then any bounded sequence in E^* has a weak* converging subsequence.*

Definition A.4.1 (Equi-integrability). Let (X, Σ, μ) be a probability space and let $\mathcal{F}$ be a subset of $L^1(\mu)$. We say that $\mathcal{F}$ is equi-integrable if for every $\varepsilon > 0$ there is some $\delta > 0$ such that for any $A \in \Sigma$ with $\mu(A) \leq \delta$ and for all $f \in \mathcal{F}$

$$\int_A |f| d\mu \leq \varepsilon.$$

Theorem A.4.2 (Dunford-Pettis). *Suppose that (X, Σ, μ) is a probability space and that $\mathcal{F}$ is a bounded subset of $L^1(\mu)$. Then $\mathcal{F}$ is equi-integrable if and only if $\mathcal{F}$ is a relatively compact subset of $L^1(\mu)$ in the weak topology.*

Bibliography

[1] P. Andries, K. Aoki, and B. Perthame. A consistent BGK-type model for gas mixtures. *Journal of Statistical Physics*, 106(5):993–1018, 2002.

[2] P. Andries, P. Le Tallec, J.-P. Perlat, and B. Perthame. The Gaussian-BGK model of Boltzmann equation with small Prandtl number. *European Journal of Mechanics-B/Fluids*, 19(6):813–830, 2000.

[3] P. Andries and B. Perthame. The ES-BGK model equation with correct Prandtl number. In *AIP Conference Proceedings*, volume 585, pages 30–36. AIP, 2001.

[4] L. Arkeryd. On the Boltzmann equation part II: The full initial value problem. *Archive for Rational Mechanics and Analysis*, 45(1):17–34, 1972.

[5] V. I. Arnold. *Mathematical methods of classical mechanics*, volume 60. Springer Science & Business Media, 2013.

[6] P. Asinari. Asymptotic analysis of multiple-relaxation-time lattice Boltzmann schemes for mixture modeling. *Computers & Mathematics with Applications*, 55(7):1392–1407, 2008.

[7] C. Bardos, F. Golse, and C. D. Levermore. Fluid dynamic limits of kinetic equations II convergence proofs for the Boltzmann equation. *Communications on pure and applied mathematics*, 46(5):667–753, 1993.

[8] C. Bardos and E. Titi. Euler equations for incompressible ideal fluids. *Russian Mathematical Surveys*, 62(3):409, 2007.

[9] A. Bastounis, T. Holding, and V. Silvestr. The Cauchy problem for the Boltzmann equation. lecture notes, 2013.

[10] J. Bell. The Dunford-Pettis theorem. lecture notes, 2015.

[11] P. M. Bellan. *Fundamentals of Plasma Physics*. Cambridge University Press, 2006.

[12] M. Bennoune, M. Lemou, and L. Mieussens. Uniformly stable numerical schemes for the Boltzmann equation preserving the compressible Navier–Stokes asymptotics. *Journal of Computational Physics*, 227(8):3781–3803, 2008.

[13] F. Bernard, A. Iollo, and G. Puppo. Accurate asymptotic preserving boundary conditions for kinetic equations on cartesian grids. *Journal of Scientific Computing*, 65(2):735–766, 2015.

[14] F. Bernard, A. Iollo, and G. Puppo. Polyatomic models for rarefied flows. submitted for publication, 2017.

[15] C. Besse, P. Degond, F. Deluzet, J. Claudel, G. Gallice, and C. Tessieras. A model hierarchy for ionospheric plasma modeling. *Mathematical Models and Methods in Applied Sciences*, 14(03):393–415, 2004.

[16] P. L. Bhatnagar, E. P. Gross, and M. Krook. A model for collision processes in gases. I. Small amplitude processes in charged and neutral one-component systems. *Physical review*, 94(3):511, 1954.

[17] C. K. Birdsall and A. B. Langdon. *Plasma physics via computer simulation.* CRC press, 2004.

[18] M. Bisi and M. J. Cáceres. A BGK relaxation model for polyatomic gas mixtures. *Communications in Mathematical Sciences*, 14(2):297–325, 2016.

[19] C. Blatter. Komplexe Analyis, Fourier- und Laplacetransformation für Elektroingenieure. Autographie, 2006.

[20] A. Bressan. Notes on the Boltzmann equation. *Lecture notes for a summer course, SISSA Trieste*, 2005.

[21] S. Brull. An ellipsoidal statistical model for gas mixtures. *Communications in Mathematical Sciences*, 8:1–13, 2015.

[22] S. Brull, V. Pavan, and J. Schneider. Derivation of a BGK model for mixtures. *European Journal of Mechanics-B/Fluids*, 33:74–86, 2012.

[23] S. Brull and J. Schneider. On the ellipsoidal statistical model for polyatomic gases. *Continuum Mechanics and Thermodynamics*, 20(8):489–508, 2009.

[24] C. Cercignani. *The Boltzmann Equation and its Applications.* Springer, 1975.

[25] C. Cercignani. *Rarefied gas dynamics: from basic concepts to actual calculations,* volume 21. Cambridge University Press, 2000.

[26] S. Chapman and T. G. Cowling. *The mathematical theory of non-uniform gases: an account of the kinetic theory of viscosity, thermal conduction and diffusion in gases.* Cambridge university press, 1970.

[27] C. K. Chu. Kinetic-theoretic description of the formation of a shock wave. *The Physics of Fluids*, 8(1):12–22, 1965.

[28] A. Crestetto, N. Crouseilles, and M. Lemou. Kinetic/fluid micro-macro numerical schemes for Vlasov-Poisson-BGK equation using particles. *Kinetic and Related Models*, 5(4):787–816, 2012.

[29] A. Crestetto, C. Klingenberg, and M. Pirner. Kinetic/fluid micro-macro numerical scheme for a two component plasma. submitted, 2017.

[30] N. Crouseilles and M. Lemou. An asymptotic preserving scheme based on a micro-macro decomposition for collisional Vlasov equations: diffusion and high-field scaling limits. *Kinetic and related models*, 4(2):441–477, 2011.

[31] A. De Cecco, F. Deluzet, C. Negulescu, and S. Possanner. Asymptotic transition from kinetic to adiabatic electrons along magnetic field lines. *Multiscale Modeling & Simulation*, 15(1):309–338, 2017.

[32] S. Dellacherie. Relaxation schemes for the multicomponent euler system. *ESAIM: Mathematical Modelling and Numerical Analysis*, 37(6):909–936, 2003.

[33] R. Denk and R. Racke. *Kompendium der ANALYSIS*. Springer, 2012.

[34] G. Dimarco, L. Mieussens, and V. Rispoli. An asymptotic preserving automatic domain decomposition method for the Vlasov–Poisson–BGK system with applications to plasmas. *Journal of Computational Physics*, 274:122–139, 2014.

[35] G. Dimarco and L. Pareschi. Numerical methods for kinetic equations. *Acta Numerica*, 23:369–520, 2014.

[36] R. J. DiPerna and P.-L. Lions. On the Cauchy problem for Boltzmann equations: global existence and weak stability. *Annals of Mathematics*, 1:321–366, 1989.

[37] R. J. DiPerna and P. L. Lions. Global solutions of Boltzmann's equation and the entropy inequality. *Archive for Rational Mechanics and Analysis*, 114(1):47–55, 1991.

[38] C. Eck, H. Garcke, and P. Knabner. Partielle Differentialgleichungen. In *Mathematische Modellierung*. Springer, 2011.

[39] R. Esposito and M. Pulvirenti. From particles to fluids. *Handbook of mathematical fluid dynamics*, 3:1–82, 2004.

[40] L. Evans. *Partial Differential Equations, Graduate Studies in Mathematics*, volume 19. American Mathematical Society, 1997.

[41] F. Filbet and S. Jin. A class of asymptotic-preserving schemes for kinetic equations and related problems with stiff sources. *Journal of Computational Physics*, 229(20):7625–7648, 2010.

[42] G. Fischer. Lineare Algebra, eine Einführung für Studienanfänger, 14. *Auflage, Vieweg*, 2003.

[43] V. Garzó, A. Santos, and J. J. Brey. A kinetic model for a multicomponent gas. *Physics of Fluids*, 1:380–383, 1989.

[44] Gerthsen, Kneser, and H. Vogel. *Physik: Ein Lehrbuch zum Gebrauch neben Vorlesungen*. Springer Berlin Heidelberg, 1977.

[45] R. T. Glassey. *The Cauchy problem in kinetic theory*. SIAM, 1996.

[46] F. Golse. From kinetic to macroscopic models, session "l'etat de la recherche" de la soc. *Math. de France, Orleans*, 1998.

[47] F. Golse. The Boltzmann equation and its hydrodynamic limits. *Evolutionary equations*, 2:159–301, 2005.

[48] M. Groppi, S. Monica, and G. Spiga. A kinetic ellipsoidal BGK model for a binary gas mixture. *EPL (Europhysics Letters)*, 96(6):64002, 2011.

[49] E. P. Gross and M. Krook. Model for collision processes in gases: Small-amplitude oscillations of charged two-component systems. *Physical Review*, 102(3):593, 1956.

[50] S.-Y. Ha, S. Noh, and S. Yun. Global existence and stability of mild solutions to the Boltzmann system for gas mixtures. *Quarterly of Applied Mathematics*, 65(4):757–779, 2007.

[51] B. B. Hamel. Kinetic model for binary gas mixtures. *The Physics of Fluids*, 8(3):418–425, 1965.

[52] D. Hänel. *Molekulare Gasdynamik: Einführung in die kinetische Theorie der Gase und Lattice-Boltzmann-Methoden*. Springer, 2004.

[53] C. V. Heer. *Statistical Mechanics, Kinetic Theory, and Stochastic Processes*. Elsevier, 2012.

[54] L. H. Holway Jr. New statistical models for kinetic theory: methods of construction. *The Physics of Fluids*, 9(9):1658–1673, 1966.

[55] J. Jeans. LXX. The persistence of molecular velocities in the kinetic theory of gases. *The London, Edinburgh, and Dublin Philosophical Magazine and Journal of Science*, 8(48):700–703, 1904.

[56] J. H. Jeans. *The dynamical theory of gases*. University Press, 1921.

[57] F. John. *Partial Differential Equations*. Springer, 1982.

[58] C. Klingenberg and M. Pirner. Existence, Uniqueness and Positivity of solutions for BGK models for mixtures. *Journal of Differential Equations*, 2017.

[59] C. Klingenberg and M. Pirner. Using a kinetic BGK model to determine transport coefficients of gas mixtures. *submitted*, 2018.

[60] C. Klingenberg, M. Pirner, and G. Puppo. A consistent kinetic model for a two-component mixture of polyatomic molecules. *submitted*, 2017.

[61] C. Klingenberg, M. Pirner, and G. Puppo. A consistent kinetic model for a two-component mixture with an application to plasma. *Kinet. Relat. Models*, 10(2):445–465, 2017.

[62] C. Klingenberg, M. Pirner, and G. Puppo. Kinetic ES-BGK models for a multi-component gas mixture. *Springer Proceedings in Mathematics and Statistics (PROMS) 236*, 2018.

[63] L. Landau. On the vibration of the electronic plasma. *J. Phys USSR*, 1946.

[64] Q. Li, C. Klingenberg, and M. Pirner. On quantifying uncertainties for the linearized BGK kinetic equation. *Springer Proceedings in Mathematics and Statistics (PROMS) 236*, 2018.

[65] A. Majda, A. Bertozzi, and A. Ogawa. Vorticity and incompressible flow. Cambridge texts in applied mathematics. *Applied Mechanics Reviews*, 55:B77, 2002.

[66] D. Matthes. Entropy Methods and Related Functional. lecture notes, 2010.

[67] M. Monteferrante, S. Melchionna, and U. M. B. Marconi. Lattice Boltzmann method for mixtures at variable Schmidt number. *Journal of Chemical Physics*, 141:014102, 2014.

[68] T. Morse. Kinetic model for gases with internal degrees of freedom. *The physics of fluids*, 7(2):159–169, 1964.

[69] B. Perthame. Global existence to the BGK model of Boltzmann equation. *Journal of Differential equations*, 82(1):191–205, 1989.

[70] B. Perthame and M. Pulvirenti. Weighted L^∞ bounds and uniqueness for the Boltzmann BGK model. *Archive for rational mechanics and analysis*, 125(3):289–295, 1993.

[71] C. E. Pico, L. O. dos Santos, and P. C. Philippi. thermal lattice Boltzmann BGK model for ideal binary mixtures. In *19th International Congress of Mechanical Engineering*, 2007.

[72] S. Pieraccini and G. Puppo. Implicit–explicit schemes for BGK kinetic equations. *Journal of Scientific Computing*, 32(1):1–28, 2007.

[73] M. Pirner. Von Mikro zu Makro: Herleitung der Eulergleichungen. Bachelorarbeit, 2012.

[74] M. Pirner. A kinetic model for mixtures with an application to plasma. Masterthesis, 2014.

[75] B. Rejeb. On the existence of conditions of a classical solution of BGK–Poisson's equation in finite time. *International Journal of Applied Mathematics*, 39(4):01, 2009.

[76] V. Rykov. A model kinetic equation for a gas with rotational degrees of freedom. *Fluid Dynamics*, 10(6):959–966, 1975.

[77] L. Saint-Raymond. Du modèle BGK de l'équation de Boltzmann aux équations d'Euler des fluides incompressibles. *Bulletin des sciences mathematiques*, 126(6):493–506, 2002.

[78] L. Saint-Raymond. *Hydrodynamic limits of the Boltzmann equation*. Springer, 2009.

[79] V. Sofonea and R. F. Sekerka. BGK models for diffusion in isothermal binary fluid systems. *Physica A: Statistical Mechanics and its Applications*, 299(3):494–520, 2001.

[80] E. Sonnendrücker. Numerical methods for the Vlasov-Maxwell equations. Lecture notes ETH Zürich, 2011.

[81] H. Struchtrup. *Macroscopic transport equations for rarefied gas flows*. Springer, 2005.

[82] S. Ukai and T. Okabe. On classical solutions in the large in time of two-dimensional Vlasov's equation. *Osaka Journal of Mathematics*, 15, 1978.

[83] C. Villani. A review of mathematical topics in collisional kinetic theory. *Handbook of mathematical fluid dynamics*, 1(71-305):3–8, 2002.

[84] C. Villani et al. Landau damping. *Notes de cours, CEMRACS*, 2010.

[85] B. von Harrach. Numerik partieller Differentialgleichungen. lecture notes, Wintersemester 2012/2013.

[86] D. Werner. *Funktionalanalysis*. Springer, 2010.

[87] S.-B. Yun. Classical solutions for the ellipsoidal BGK model with fixed collision frequency. *Journal of Differential Equations*, 259(11):6009–6037, 2015.

www.ingramcontent.com/pod-product-compliance
Lightning Source LLC
Chambersburg PA
CBHW081102220326
41598CB00038B/7197